AF517119

Plant Protection in Horticulture

Volume 2

Fungal Diseases and their Management

OTHER VOLUMES

Plant Protection in Horticulture

Vol. 1 : Insect, Mite and Vertebrate Pests and their Management
ISBN: 978-81-7233-615-8

Vol. 3 : Bacterial and Viral Diseases and their Management
ISBN: 978-81-7233-617-2

BY THE SAME AUTHOR

- Diseases of Horticulture Crops: Nematode Problems and their Management ISBN: 978-81-7233-543-4
- Integrated Nematode Management in Horticultural Crops ISBN: 978-81-7233-523-6
- Organic Farming for Sustainable Horticulture: Principles and Practices ISBN: 978-81-7233-501-4

Plant Protection in Horticulture

Volume 2
Fungal Diseases and their Management

P. Parvatha Reddy
Former Director and Former Head
Division of Entomology & Nematology
Indian Institute of Horticultural Research
Bangalore-560 089

Published by:

SCIENTIFIC PUBLISHERS (INDIA)
5A, New Pali Road, P.O. Box 91
Jodhpur 342 001, India
E-mail: info@scientificpub.com
Website: www.scientificpub.com

Print:2025

ISBN: 978-81-72336-18-9
eISBN: 978-93-87913-00-4

Printed in India

Affectionately dedicated to
My brothers

Y.S. Parwath Reddy
P. Hema Reddy
P. Mallikarjuna Reddy
P. Basava Reddy
P. Chandrasekhara Reddy

PREFACE

Horticulture in India is fast emerging as a major commercial venture, because of higher remuneration per unit area and the realization that consumption of fruits and vegetables is essential for health and nutrition. In the last one decade, export potential of horticultural crops has significantly increased attracting even multinationals into floriculture, processing and value added products.

During 2005-06 coverage of area under various horticultural crops in the country was 19.237 million hectares (which is about 13.08% of the total cultivated area) with production of 185.207 million MT. The horticulture sector contributed around 28% of GDP in agriculture. India is the largest producer in the world of cashew nuts, coconuts, tea, ginger, turmeric and black pepper. India is the second largest producer of fruits and vegetables (after China) with a production of 58.74 and 109.05 million MT and contributes 10.9 and 11.9% share in global fruit and vegetable production, respectively. India is the second largest producer of flowers after China. About 126,000 hectare of area is under floriculture, producing 694,000 million tonnes of loose flowers annually (2005-2006). The production of cut flowers has increased to 2,762 million numbers. India is still a marginal player in the world floriculture trade indicating strong potential that can be exploited in this sector. India is the third largest producer of coconut, fourth largest producer and consumer of rubber and sixth largest producer of coffee in the world.

Productivity of horticultural crops in India is relatively low compared to other countries. Of the several factors responsible for lower productivity of horticultural crops, fungal diseases are considered as important limiting factors. Some disease problems are of national importance which causes significant losses to horticultural crops. Diseases of horticultural crops continue to cause losses of about 10% of the crop yields worth more than Rs. 15,000 crores annually. More than 9,600 MT of technical grade fungicides are used annually to manage the diseases in India.

The information on fungal diseases of horticultural crops (fruits, vegetables, plantation, spice, tuber, ornamental, medicinal and aromatic crops) is very much scattered. There is no book at present which comprehensively and exclusively deals with the above aspects on horticultural crops. The present book deals with integrated disease management in horticultural crops in detail using regulatory, physical, cultural, chemical, biological, host plant resistance and integrated methods. The book is extensively illustrated with excellent quality photographs enhancing the quality of publication. The book is written in lucid style, easy to understand language along with adoptable recommendations involving eco-friendly components of IDM.

In view of greater emphasis being given by the Government of India to horticulture by establishing horticulture mission and exclusive State Horticultural Universities, there is an urgent need for good text books to teach courses on different disciplines of horticulture. In this context the present book is of immense help to all those concerned with diseases of horticultural crops.

This book is a practical guide to practicing farmers of horticultural crops. Further, it is a useful reference to policy makers, research and extension workers and students. The material can also be used for teaching under-graduate and post-graduate courses. Suggestions to improve the contents of the book are most welcome (E-mail: reddy_parvatha@yahoo.com).

I profusely thank Dr. K. Srinivasan, Chief Agricultural Economist, TAFE, Chennai; Dr. N. Ramachandran, Indian Institute of Horticultural Research, Bangalore; and Dr. Dr. V.K. Das and Dr. D.K. Ghosh, National Research Centre for Citrus, Nagpur; who have kindly made available many of the photographs included in this book. The Publisher, Scientific Publishers (India), Jodhpur, deserves commendation for their professional contribution.

Bangalore
August, 15th, 2009

P. Parvatha Reddy

CONTENTS

Chapter 1

HORTICULTURAL CROPS

1.1. Importance of Horticultural Crops

Agriculture is the mainstay of the Indian economy, and horticulture is a crucial component thereof. Horticulture development has been accorded high priority in the XIth Five Year Plan. The impact of enhanced investment in horticulture has been highly encouraging in terms of vastly improved quality production and export potential-an increase from 96.1 million tonnes in 1990-91 to 185.2 million tonnes during 2005-06 (Bijay Kumar, 2007) (Table 1.1).

Table 1.1. Area, production and productivity of horticultural crops in India (2005-06) (Bijay Kumar, 2007)

Horticultural crops	Area (million ha)	Production (million MT)	Productivity (MT/ha)	% Share of hortil. crops
Fruits	5.510	58.740	10.7	31.74
Vegetables	7.164	109.050	15.2	58.93
Flowers-loose	0.126	0.694	---	0.38
Flowers-cut	---	2762.000*	---	---
Mushrooms	---	0.040	---	0.03
Spices	2.858	5.108	1.8	2.76
Plantation crops	3.283	11.263	3.4	6.09
Nuts	0.121	0.134	1.1	0.07
Aromatic & Medicinal	0.266	0.220	---	---
Total	19.237	185.207	9.6	---

*Million numbers

India is the second largest producer of fruits next only to China contributing 10.9% of the total world production. Total production of fruits during 2005-06 has been estimated at 58.74 million MT from 5.51million ha. (Bijay Kumar, 2007). India is the largest producer of mango, banana, and papaya. India produces 41.7% of the world mangoes, and 25.7% of the bananas. In grapes, India has recorded the highest productivity per unit area in the world. The per capita consumption of fruits has increased from 40 to 85g/day within the last one decade.

India is the second largest producer of vegetables next only to China contributing 11.9% of the total world production. Vegetables occupy an area of 7.164 million ha with a production of 109.05 million tonnes (Bijay Kumar, 2007). India produces 13.6% of the world onions. In vegetables, India occupies the first position in the production of peas and okra, second in onion, brinjal, cabbage, cauliflower and third in potato. The per capita consumption of vegetables has increased from 96 to 175 g/day within the last one decade.

Floriculture is estimated to cover an area of 126,000 ha with production of 694,000 MT of loose flowers. The area under cut flowers has increased significantly in recent years and the production has reached 2762 million numbers.

India has emerged as the largest producer of spices, tea, cashew and areca nut. Coconuts sustain an area of 2.029 million ha with a production of 101.90 million tonnes. Areca nut is cultivated in 0.381 million ha with a production of 0.483 million tonnes. Cashew nuts are grown in an area of 0.843 million ha with a production of 0.579 million tonnes. The annual production of spices is 5.108 million tonnes from 2.858 million ha.

Horticulture has emerged as an important sector for diversification of agriculture. Presently, horticulture has established its credibility in improving income through increased productivity, generating employment and in enhancing exports besides providing protective food, household nutritional security, precious raw material for various processing and ancillary industries and finally in enhancing the aesthetic aspects of life. The focused attention on investment in horticulture during the last decade has been rewarding in terms of increased production and productivity of horticultural crops with manifold export potential. Horticulture has played a significant role in women empowerment, providing employment opportunities through mushroom cultivation, floriculture, processing, nursery raising, vegetable seed production etc.

1.2. Export of Horticultural Produce

Horticulture in India is fast emerging as a major commercial venture, because of higher remuneration per unit area and the realization that consumption of fruits and vegetables is essential for health and nutrition. In the last one decade, export potential of horticultural crops has significantly increased attracting even multinationals into floriculture, processing and value added products.

The present annual growth rate of horticulture is more than 6.5% which contributes 24.5% from the mere 8% of area to the GDP and 54.55% to the export earnings in the agriculture sector (Table 1.2).

Table 1.2. Export of horticultural produce from India (2005-06) (Bijay Kumar, 2007)

Horticultural product	Export quantity (MT)	Value (Rs. Crores)	% Share of value
Floriculture & seeds			
Floriculture	35186.00	299.41	7
Fruit & vegetable seeds	7472.9	92.83	2
Total	42658.90	392.24	
Fruits & vegetables			
Fresh onions	960507.32	708.16	18
Other fresh vegetables	217280.59	267.69	6
Walnuts	5256.56	114.47	3
Fresh mangoes	69606.60	128.11	3
Fresh grapes	54049.87	214.61	5
Other fresh fruits	158339.26	225.67	5
Total	1465040.20	1658.71	
Processed fruits & vegetables			
Dried & preserved vegetables	124495.71	364.11	9
Mango pulp	134613.20	364.24	9
Pickle & Chutneys	135381.85	260.98	6
Other processed fruits and vegetables	107335.24	370.21	27
Grand Total	943568.59	2454.60	

The major destinations for export of different horticultural products and potential international markets for fresh and processed products are presented in Table 1.3.

Table 1.3. Major destinations for export of different horticultural products (Chamber, 2006)

Horticultural crops	Major destinations for export
Fruits	UAE, Saudi Arabia, Sri Lanka, Bangla Desh, The Netherlands, USA, Malaysia
Fresh grapes	UK, UAE, The Netherlands, Bangla Desh, Germany, Sri Lanka
Pomegranate	UAE, Saudi Arabia, Sri Lanka, Kuwait
Fresh guava	Saudi Arabia, Netherlands, USA, UAE, Kuwait
Gooseberries	UAE, Qatar
Mango pulp	UK, Canada, Bangle Desh, Bahrain, Netherlands, Japan, Saudi Arabia
Mango slices	UK, Saudi Arabia, Japan, USA
Dried grapes	UK, Malaysia, Netherlands, UAE

Mango pickles	Saudi Arabia, UAE, UK, Canada
Mango chutney	Oman, USA
Guava jam & jelly	UK, Germany, Hong Kong, USA, Netherlands, Saudi Arabia
Dried tamarind	Egypt, UAE, Pakistan, Saudi Arabia, UK
Dried figs	Nepal, Bangla Desh
Vegetables	UAE, UK, Saudi Arabia, Sri Lanka, USA, Singapore
Fresh vegetables	UAE, Saudi Arabia, Sri Lanka, USA
Fresh onion	UAE, Singapore, Sri Lanka, Bangla Desh, Malaysia
Fresh tomato	Bangle Desh, Nepal, Sri Lanka
Gherkin	France, USA, Spain, Belgium
Dried vegetables	UAE, Sri Lanka, USA, UK, Singapore
Dehydrated onion	Germany, Romania, Sri Lanka, USA, Malaysia, Netherlands, UK
Vegetable seeds	USA, The Netherlands, France, Italy, Japan, Pakistan
Floriculture	USA, Japan, The Netherlands, UK, Italy

Horticulture today is not merely a means of diversification but forms an integral part of food and nutritional security, as also an essential ingredient of economic security. Adoption of horticulture both by small and marginal farmers brought prosperity in many regions of the country of which Maharashtra, Karnataka and Andhra Pradesh are the prime examples. A Golden Revolution through significant production increases in horticultural crops is already on the horizon.

Since India is now a signatory member of WTO, it will require its produce and products to be competitive both in the domestic and export markets. This would call for the use of improved HYV/F1 hybrids, recommended agro-techniques and plant protection measures, hi-tech intervention and scientific post-harvest management.

After the success of 'Green Revolution' in agriculture, India is now moving towards the **'Golden Revolution'** comprising fruits, vegetables and flowers. Commercialization of horticulture along with post harvest management, processing and marketing assume great significance and add value not only in the context of farmers' income, but also its contribution to the agriculture GDP. Commercialization of horticulture is largely dependent upon use of appropriate technology and market oriented production.

Chapter 2

FRUIT CROPS

A large variety of fruits are grown in India. Of these, banana, citrus, pineapple, papaya, sapota, jack fruit and avocado (tropical fruits); mango, grape, guava, litchi and loquat (subtropical fruits); apple, pear, peach, plum, apricot, almond and walnut (temperate fruits) and amla, ber, pomegranate, annona, fig and phalsa (arid zone fruits) are important. The leading fruit growing states are Maharashtra, Andhra Pradesh, Tamil Nadu, Kerala, Gujarat, Karnataka, Bihar, Uttar Pradesh, West Bengal and Orissa.

India is the second largest producer of fruits next only to China contributing 10.9% of the total world production. The present area under fruit crops is estimated to be about 5.510 million hectares with an estimated production of 58.740 million tonnes (Tables 2.1 and 2.2). Fruit crops and nuts contribute 31.81% share to the total production of horticultural crops. The nut crops are grown in 0.1209 million hectares with an annual production of 0.1335 million tonnes (Table 2.3). India is the largest producer of mango, banana and papaya. India produces 40.1% of the world mangoes, 24.5% of the bananas and 28.8% of world papaya. In grapes, India has recorded the highest productivity per unit area (25.4 tonnes/ha) in the world. The per capita consumption of fruits has increased from 40 to 85 g/day within the last one decade.

Table 2.1. Major fruit producing countries of the world (Bijay Kumar, 2007)

Country	Area (in '000 ha)	Production (in '000 MT)	Productivity (in MT/ha)	% share of world production
China	10456	89355	8.5	16.7
India	5510	58740	10.7	10.9
Brazil	2386	36952	15.5	6.9
USA	1215	28144	23.2	5.2
Italy	1219	18134	14.9	3.4
Spain	1806	15537	8.6	2.9

Indonesia	1167	15406	13.2	2.9
Mexico	1128	14903	13.2	2.8
Iran	1354	13848	10.2	2.6
Philippines	1073	13110	12.2	2.4
Turkey	1074	12998	12.1	2.4
Others	24781	219465	8.9	40.9
Total	53170	536591		

Table 2.2. Area, production and productivity of fruit crops in India (2005-06) (Bijay Kumar, 2007)

Fruit crop	Area ('000 ha)	Production ('000 MT)	Productivity (MT/ha)	% share of India production
Apple	235	1756	7.5	3.0
Banana	565	18702	33.1	31.8
Citrus	749	6326	8.4	10.8
Grapes	64	1631	25.4	2.8
Guava	182	1823	10.0	3.1
Litchi	63	381	6.0	0.6
Mango	2021	12538	6.2	21.3
Papaya	73	2317	31.7	3.9
Pineapple	85	1353	15.8	2.3
Pomegranate	116	849	7.3	1.4
Sapota	140	1117	8.0	1.9
Others	1217	9947	8.2	16.9
Total	5510	58740	10.7	

Table 2.3. Area, production and productivity of nut crops in India (2005-06) (Bijay Kumar, 2007)

Nut crop	Area ('000 ha)	Production ('000 MT)	Productivity (MT /ha)
Almond	21.1	14.6	0.7
Walnut	99.8	118.9	1.2
Total	120.9	133.5	

India has been exporting fruits and fruit products since several decades. However, in the recent past export promotional activities received greater attention (Table 2.4). Mango and grapes are the main items in fresh fruit export and some other fruits like pomegranate, sapota, banana, citrus, litchi, apple and strawberry also entered fresh fruit export market in limited quantities.

Table 2.4. Export of fruit produce from India (2005-06) (Bijay Kumar, 2007)

Fruit products	Quantity (in MT)	Value (in Rs. Million)
Fruit & vegetable seeds	7472.90	928.3
Walnuts	5256.56	1144.7
Fresh mangoes	69606.60	1281.1
Fresh grapes	54049.87	2146.1
Other fresh fruits	158339.26	2256.7
Mango pulp	134613.20	3642.4
Pickle & Chutneys	135381.85	2609.8
Other processed fruits & vegetables	107335.24	3702.1

The major potential international markets for fresh and processed fruit products are presented in Table 2.5.

Table 2.5. Potential international markets for fresh and processed fruits

Fresh & processed fruits	Potential international markets
Fresh grapes	UK, Netherlands, Sri Lanka, UAE
Pomegranate	UAE, Saudi Arabia, Sri Lanka, Kuwait
Fresh guava	Saudi Arabia, Netherlands, USA, UAE, Kuwait
Gooseberries	UAE, Qatar
Mango pulp	UK, Canada, Bangla Desh, Bahrain, Netherlands, Japan, Saudi Arabia
Mango slices	UK, Saudi Arabia, Japan, USA
Dried grapes	UK, Malaysia, Netherlands, UAE
Mango pickles	Saudi Arabia, UAE, UK, Canada
Mango chutney	Oman, USA
Guava jam & jelly	UK, Germany, Hong Kong, USA, Netherlands, Saudi Arabia
Dried tamarind	Egypt, UAE, Pakistan, Saudi Arabia, UK
Dried figs	Nepal, Bangla Desh

Disease Scenario in Fruit Crops

Of the several factors responsible for lower productivity of fruits, fungal diseases constitute important limiting factors. It is estimated that the fruit production could be increased at least by 26% if one protects the crops from various diseases affecting them. Many of the diseases in orchards lead to complete failure of the crop. Citrus decline and powdery mildew; guava wilt; downy mildew, powdery mildew and anthracnose of grapevine; Panama wilt of banana and stem rot of papaya are considered as national problems in India.

2.1. BANANA

Globally, banana (*Musa* spp.) is one of the maximum distributed fruit crop grown in more than 130 countries in 4.306 million hectares with an annual production of 76.436 million tonnes (Table 2.6). It is the fourth most important food crop after rice, wheat and maize. It is a dessert fruit for millions and is used in different regions of the world as staple food. As a diet, banana is filling, easy to digest, nearly fat free, rich source of carbohydrate with calorific value of 67 calories/100g and is free from sodium, making it salt-free diet and also has zero cholesterol. Hence, regular use of it is also considered to be good for heart. Banana also contains several minerals such as Calcium, Magnesium, Potassium and Phosphorus. Banana fruit is particularly rich in Vitamin-C and contains significant amount of other vitamins such as Vitamin-A.

Table 2.6. Major banana producing countries of the world (Bijay Kumar, 2007)

Country	Area (in '000 ha)	Production (in '000 MT)	Productivity (in MT/ha)	% share of world production
India	565	18702	33.1	24.5
Brazil	491	6703	13.6	8.8
China	274	6670	24.4	8.7
Philippines	418	6298	15.1	8.2
Ecuador	221	6118	27.7	8.0
Indonesia	315	5178	16.4	6.8
Coasta Rica	41	2353	57.2	3.1
Mexico	77	2250	29.2	2.9
Thailand	141	1865	13.2	2.4
Colombia	65	1765	27.2	2.3
Others	1698	18534	10.9	24.2
Total	4306	76436	33.1	

Banana is an important fruit crop having great socio-economic significance in India as it supports livelihood of millions of people. India is the largest producer of banana in the world with an annual production of 18.702 million tonnes from an area of 0.565 million ha (Table 2.7). Although, banana occupies 10.3 per cent of the area under fruits it contributes 31.8 per cent of the total fruit production in the country. The national productivity of banana is very impressive (33.1 tonnes per ha). Its productivity in commercial zones especially in Maharashtra, Gujarat, Tamil Nadu, Karnataka, Andhra Pradesh and Bihar is highly impressive and comparable to the productivity in other countries.

Table 2.7. Area, production and productivity of leading banana growing states of India (2005-06) (Bijay Kumar, 2007)

State	Area (in '000 ha)	Production (in '000 MT)	Productivity (in MT/ha)
Tamil Nadu	94.6	4647.6	49.1
Maharashtra	73.2	4608.5	63.0
Gujarat	49.2	2498.8	50.8
Andhra Pradesh	61.1	1528.7	25.0
Karnataka	56.4	1399.1	24.8
Bihar	28.0	920.0	34.3
Kerala	56.2	443.1	7.9
Assam	47.0	699.4	14.9
Madhya Pradesh	15.4	615.6	40.0
West Bengal	27.8	544.9	19.6
Others	56.2	756.9	13.5
Total	565.1	18701.9	33.1

Research and development efforts made in the last two decades have resulted in commendable increase in productivity (144 per cent in the last decade) and overall production of banana in the country. The improved production technologies include higher plant densities (4500 plants per ha), improved crop geometries coupled with adoption of drip irrigation, more judicious fertilization management through balanced use of organic and inorganic fertilizers, appropriately timed split application of inorganics, use of proper sized rhizomes and well timed plantings, commercial production of *in vitro* plants and development of effective integrated pest management strategies. By adoption of these improved technologies, productivity of several individual farmers in different parts of the country has resulted in substantial increase (60 to 65 tonnes/ha).

2.1.1. Panama Wilt, *Fusarium oxysporum* f. sp. *cubense*

Panama wilt is one of the most devastating diseases of banana in the world. The disease is prevalent in Australia, Coasta Rica, Hawaii, India, Jamaica, Panama, South America, Surinam and West Africa. In India, it became widespread in Tamil Nadu, Kerala, Karnataka, Bihar and Assam. The disease is also prevalent in West Bengal, Maharashtra and Andhra Pradesh, especially where Rasabale, Amritpani, Malbhog and Martban cvs. belonging to Rasthali group are grown.

i) Symptoms: The entry of the fungus is facilitated by root damage caused by the nematodes (*Radopholus similis, Meloidogyne incognita*, etc.). The fungus blocks the vascular system and causes wilting (Fig.2.1). The infected plants show characteristic yellowing of leaf blades developing as a band along the margin and spreading towards midrib. The leaf wilts and the petiole buckles. The leaf hangs between the pseudo stem while the middle of lamina is still green. All leaves eventually collapse, whereas the petioles join the pseudo stem and die. Often the emerging heart leaf gets affected. Four to six weeks after the appearance of first symptoms, only the pseudo stem with dead leaves hanging round it remains. Young and old plants show dwarfing or stunting. When an affected rhizome is cut transversely, the disease is seen localized in the vascular strands. Individual strand appears yellow. Red or brown dots or streaks are also seen. The cut stem smells like rotten fish. The suckers growing out of diseased corms wilt and eventually the whole mat dies.

Fig. 2.1. Panama wilt affected banana plant.

ii) Epidemiology: Light textured loam and sandy loam soils which are acidic favours the disease development. Such soils are referred as conducive soils for fusarial infection. Soil infestation by nematodes (*Radopholus similis, Meloidogyne incognita*, etc.) predisposes the plant to infection. The pathogen is soil-borne and survives for long periods in soil as chlamydospores (even up to 20 years in the absence of its host). Ramakrishnan and Damodaran (1956) found that liming of soil reduced the survival period to two months. The primary spread of the disease is through infected rhizomes and secondary spread is through irrigation water. Continuous cultivation of banana in the same field results in abundant build up of inoculum in the soil.

iii) Management

- Flooding of wilt affected areas for 6 months has been reported to give satisfactory results. Use of healthy planting material, removal of infected plants at first sight and avoidance of root injury through intercultural operations helps in reducing the disease incidence.

- Crop rotation with rice or sugarcane helps in reduction of the disease incidence. Improvement in drainage is also recommended.
- Application of the capsule filled with *Pseudomonas fluorescens* at 60 mg/capsule is recommended at 2, 4 and 6 months after planting.
- The modified Panchagavya mixture at 10^1 dilution should be applied (soil drenching) at 2-3 litres/plant.
- Application of the capsule filled with carbendazim 50 WP at 60 mg/capsule is recommended at 2, 4 and 6 months after planting. One capsule is applied to each plant in a hole made at an angle of 45° diagonally in the diseased corm. If the capsule is not available, the hole may be injected with 10 to 15 ml of 2% carbendazim solution. Soil drenching with 0.01% vapam or 0.15% thiram helps in managing the disease. Dipping of banana suckers in 0.1% carbendazim before planting followed by bimonthly soil drenching starting from 6 months after planting gives good control.
- Banana cvs. Giant Cavendish, Lacatan, Rajavazhai, Peyladen, Moongil, Poovan and Vamanakeli are reported to be resistant to wilt.

2.1.2. Sigatoka Leaf Spot, *Cercospora musae*

The disease mostly attacks leaves and has been reported to cause economic losses to the banana crop. It has been reported to cause severe losses to banana crop in Java, Guyana, Honduras, Surinam, Tanzania, Fizi, Uganda, Colombia, Coasta Rica, Guatemala, Haiti, India, Jamaica, Mexico, Panama and Venezuela. In India, it is distributed in Assam, Andhra Pradesh, Bihar, West Bengal, Kerala, Tamil Nadu, Karnataka, Maharashtra and Gujarat states.

i) Symptoms: The initial symptoms seen in the field are small lesions on leaves which are pale yellow or greenish yellow streaks that appear on both sides of leaves parallel to the leaf veins. The leaf spots on leaves increase in size to form dark brown to black linear oblong areas (Fig. 2.2). Center of the spots eventually dry out, becoming light gray, but a narrow, dark brown to black border persists giving an eye spot appearance. Rapid drying and defoliation is the characteristic feature of this disease. In severe cases, the petiole collapses and the leaf hangs down from the pseudo stem. The infection may cause complete failure of maturity of the bunch. In some cases, premature ripening of fruits is also reported.

ii) Epidemiology: During rainy season, spread of the disease is very rapid. Both conidia and ascospores can cause infection, but conidia predominate for most of the period. Conidia are produced in wet weather or in dew on the surface of lesions and are dispersed by water droplets to other leaves, particularly to the leaves of suckers. Ascospores are borne on older leaves in perithecia sunken into the tissues of the leaves. The disease is favoured by warm temperature (23°-25°C), rainy or humid weather, close planting, heavy weed or grass cover and failure to remove the suckers.

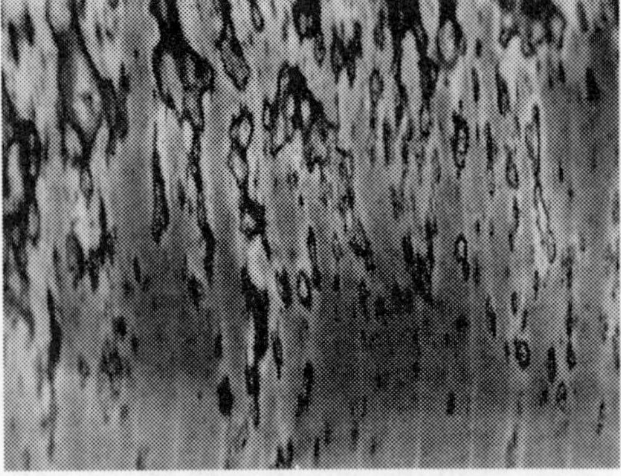

Fig. 2.2. Sigatoka leaf spot on banana (Courtesy, K. Srinivasan, TAFE, Chennai).

iii) Management

- Removal and destruction of infected leaves, spraying 1% Bordeaux mixture mixed with 2% linseed oil, providing improved drainage, good weed control, removal of suckers and correct spacing helps in reducing the disease incidence.
- Spraying 0.1% topsin-M or prochloraz or 0.1% carbendazim or 0.2% chlorothalanil or 0.15% kitazin 3-4 times at fortnightly interval was quite effective in field trials.

2.1.3. Anthracnose, *Gloeosporium musarum (Colletotrichum musae)*

Anthracnose, also called as black rot or ripe rot, is one of the most important and widely distributed diseases of ripening or ripe bananas. It is a serious disease in the entire banana growing regions, especially in Bihar, Karnataka and Tamil Nadu. Almost all the banana varieties are susceptible but the severity of disease may vary. It attacks the plants at all stages of development.

i) Symptoms: The fungus attacks the young banana fruits usually at distal end. The skin of the fruit turns black, shrivels and covered with characteristic pink acervuli. Finally the whole finger is affected (Fig. 2.3). Later the disease spreads and affects the whole bunch resulting in premature ripening and shriveling of fruits which are covered with pink spore masses. Infected fruits turn black and rot.

ii) Epidemiology: The disease is favoured by high atmospheric temperature and humidity, wounds and bruises caused in the fruit and susceptibility of the cultivar. The optimum temperature for germination of conidia is 30°C and for disease development on ripe fruits is 30°-35°C. The conidia germinate in RH ranging from 85.7 to 100%. The disease is more abundant in rainy season (June to September). The spread of the disease is by air-borne conidia and by numerous insect pollinators that visit banana flowers.

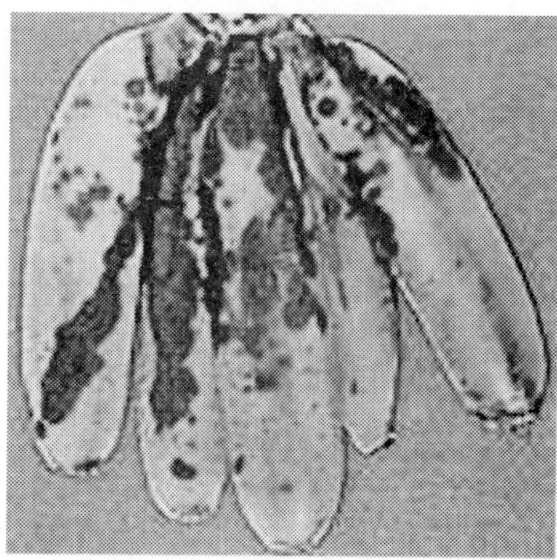

Fig. 2.3. Anthracnose on banana fruits (Courtesy, K. Srinivasan, TAFE, Chennai).

iii) Management

- In the field, the distal bud should be removed when all the hands have opened to prevent infection.
- Protective spray of young bunches with 1% Bordeaux mixture controls the disease.
- Pre-harvest spray with 0.2% prochloraz or 0.1% carbendazim or 0.2% chlorothalanil or 0.15% kitazin 3 to 4 times at fortnightly interval is highly effective (Rawal and Ullasa, 1989).
- Post-harvest dipping of fruits in 440 ppm mycostatin or 100 ppm aureofungin or 400 ppm carbendazim or 1000 ppm benomyl prevents fruit rot.
- Banana bunches should be harvested at correct stage of maturity and should be stored at 10°C.
- Banana cvs. Kothia and Muthia are reported to be fairly resistant (Thakur and Mishra, 1966).

2.1.4. Cigar End Rot, *Verticillium theobromae*

The disease is widespread in banana growing areas and can cause serious fruit loss. It is largely confined to Australia, Egypt, India and West Africa. In India, it has been reported from Andhra Pradesh and Madhya Pradesh.

i) Symptoms: The pathogen attacks both mature and immature fruits. In some bunches only a few fruits are affected, while in others all the fingers show the disease symptoms. The initial infection in the perianth slowly spreads along the finger and cause blackening, shrinkage and folding of the skin tissues. The corrugated diseased portion becomes covered with powdery conidiophores and conidia of the fungus. It resembles the ash on a cigar end and hence the name cigar end disease (Fig. 2.4). Normally less than 2 cm length of fruit is rotted. But in severe infections, even one-third of the fruit length show rotting from the end. The internal pulp is reduced to dry and fibrous condition known as 'dry rot'.

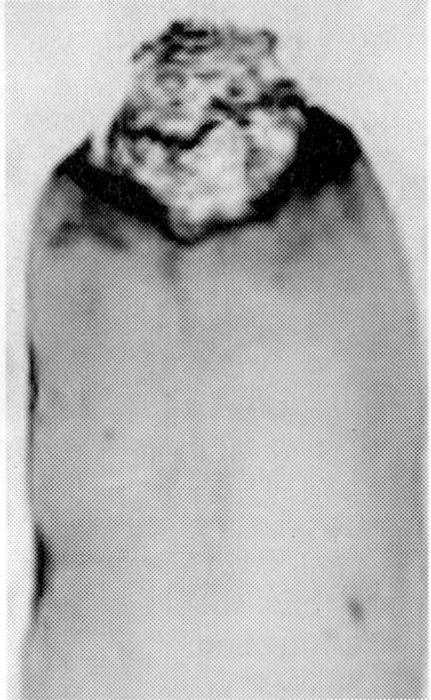

Fig. 2.4. Cigar end rot on banana fruit (Courtesy, K. Srinivasan, TAFE, Chennai).

ii) Management

- Eradication of other *Musa* spp., *Bamboos* spp., *Heliconia bethel* and *H. resiliencies* from banana plantation reduces the risk of spread of infection.
- Removal of perianth and pistil by hand immediately after the fruits are formed is effective in disease control. Pistils should be removed 8-11 days after bunch emergence.
- Young bunches should be opened up to the light and air and the bracts which remain attached to the bunch should be removed especially during wet weather.
- The plantations should have enough aeration by avoiding over-crowding of plants. Further, improved sanitation helps in the reduction of disease incidence.
- The disease can be effectively managed by placing polythene sheets over the stems before the hands emerge.
- The bunches may be sprayed with 0.25% copper oxychloride along with wetting agent at 0.5 to 1.0 ml/litre of spray fluid.

2.1.5. *Botryodiplodia* Fruit Rot, *Botryodiplodia theobromae*

This is one of the common and most severe banana storage pathogen. It is distributed in Uttar Pradesh, Delhi, Maharashtra and Andhra Pradesh.

i) Symptoms: The fungus is a wound parasite. Infection originates in or immediately below the decayed perianth or style and spreads uniformly along the fruit causing a progressive brownish black discolouration. Generally, rotting of the pulp is comparatively faster than that of the skin. In maturing or matured fruits, two thirds of whole fruit may be affected. The skin becomes black, soft and wrinkled. Under moist conditions, white or light grey cottony mycelium may be seen on the affected area. Later, the skin gets encrusted with pycnidia. While maximum damage of fruit occurs at 25-30°C, it is minimum at 7°C and below.

ii) Management: The decay can be prevented by storing the banana bunches at 10°C.

2.2. CITRUS

Citrus fruits (*Citrus* spp.) rank third in area and production after banana and mango with an estimated production of 6.323 million tonnes from an area of 0.749 million hectares (Table 2.8). There are many commercially grown *Citrus* spp. such as mandarins (*Citrus reticulata*), sweet orange (*C. sinensis*), acid lime (*C. aurantifolia*), lemon (*C. limon*), grape fruit (*C. paradisi*), pumello (*C. grandis*), etc.

Table 2.8. Area, production and productivity of leading citrus growing states of India (2005-06) (Bijay Kumar, 2007)

States	Area (in '000 ha)	Production (in '000 MT)	Productivity (in MT/ha)
Andhra Pradesh	166.2	2234.9	13.4
Maharashtra	248.8	1469.0	5.9
Punjab	26.0	365.4	14.0
Gujarat	30.5	303.3	9.9
Madhya Pradesh	16.7	266.9	16.0
Rajasthan	12.7	237.7	18.7
Karnataka	11.0	231.5	21.0
Orissa	26.5	205.3	7.7
Manipur	27.0	167.7	6.2
Assam	18.7	159.5	8.5
Uttarakhand	26.1	122.9	4.7
Bihar	16.8	112.3	6.7
Others	121.6	449.7	3.7
Total	748.8	6323.2	8.4

2.2.1. Damping-Off, *Phytophthora nicotianae, P. citrophthora, P. palmivora, Rhizoctonia solani, Pythium* spp.

Damping off of seedlings in nursery bed is wide spread problem in citrus industry. The disease frequently occurs in field nurseries where maintenance of sanitary measures is difficult. More than 20% seedling mortality has been observed in Central India due to this disease (Naqvi, 2000).

i) Symptoms: Necrosis of tissue and typical damping off of seedlings occur due to fungal infection just above the soil level. The seeds/soil infested with the pathogen results in pre-emergence rot of seeds and post-emergence damping off of seedlings. In infested seed beds, the mortality of seedlings occurs in patches. The seedling mortality increases where excessive soil moisture accompanies the favourable temperature for the pathogen. Pathogens survive in soil either through saprophytic growth (*R. solani*) or production of resistant structures such as chlamydospores or oospores (*Phytophthora* spp.). Seedlings become tolerant to *R. solani* infection on maturity of first true leaf. Flood irrigation in flat bed system spreads the pathogen from one bed to other. However, infection of seedlings with *Phytophthora* spp. in primary nursery beds perpetuates and causes further losses to seedlings in secondary nursery beds. The budded plants show stunted, chlorotic growth with development of poor feeder roots.

ii) Management: Sanitary measures to exclude the pathogen from seeds and soil are of utmost importance to avoid disease establishment. The seeds should be extracted from healthy fruits picked up from resistant rootstock tree from more than 1.5 m height to avoid *Phytophthora* infection. Seeds should be immersed in agitating hot water maintained at 51.7°C for 10 min to disinfect *Phytophthora* spp. For storage, seed should be treated with 8-hydroxyquinoline for 3 min. Raised seed beds should be either solarized or fumigated only with available fumigants like basamid (dazomet) granules at 50-60 g/cubic metre well in advance. Nursery should be away from citrus plantation and operational equipments for nursery should be separate to avoid introduction of pathogens. To avoid the risk of introduction of pathogens through wind, soil and workers in field nurseries, seedlings should be raised in portrays in greenhouse. It is advisable to use metalaxyl spray just after seed sowing to avoid *Phytophthora* infection, if any, present in trays or nursery beds.

2.2.2. Foot Rot, Root Rot, Crown Rot, Gummosis, Leaf Fall and Fruit Rot, *Phytophthora palmivora, P. nicotianae* var. *parasitica*

The disease seems to occur especially in the high rainfall areas. Its prevalence has been reported in south India, Maharashtra, Gujarat, Punjab and Assam states. *Phytophthora nicotianae* var. *parasitica* is widespread in Assam, while *P. palmivora* is prevalent throughout India.

i) Symptoms: Profuse gumming on the surface of the attacked bark is the main symptom (Fig. 2.5). When gumming occurs on the stem, droplets of gum trickle down the stem. The bark gradually turns brown to dark brown and develops longitudinal cracks. A thin layer of wood tissue is also affected. When gumming starts close to the soil, the disease spreads to the main roots and then around the base of the trunk. As a result of severe gumming, the bark becomes completely rotten and the tree dies owing to girdling effect. The trees usually blossom heavily and die after the fruits mature. In such cases, the disease is called as foot rot or collar rot. The pathogen produces symptoms of decline through rotting of the rootlets, girdling of the trunk and dropping of the blighted leaves. The fruits lying on the ground are liable to invasion by the pathogen and develop brown rot.

Fig. 2.5. Gummosis on main stem, foot rot and decline of Nagpur mandarin tree (Courtesy: A.K. Das, NRCC, Nagpur).

Leaf fall and fruit rot phase of the disease is severe on mandarin oranges in heavy rainfall areas of South India. Quick shedding of leaves is the earliest symptom. The infection starts as water-soaked lesions at the leaf base. By the time the lesions extend to the whole leaf, the leaf drop off. The infection may spread to young twigs and fruits of all stages. The affected fruits show water-soaked patches on rind and subsequently drop off and rot.

ii) Epidemiology: Severe occurrence of the disease is noticed in sweet oranges, acid lime and lemon. Heavy soil, high water table, high soil moisture, soil pH of 5.4 to 7.5 and temperature of 25° to 28°C are conducive for disease development. Low grafting, deep planting and nearness of bud union to ground level increase the chances for soil-borne infection. The fungi survive on fallen fruits, twigs, leaves and in cracks of the tree and spread by irrigation water, rain splashes, wind and insects to stems, leaves and fruits.

iii) Management

a) Cultural Methods: Planting material should be selected from *Phytophthora*-free certified nurseries and with high budding (above 25 cm height from soil level). While planting, care should be taken to keep bud union as high as possible from the soil level so that irrigation water should not touch the scion. Soil should be kept well drained. Double ring method should be adopted for irrigation and fertilization to keep the trunk of the tree and the area around the trunk as far as possible dry. Flood irrigation and stagnation of water for longer periods in the basin should be avoided. Injuries to trunk and root system by farm operations should be avoided (Naqvi, 2001).

b) Chemical Methods: Copper fungicides are effective in controlling foot rot and gummosis of citrus if used at the correct time. Bordeaux paste should always be applied before the onset of monsoon and after the monsoon on tree trunk as prophylactic measure. Foot rot or gummosis affected portions should be scraped out with sharp knife taking care not to damage the wood before application. Generally, citrus growers in Central India apply Bordeaux paste in winter season after harvesting the crop.

Availability of systemic fungicides like metalaxyl (Ridomil MZ 72) and fosetyl AL (Aliette) has given the citrus growers additional and more effective options for the management of *Phytophthora* diseases. Due to systemic in nature and long term persistence of these fungicides in plant system, the foliar spray and soil drench treatments or pasting the affected trunks are effective for disease management. Foliar sprays + soil drench of 0.2% a.i. metalaxyl or foliar sprays of 0.2% a.i. fosetyl at 45 days interval before onset of monsoon were effective for the management of gummosis, foot and root rot diseases and in increasing the feeder root density by more than 50% of the treated plants (Naqvi, 1994).

c) Biological Methods: Certain strains of *Pseudomonas putida* were found to reduce *Phytophthora* population up to 96% in greenhouse. This bioagent when tried in field with seedlings reduced 41-66% population of *P. citrophthora* and controlled 22% root infection for 3 months (Menge, 1993). *Trichoderma harzianum* has been advocated to have a potent antagonistic action against *Phytophthora* root rot of Coorg mandarin when applied along with coffee waste, poultry manure and FYM (Sawant and Sawant, 1989). *Myrothecium roridum* also inhibited *Phytophthora* infection in citrus (Tuset *et al.*, 1990).

d) Resistant Rootstocks: Citrus rootstocks such as Swingle Citrumelo, Trifoliate Orange USA, Shaub Rough Lemon, Jallandari Katti, Methi Tulia, Sun Chu Sha, Severinia and Cleopatra Coorg were found to be tolerant against root rot caused by *P. nicotianae*, while Rangpur Abohar, *C. macrophylla*, Souranthan, Rangpur Lime Kirumakki, Sour Orange, *C. volkameriana* and Rangpur Lime Poona were moderately resistant/ susceptible (Naqvi, 2002).

e) Integrated Methods

- Use of disease-free planting material, avoiding water stagnation in contact with susceptible portions of the bark above the bud union by providing good drainage, avoiding excess irrigation, collection and burning of diseased leaves and fruits and avoiding injury to plants reduce the disease incidence.
- Adopt ring system of irrigation by heaping soil around the base of plants to prevent direct contact of water with the trunk.

- Use of resistant rootstocks such as Trifoliate orange, Troyer citrange, Rangpur lime and Cleopatra mandarin is effective.
- Scrape off the affected parts and apply Bordeaux paste. Drench the soil around the tree covering 1 m diameter area with 20 litres of 1% Bordeaux mixture/tree at an interval of 1 month from July to September. Healthy trees should be protected by pasting the trunk with Bordeaux paste up to a height of 50-75 cm above the ground level once in a year before onset of monsoon.
- Spraying with 1% Bordeaux mixture alone or with tin sulphate or 0.3% difolatan or 0.2% metalaxyl-MZ or 0.1% fosetyl-AL (Aliette) is effective in the management of the disease.
- Soil application of *Trichoderma* spp. along with organic matter in the ratio of 1: 40 at 2 kg/plant or FYM enriched with *Trichoderma* spp. at 2 kg/plant and application of the mycorrhiza, *Glomus fasciculatum* is effective against the pathogen.

2.2.3. *Fusarium* Root Rot and Dry Rot, *Fusarium solani*

The destruction of root system of citrus by *Fusarium solani* is under estimated when compared with the diseases caused by *Phytophthora* spp. The disease has been reported to cause serious damage to citrus in North Arcot district of Tamil Nadu and Chittoor and surrounding areas of Andhra Pradesh. The disease causes sporadic losses in Central India, North Eastern Region and in Nepal (Ghosh and Singh, 1993).

i) Symptoms: Dry root rot affects larger roots and trunk below the bud union without oozing any gum. It is characterized by moist decay of bark in early stages and dry shredded cankers at later stage with hard, dried dark brown to grey colour of wood. The disease may progress in wood for many years with the only symptom being a slight wilting under dry conditions. Chronic rotting of feeder roots and scaffold roots by *F. solani* is often associated with gradual decline of the canopy. The affected feeder roots are characterized by cortical sloughing exposing the light coloured stele in the centre of the root. Canopy symptoms include wilting (leaf curl), defoliation and die-back of twigs. The leaves turn dull green in contrast to the lush green appearance of leaves on healthy trees.

ii) Management

a) Cultural Methods: Faulty cultural practices like damaging roots and trunk during farm operations, water logging in orchard favour the disease development and should be avoided. At early stage of the disease, removal of one or two infected roots when tap root and crown are still in good condition can control further disease development. Providing good drainage and soil aeration helps in reducing the incidence of the disease. Soil solarization also helps in the management of the disease.

b) Chemical Methods: Soil treatment with benzimidazole fungicides provides effective management of dry root rot.

c) Biological Methods: Biocontrol with *Trichoderma* spp. and avirulent strains of *F. oxysporum* would be long term solutions for the disease.

2.2.4. Powdery Mildew, *Acrosporium tingitanium* (Syn. *Oidium tingitanium*)

The powdery mildew disease is a major cause for decline and death of mandarin plantations in Jumpui hills of Tripura and Mezoram states. The Khasi mandarin plantations from a number of locations in Jumpui hills have virtually vanished due to repeated attack of the pathogen on new flushes. The disease is prevalent in India, Nepal, Sri Lanka, Java and Philippines on almost all citrus cultivars. In India, it is distributed in sub-mountain tracts of Kodagu, Wynad and Shevroy hills in Southern India causing severe losses to mandarin and sweet orange plantations. It has also been reported from North Eastern states, Darjeeling, Sikkim and Bhutan. In Southern India, the disease is severe from October to March.

i) Symptoms: The pathogen attacks all aerial parts of the citrus plants in nurseries and orchards. White powdery patches of mildew appear on upper part of young leaves and on twigs (Fig. 2.6). The leaf tissue at infection court in initial stages of infection turns darker watery green than normal and later becomes yellow. In severe cases, infected leaves and premature fruits drop off and twigs show die-back symptoms. The attack of powdery mildew drastically reduces the vigour and yield of the plants.

Fig. 2.6. Powdery mildew on citrus leaves (Courtesy: K. Srinivasan, TAFE, Chennai).

ii) Epidemiology: High humidity and cloudy weather favour the disease development and spread. The fungus over winters as dormant mycelium in the buds and with the onset of favourable conditions it multiplies and causes the disease. The conidia are formed in chains and disseminate by wind.

iii) Management

a) Cultural Methods: Water shoots should be pruned regularly which are supposed to catch early infection.

b) Chemical Methods: The disease can be effectively controlled by spray applications of tridemorph, triadimefon, dinocap and benzimidazole fungicides at first sign of attack and should be repeated at 10-day interval (Narasimhan *et al.*, 1984). Dusting of sulphur at 8-day interval is also recommended for the control of the disease effectively.

2.2.5. Anthracnose, *Colletotrichum gloeosporoides*

Anthracnose, also known as wither tip, in recent years is responsible for decline in lime production in north India. The disease is distributed in Punjab, Maharashtra, Madhya Pradesh, Uttar Pradesh, Assam, Rajasthan and South India.

i) Symptoms: The pathogen causes leaf spot and fruit rot (Fig. 2.7). Wither tip is characterized by shedding of leaves and die-back of twigs. Leaves show light green spots which turn brown. On dead twigs, black dot-like acervuli appear in concentric rings. The stem end infection of immature fruit result in fruit drop. In severe cases, branches show die-back and the tree dies in few years. Dead parts of the twigs assume silver gray appearance. Malta oranges and grapefruits suffer much more damage.

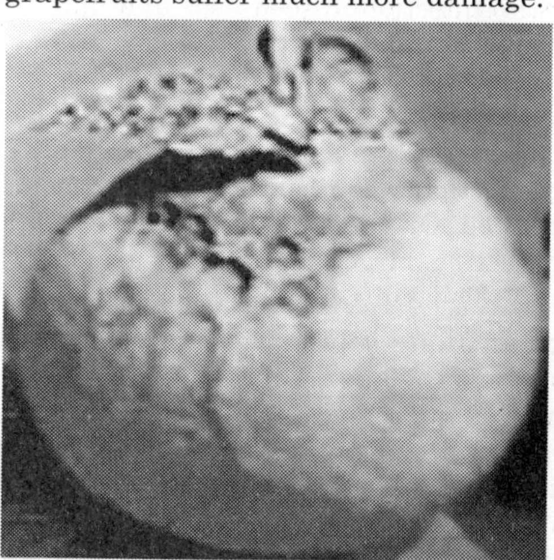

Fig. 2.7. Anthracnose on citrus (Courtesy: K. Srinivasan, TAFE, Chennai).

ii) Epidemiology: *Colletotrichum gloeosporoides* remains in a dormant stage in the dead twigs and branches. This leads to the twig blight. The infection hyphae produced by appresoria remains latent even after the fruit matures and produces anthracnose in oranges and grapefruits if the peel is injured or fruits are over matured. The conidia of the fungus are produced on dead twigs of the mother plant and dispersed by rain splashes to developing fruits.

iii) Management

- The disease can be minimized by proper irrigation, manuring, cultural practices and plant protection measures.
- Dead twigs should be pruned and destroyed during winter and sprayed with 1% Bordeaux mixture in February, March and September to reduce fruit drop. Cut ends should be protected with Bordeaux paste.
- Trees should be sprayed with 0.1% bavistin or 0.2% captafol 3 times after pruning. Prochloraz and benomyl gave the best results in controlling this disease during storage in Coorg mandarin (Ullasa and Rawal, 1983).

2.2.6. Scab, *Elsinoe fawcetti*

Scab is widely prevalent on a number of citrus cultivars in many countries where temperature and rainfall conditions are favourable for infection and causing considerable damage in nurseries and orchards. It is one of the common diseases of sour orange, rough lemon, tangelo and pummelo. It has also been recorded on sweet oranges from Karnataka (Rawal and Ullasa, 1988). The disease is distributed in Andhra Pradesh, Assam, Karnataka, Uttar Pradesh and West Bengal. Generally it is prevalent in nursery. In Assam, the scab caused up to 55% spoilage of citrus fruits.

i) Symptoms: Scab infection in seed-beds and nurseries causes stunting of seedlings and wrinkles and distortion of leaves. The pathogen attacks leaves, twigs and fruits. The lesions on leaves become sharply defined pustular elevations generally on the underside. The opposite surface corresponding to the warty growth shows a circular depression with a pink to red centre. The twig infection also results in similar lesions and the affected twigs are ultimately killed. On the fruits scabby or corky lesions develop and they soon coalesce affecting larger areas on the fruits. Infection on fruits deteriorates the quality. The fruits may be dropped.

ii) Epidemiology: Young leaves are highly susceptible. Fruits remain susceptible up to three months after fruit set. Wet weather for 2.0 to 2.5 hr is essential for germination of conidia and fresh infection. The conidia are formed at temperatures ranging from 7 to 30°C with 66 to 100 % RH. The pathogen infects tissues only when the surface is wet at optimum temperature of 16 to 23°C. The disease is severe under low temperature and high humidity conditions. The pathogen perpetuates and survives in off-season in scab pustules as ascospores. The fungus spreads through conidia.

iii) Management

a) Cultural Methods: Infected nurseries become primary source of inoculum to any new area. Disease-free plants should be selected to new plantation to avoid introduction of inoculum. Overhead sprinklers should not be used. The diseased leaves, twigs and fruits should be collected and destroyed to minimize the inoculum load and reinfection.

b) Chemical Methods: The disease coincides with the emergence of new flush. Prophylactic sprays of captafol, benomyl, methyl thiophanate and carbendazim during new flush provide satisfactory control of the disease. Copper fungicides if applied as protectants after bloom gives good control.

c) Integrated Methods

- The diseased leaves, twigs and fruits should be collected and destroyed.
- Spraying with 1% Bordeaux mixture gives effective control.
- Spraying with 0.3% copper oxychloride or 0.2% difolatan or 0.2% chlorothalanil at 15 days interval is effective in the management of the disease.

2.2.7. Pink Disease, *Pellicularia salmonicolor* (Syn. *Corticium salmonicolor*)

The disease is very destructive on all citrus cultivars in tropical high rainfall areas. The infected trunk, limbs and twigs turn pink in colour due to pink colour of mycelial covering. Hence, the disease is named 'Pink disease'. In India, it causes severe damage to citrus plantations in Southern zone and North Eastern zone

i) Symptoms: The disease appears during or just after rains. The bark of trunk and limbs of mature trees is attacked in humid conditions. The infected portion is rapidly covered with white silvery mycelium, which later turns to pink. The infected portion of bark is killed and fungus invades into the wood to impede translocation in vascular tissue. The infected plants die when complete girdling of trunk and limbs takes place. The pathogen also attacks surrounding plantations of rubber, cacao, coffee, mango, jackfruit, etc.

ii) Epidemiology: The fungus perpetuates in the callus formed around the decayed bark during dry season. In wet season, all the stages (mycelium, pustules, stromatic tissue, and spores) are capable to spread the disease. The formation of basidiospores has been reported on collateral hosts but not on citrus.

iii) Management

a) Cultural Methods: The infected and damaged portions of plants should be pruned and burnt. This operation should also be done on other collateral hosts near the orchard. Cut ends and scrapped portion of plants should be covered with Bordeaux paste.

b) Chemical Methods: Benzimidazole fungicides may be sprayed in field to control the disease.

2.2.8. Soft Rot, *Aspergillus niger*

i) Symptoms: *Aspergillus* rot is an important post-harvest disease of citrus in India. *A. niger* invades through wounds/injuries and causes soft rot. The pathogen infects lemon, lime, sweet orange and mandarin. The disease has been reported on citrus from Americas, South Africa, Mediterranean countries and most particularly in India (Snowdon, 1990). Pre-harvest infection may also occur if the fruit is injured by thorns in case of acid lime during windy weather. The infected area of the fruit turn pale yellow with characteristic halo leaving the shrink intact cuticle on the macerated peel. In advanced stage, infected surface is covered with black mass of spores that disseminate to contaminate the whole lot. The losses due to this fungus were recorded 4-6% in certain consignments of Nagpur mandarin sent from Nagpur to New Delhi during 1991-92. Nagpur mandarin was found comparatively more susceptible to *Aspergillus* rot in green stage than colour break or orange coloured fruit (Naqvi, 1992).

ii) Management: The rot can be minimized by avoiding the injury to fruits, sorting and discarding the infected fruits and by storing fruits at low temperature in baskets or in open containers. In lemons treated with *Bacillus subtilis* (AF 1) cells, only 3-4% fruits showed development of soft rot in comparison to 100% in control.

2.2.9. *Pencillium* Rot, *Pencillium digitatum, P. italicum*

Green mold (*P. digitatum*) and blue mold (*P. italicum*) decay are important post-harvest diseases which occur in all citrus growing areas and often constitute the predominant type of decay (Gardner *et al.*, 1986). Post-harvest losses of citrus fruit caused by these pathogens can account for more than 90% of all post-harvest losses in semi-arid production areas of the world (Eckert and Eaks, 1989).

i) Symptoms: A soft water-soaked area is developed at the infection site in both the diseases. Coloured spore mass is developed at the centre of the lesion surrounded by broad band of white mycelial growth in green mold infection, whereas white mycelial growth around the spore mass of blue mold is usually not more than 2 mm wide. Both the pathogens occur frequently, but green mold grows faster at moderate temperature and contaminates the fruit. The spores of green mold are unable to infect healthy uninjured adjacent fruits while the blue mold develops nesting onto uninjured healthy fruits and may cause serious damage.

ii) Epidemiology: The spores of the pathogens are air-borne. From the infected fruit in the packing case, the infection spreads to other fruits through wounds or through stem ends but also via sound part of the rind. Stem end was the common entry point for both the pathogens. Oil glands of citrus fruits can be a favourable avenue for infection. Fruits weakened by long cold storage decay more rapidly than fresh fruits.

iii) Management: Dipping of the fruits either in 2% bleaching powder for 5 min or in 0.2% Flit 406 for 10 min and storage at 2-4°C was effective against both the pathogens (Tandon, 1961). Treatment of fruits with 0.1% TBZ incorporated in Waxol-0-12 was effective under refrigeration (Subramanyam *et al.*, 1973). Boric acid at 4% was also effective against green mold. Imazalil WP and EC formulations were highly effective and showed a decay reduction index of 76.2 to 96.7 during a storage period of 36 days.

2.2.10. Diplodia Stem End Rot, *Diplodia natalensis*

i) Symptoms: This is one of the most important fungal rots in the packing houses and transit. The earliest symptom is softening of the rind around the bottom. The affected area increases rapidly and becomes brown. The rot may develop at the styler end as a result of rapid progression of internal decay. The tendency of decay at both ends is a characteristic symptom of *Dipolodia* stem end rot.

Peak decay occurs 2-3 weeks after harvest. *Dipolodia* is a major cause of stem end rot in fruits degreened with ethylene.

ii) Management: Post-harvest treatment of fruits with 6-8% borax solution or 300-400 ppm microstatin is recommended for the management of the disease.

2.2.11. *Phomopsis* Stem End Rot, *Phomopsis citri*

i) Symptoms: This is also one of the most important fungal rots (along with *Dipolodia* stem end rot) in the packing houses and transit. Fruits may decay at any time during the 21 days period from harvest to consumption. The rot starts around the bottom, producing at first a slight softening of the rind. The affected area increases rapidly and the fruit develops a pliable tan or brown rot. Occasionally, fruits may be attacked while still on the tree but usually when they are injured or overripe. After harvest, the fungus enters the abscission area. Symptoms of infection can be seen on the rind 2-4 days after the fungus has gained entry.

ii) Management : The disease can be checked in the packing house by fungicidal treatment of fruits with thiobendazole or diphenyl. A substantial reduction in the development of stem end rot can be obtained by pre-harvest spray with benomyl.

2.2.12. Brown Rot, *Phytophthora nicotianae*

This is a severe disease of mandarin oranges in heavy rainfall areas of Kodagu, Wynad and parts of Pulney hills between May and October.

i) Symptoms: The pathogen becomes active soon after the start of the monsoon on fruits of all stages. The affected fruits show water-soaked patches on rind which subsequently drop off and rot. The surface of the dropped fruits gets covered by cottony fungal growth and rotten fruits emit foul odour. The mycelium of the fungus survives within the infected stem and spreads through sporangia and zoospores. Apparently healthy fruits carry infection and develop fruit rot during storage.

ii) Management: A single application of 1% Bordeaux mixture can prevent the onset of the disease but should be repeated if the symptoms are observed during August or September.

2.2.13. *Alternaria* Fruit Rot, *Alternaria citri*

Black rot has been noticed in Punjab, West Bengal, Maharashtra, Rajasthan and Madhya Pradesh.

i) Symptoms: On naturally infected fruits, small, discoloured spots appear which turn brown covering an area of 2-3 cm. Sometimes, the entire fruit may be girdled and the fruit may drop. The temperature range for development of the disease is between 15° and 35°C with a peak between 25° and 35°C (Agarwal and Hasija, 1967). External injury to rind of Malta orange was necessary for infection and disease development which increased with the extent of injury (Chand *et al.*, 1967). The fungus reduced total soluble solids and reducing sugars and increased acidity in the affected fruits (Cheema *et al.*, 1976).

ii) Management: Phaltan at 5 ppm and Bordeaux mixture (4-4-50) were effective against the fruit rot. Karathane and Kirti-copper each at 0.1% as pre-harvest spray can also check the disease development (Chand *et al.*, 1967).

Management of Post-harvest Diseases

a) Pre-harvest Sprays: *Colletotrichum gloeosporoides, Pencillium citri* and *Botryodiplodia theobromae* are the major pathogens responsible for twig blight, anthracnose, flower and fruit infections in citrus orchards. The dried/ dead twigs on the bearing trees act as reservoir of inoculum for infection of developing fruits. The spores dispersed by rain splash infect the fruits from fruit set to harvest stage leading to pre-harvest fruit drop and incipient pre-harvest infections (Eckert and Ogawa, 1985; Naqvi, 1993a, 1999a). Pruning of dead twigs from citrus trees (Naqvi, 1993a) and removing fallen fruits from the grove reduce the inoculum load of pre-harvest pathogens responsible for post-harvest fruit rots. The inoculum load and pre-harvest infection of citrus can effectively be minimized by three pre-harvest sprays of benzimidazole fungicides at 15 days interval prior to harvest. These pre-harvest sprays could reduce pre-harvest fruit drop of Nagpur mandarin to 9.94% as against 22.26% in control (Naqvi, 1993a,b) and also control more than 70% fruit decay after harvest up to three weeks of storage at ambient conditions (Naqvi, 1997).

b) Post-harvest Treatments: Post-harvest fungicide treatment through washing, drenching and waxing either manually or on mechanical packing line, the losses up to 80% can be curtailed for three weeks storage/ transport/marketing of Nagpur mandarin fruits at ambient conditions without any hazardous residual effect and quality deterioration. The losses from 18-24% could be brought down to 3-4% or even less when fruits were handled smoothly and packed in well cushioned adequately ventilated corrugated boxes. Under refrigerated conditions, such treated fruits can be stored up to 40-45 days. Reduced post-harvest decay and enhanced shelf life fetch about 27% additional income to the grower/trader (Naqvi, 1999b). This technology can be used to exploit the export potential of citrus.

Post-harvest treatments like washing, clean water rinse, uniform fungicide application, drying, grading and packing can be performed efficiently on mechanical packing line. After removal of trashes on conveyor belt, the fruits are washed within 1-2 days after harvest with soda ash (4-6% Sodium carbonate) to inactivate the spore load on fruits. The soda ash washing cleans up to 70% spore load of *Pencillium* from the fruit surface (Smilanick and Margosan, 1996). The fruits are further rinsed in clean water to remove the soda ash residues. The excess water is removed with the help of spongy rollers. The cleaned fruits are further treated with wax and fungicides either in the form of foam or through spray. A uniform wax and fungicide coating on fruit surface can easily be done through specially designed brushes on roller which helps in the control of post-harvest infections and weight loss of fruits. Wax application also provides an attractive shine to the treated fruits. This coating is immediately dried in drying tunnel and then fruits are graded by mechanical or electronic graders. The graded fruits can be packed manually or mechanically as per their grades. The fruits are packed in telescopic corrugated boxes with the information about the produce printed on the boxes for distant transport/ storage/ marketing.

Table 2.9. Calendar of IPM activities in citrus

Month/s	Target pest	IPM activities
January	Gummosis	Scrape out gum oozing portion with a sharp knife from the infected tree trunk and apply metalaxyl or fosetyl-AL paste.
March	*Phytophthora* foot rot	Spray 2.5 g metalaxyl MZ 72 or 2.5 g fosetyl AL in 1 litre of water on the tree covering the whole plant canopy.
April	Powdery mildew	The dead wood on trees should be pruned about 2 cm below the dead portion followed by spraying with 1 g carbendazim in 1 litre of water.
May	*Phytophthora* foot rot	Apply fresh Bordeaux paste on the tree trunk up to the height of 60 cm by brush. Spray and drench with 2.5 g metalaxyl MZ 72 or 2.5 g fosetyl AL in 1 litre of water on the tree covering the whole plant canopy.
	Gummosis	Scrape out gum oozing portion with a sharp knife from the infected tree trunk and apply metalaxyl or fosetyl-AL paste.
June	*Phytophthora* foot rot	Apply fresh Bordeaux paste on the tree trunk up to the height of 60 cm by brush.
	Canker	The affected leaves and branches should be pruned and burnt. Spray 200 g copper oxychloride along with 6 g streptocycline in 60 litres of water before the onset of monsoon. Repeat the spray 3-4 times at interval of 30 days.

July	Canker	The affected leaves and branches should be pruned and burnt. Spray 200 g copper oxychloride along with 6 g streptocycline in 60 litres of water before the onset of monsoon. Repeat the spray 3-4 times at interval of 30 days.
August	Canker	The affected leaves and branches should be pruned and burnt. Spray 200 g copper oxychloride along with 6 g streptocycline in 60 litres of water before the onset of monsoon. Repeat the spray 3-4 times at interval of 30 days.
September	Canker	The affected leaves and branches should be pruned and burnt. Spray 200 g copper oxychloride along with 6 g streptocycline in 60 litres of water before the onset of monsoon. Repeat the spray 3-4 times at interval of 30 days.
October	*Phytophthora* foot rot	Apply fresh Bordeaux paste on the tree trunk up to the height of 60 cm by brush. Spray with 2.5 g metalaxyl MZ 72 or 2.5 g fosetyl AL in 1 litre of water on the tree covering the whole plant canopy.
	Powdery mildew	The dead wood on trees should be pruned about 2 cm below the dead portion followed by spraying with 1 g carbendazim in 1 litre of water.
November	Gummosis	Scrape out gum oozing portion with a sharp knife from the infected tree trunk and apply metalaxyl or fosetyl-AL paste
December	Powdery mildew	The dead wood on trees should be pruned about 2 cm below the dead portion followed by spraying with 1 g carbendazim in 1 litre of water.
	Gummosis	Scrape out gum oozing portion with a sharp knife from the infected tree trunk and apply metalaxyl or fosetyl-AL paste.

2.3. SAPOTA

It is cultivated in 0.140 million ha with annual production of 1.117 million tonnes (Bijay Kumar, 2007) (Table 2.10).

Table 2.10. Area, production and productivity of leading sapota growing states of India (2005-06) (Bijay Kumar, 2007)

States	Area (in '000 ha)	Production (in '000 MT)	Productivity (in MT/ha)
Gujarat	62.5	255.8	4.1
Maharashtra	22.8	249.8	11.0
Karnataka	24.9	235.7	9.5
Tamil Nadu	6.0	150.3	25.0
Andhra Pradesh	15.4	153.6	10.0
West Bengal	4.2	49.0	11.8
Orissa	3.3	14.3	4.4

Haryana	0.5	3.7	7.4
Tripura	0.1	1.4	11.3
Pondicherry	0.1	1.0	10.0
Others	0.3	2.5	8.3
Total	140.1	1117.1	8.0

2.3.1. Leaf Spot, *Phaeophleospora indica*

This disease was first noticed in 1968 at Dharwad, North Karnataka. The disease is widespread and serious in South Karnataka also.

i) Symptoms: The pathogen produces numerous small, circular, pinkish to reddish brown, conspicuous leaf spots with whitish centres on maturity (Fig. 2.8). The diseased leaves turn yellow and shed prematurely. An appreciable reduction in yield due to defoliation has also been reported.

Fig. 2.8. Leaf spot on sapota (Courtesy: K. Srinivasan, TAFE, Chennai).

ii) Epidemiology: The maximum disease incidence in Karnataka is observed during October to December when the humidity is high. The fungus grows best at 25°C and 90% relative humidity. If rainy season extends, the defoliation may occur in August and continue up to December.

iii) Management

- Mulching of defoliated leaves followed by spraying with fungicides is more effective than spraying alone.
- Monthly spraying of 0.2% Diathane Z-78 or 0.3% Lonacol followed by 0.5% Blitox and 0.2% Cuman L proved effective (Sohi and Sridhar, 1973).
- Growing of susceptible sapota cvs. like Cricket Ball and Kinhabarthi should be avoided in new plantations.

2.3.2. Leaf Spot, *Glomerella cingulata*

The disease is reported from Karnataka.

i) Symptoms: The disease is manifested by small irregular spots on both leaf surfaces. The centre of the spot turns grayish white to dark olive with numerous acervuli of the pathogen. In severe cases, the leaves turn yellow and drop.

ii) Management

- Spraying of 0.1% carbendazim or 0.2% Topsin-M or 0.2% mancozeb or 0.2% chlorothalanil at 3 weekly intervals controlled the disease.

2.3.3. Leaf Blight, *Pestalotiopsis* spp.

The disease is caused by *Pestalotiopsis versicolor* in Kerala, by *P. scirrofaciens* in Karnataka and by *Fusicoccum sapoticola* in Maharashtra.

i) Symptoms: The disease first appears as numerous small, reddish brown specks on leaf lamina which gradually emerge to form more or less circular spots 1-3 mm in diameter. Fully mature spots have grayish centre and dark brown to reddish margins. Minute black fruiting bodies of the fungus are also seen in these spots.

ii) Management: Spraying with 0.2% zineb or 0.25% mancozeb plus 0.2% copper oxychloride at 3 weekly intervals was found effective.

2.3.4. Fruit Rot, *Phytophthora* sp.

i) Management: Clean the garden and burn the fallen leaves, fruits, etc. Drain out excess water. Spray 1% Bordeaux mixture.

2.3.5. Dry Root Rot/Wilt, *Fusarium solani*

i) Symptoms: Infected plants initially show yellowing of leaves. When the disease is severe, the infected stems produce brown to black colour lesions along with disintegration of tissues at soil level resulting in death of affected plants.

ii) Epidemiology: The disease is more severe during February to April when the dry humid weather conditions prevail.

iii) Management: Soil drenching with 0.1% carbendazim plus 0.3% copper oxychloride followed by soil application of *Trichoderma harzianum* helps in managing the disease.

2.3.6. Flat Limb, *Botryodiplodia theobromae*

i) Symptoms: Branches of affected trees become flattened and twisted showing rough marks (Fig. 2.9). Leaves become thin and yellow. Foliage and fruits tend to fall prematurely and fruits remain small and dry.

ii) Management: Spraying with 0.2% mancozeb helps in reducing the disease incidence. The spray should be repeated once in a fortnight on the leaves and fruits.

Fig. 2.9. Flat limb on sapota (Courtesy: K. Srinivasan, TAFE, Chennai).

2.4. PAPAYA

Papaya (*Carica papaya*) is one of the most important fruit crops valued for its rich source of vitamins and nutrient content. It is native to tropical America and its place of origin is said to be in Southern Mexico and Costa Rica. It is believed to have been introduced into India during the 16th century. It is grown both in tropical and sub-tropical countries of the world like Brazil, Australia, USA, Taiwan, Puerto Rico, Peru, Pakistan, Bangladesh, Malaysia, India and various parts of Central and South Africa. The ripe papaya fruits are used for table purpose, raw fruits are cooked and used as vegetable. Immature fruits are used for the extraction of proteolytic enzyme 'papain', which is mainly used as a drug in digestive disorders and in the treatment of ulcers and diphtheria. It has also varied industrial uses viz., in tanning industry, degumming of silk, etc.

India is the largest producer of papaya in the world. Papaya is grown in 73,000 ha with annual production of 2.317 million tonnes (Table 2.11). Productivity of papaya is the highest (31.7 t/ha) (Table 2.12) among the fruit crops, which has attracted the growers for its commercial cultivation resulting in the spectacular increase in area and production in last few decades. In the last decade, phenomenal increase in area and production is recorded from Southern states owing to demand of fruits for industrial use.

Table 2.11. Major papaya producing countries of the world (2005-05) (Bijay Kumar, 2007)

Country	Area (in '000 ha)	Production (in '000 MT)	Productivity (in MT/ha)	% share of world production
India	73	2317	31.7	28.8
Brazil	33	1574	48.3	19.5
Nigeria	99	834	8.4	10.4
Mexico	19	709	37.5	8.8

Indonesia	24	549	22.8	6.8
Ethiopia	13	259	20.5	3.2
Congo	13	216	17.3	2.7
Peru	14	171	12.5	2.1
Philippines	9	147	15.5	1.8
Colombia	6	138	24.7	1.7
Others	81	1142	14.0	14.2
Total	383	8055		

Table 2.12. Area, production and productivity of leading papaya growing states (2005-06) (Bijay Kumar, 2007)

States	Area (in '000 ha)	Production (in '000 MT)	Productivity (in MT/ha)
Andhra Pradesh	12.9	1009.2	78.2
Gujarat	7.7	323.0	41.8
West Bengal	9.5	263.7	27.7
Karnataka	3.5	230.1	66.1
Assam	8.1	130.5	16.1
Chhattisgarh	4.2	119.1	28.2
Kerala	18.9	85.5	4.5
Tamil Nadu	0.3	53.7	182.0
Madhya Pradesh	0.7	19.0	27.0
Others	7.3	83.4	11.4
Total	73.1	2317.2	31.7

On an average, 1000 g of ripe papaya fruit contains 2500 I.U. of Vitamin A, 8 I.U. of Vitamin B, 33 Sherman units of Vitamin B2 and 70 mg of Vitamin C. Besides, it contains papain, a proteolytic enzyme which acts as papsin and is useful in digestive disorders.

2.4.1. Damping-Off, *Pythium aphanidermatum, Phytophthora parasitica*

i) Management: Seedlings should be raised in well drained nursery soil. Diseased plants should be removed and burnt. Soil drenching with 1% Bordeaux mixture reduce the disease incidence. Soil treatment with neem cake along with *T. harzianum* gave good germination and seedling stand.

2.4.2. Stem Rot/Collar Rot/Foot Rot, *Pythium aphanidermatum*

The disease also occurs as root rot or collar rot. It is widespread in Africa, Hawaii, India and Sri Lanka. In India, it is a very serious disease prevalent throughout the country and may lead to complete failure of the crop if it appears in the early stages of plant growth.

i) Symptoms: The typical stem rot is most common in 2-3 year old trees. The symptoms appear as water-soaked patches on the stem at ground level which enlarge and girdle the base of the stem. The affected tissues turn brown, black and rot. The terminal leaves turn yellow, wilt and drop acropetally. Fruits, if formed, also shrivel and drop off. The entire plant topples over and dies because of the disintegration of parenchymatous tissues. The affected roots deteriorate. The pathogen also causes post-emergence damping-off.

ii) Epidemiology: The disease usually appears during rainy season and the severity depends upon the intensity of rainfall and temperature. One week old plants are more susceptible than one year old plants. The seedlings raised in the infected soil carry the pathogen to the field. Under favourable conditions, the disease is able to wipe off the entire plantation within one season and makes the soil unfit for planting. Optimum temperature of 36°C is favourable for the disease development. The fungus survives in the form of oospores in the soil.

iii) Management

- Maintain good drainage system in the orchard.
- Seed treatment with captafal at 3 g/kg reduces the damping off phase.
- Remove the diseased plants and destroy them.
- Apply Bordeaux paste on the affected plant parts. Spraying with 6-6-50 Bordeaux mixture or 0.3% copper oxychloride or 0.25% metalaxyl MZ or 0.2% captan with sticker on stems and drenching the soil around the stem repeated at 15-20 days interval gave good stand of the crop.
- Disease intensity was minimum in *Glomus fasciculatum* inoculated plants (6.25%) followed by *Sclorocystis dussii* (25%). Plant mortality was least in *G. fasciculatum* + vermicompost (25%) followed by *G. fasciculatum* (31.35%). The least infection was recorded in *G. fasciculatum* + vermicompost (7.01%) which was at par with *S. dussii* (7.74%) and *S. dussii* + vermicompost (8.27%).

2.4.3. Anthracnose, *Gloeosporium papayae (Colletotrichum gloeosporoides), Colletotrichum papayae*

The disease is prevalent in all papaya growing regions and becomes more prominent after reaching the consumer.

i) Symptoms: The spots on fruits first appear as dark brown superficial discolouration of the skin which develops into circular, slightly sunken areas of 1 to 3 mm in diameter (Fig. 2.10). The lesions gradually coalesce and produce sparse mycelium on the margins of the spots. Under humid conditions, an encrustation of salmon pink spores is released. Early stage infection of fruits results in mummification and deformation. Necrotic spots are also produced on leaves and stems. Erupted acervuli in concentric rings are seen on the petioles. Although the disease usually appears on the ripening fruits, but the pathogen occasionally infect in green stage causing small lesions. Lesions enlarge very slowly and rarely larger than 12 mm in diameter as long as the fruit remain green.

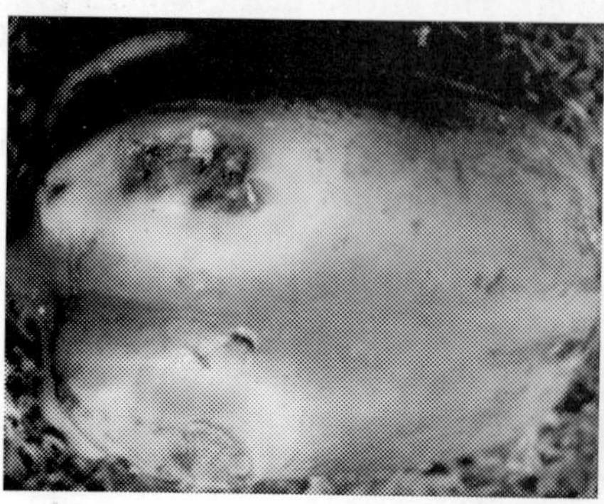
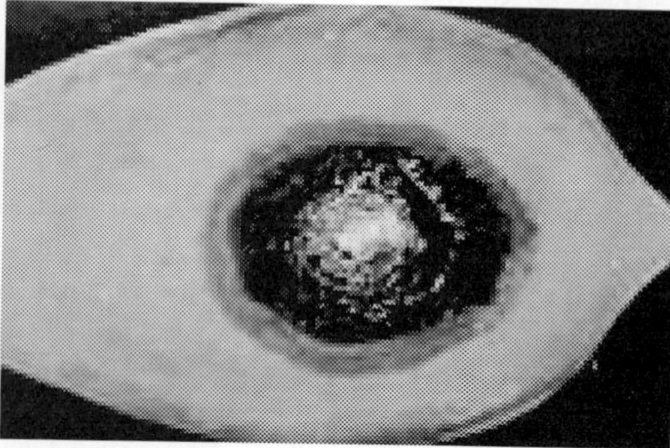

Fig. 2.10. Anthracnose on papaya fruit.

ii) Epidemiology: The severity of the disease on foliage is increased under conditions of excessive moisture. Old leaves are generally more susceptible, lesions develop more slowly on the immature fruits than on mature fruits. The incipient infection is carried by the fruits from the field. The disease is spread through wind-borne conidia and by rain splashes.

iii) Management

- Removal of infected leaves is essential for sanitation.
- Hot water treatment of fruits at 45°C for 30 min protects the fruits from the disease.
- Spraying of 0.25% mancozeb or 0.1% thiophanate methyl or 0.2% chlorothalanil at 10-15 days interval helps to manage the disease.
- Two sprays either with 0.1% carbendazim at 45 days interval or with 0.2% Daconil at 15 days interval were effective in reducing the disease incidence (Rawal *et al.*, 1983).

2.4.4. Powdery Mildew, *Oidium caricae*

The disease is distributed in Delhi, Maharashtra and Karnataka.

i) Symptoms: The fungal patches of whitish material grow superficially usually on the undersurface of the leaves drawing nutrients from the cells of the leaf through haustoria. Yellow or green pale blotches, mostly near the veins at the infection sites are observed on the upper surface of leaves. The fungal mycelium growing on the undersurface of leaves produces chains of spores which are carried by wind to the healthy leaves. These spores germinate and send haustoria into the leaf, develop a fungal vegetative body and complete the life cycle. The fungus is strictly an obligate parasite and produces spores on living plants. In severe cases, the fungus infects fruits also. The pathogen also attacks the stem of young seedlings when grown under reduced light conditions and produce typical powdery growth resulting in death of top portions of the seedlings.

ii) Epidemiology: The maximum disease incidence was recorded during September to November with a peak in October, under Bangalore conditions. The disease was more severe when the atmospheric temperature was between 16.4 and 22.9°C, relative humidity between 65 and 86% with 6.2 sunshine hours. The yield loss was directly proportionate to the disease incidence (Rawal, 1986).

iii) Management

- The disease is effectively controlled by spraying with 0.3% wettable sulphur at 10 days interval.
- The systemic fungicides such as Bayleton, carbendazim and methyl thiophanate all at 0.1% sprayed at monthly interval were more effective than other fungicides (Rawal *et al.*, 1981).

2.4.5. Blight, *Phytophthora nicotianae* var. *parasitica*

i) Symptoms: The pathogen infects the stem primarily in the region of fruit production at fruit or leaf scars. The infected area enlarges and often completely girdles the stem of young tree resulting in wilting of the top of the plant which eventually dies. The fruit may become infected at any stage while it hangs on the tree. It shrivels, turn dark brown and falls to the ground. These mummified fruits are a reservoir of the fungus and constitute a source of inoculum for further spread of the disease.

ii) Epidemiology: The fungus produces a whitish mass on rotting fruits and stems in which large quantities of sporangia are found. These sporangia are carried by rain splashes or by wind to the healthy parts of the plant where they germinate only in the presence of water and produce large number of motile swimming zoospores which infect plant and initiate the disease. These zoospores may invade uninjured leaf, stem or fruit tissues. As the fungus grows, it causes death of host cells and typical water-soaking rot or so called blight occurs. The optimum temperature for development of the disease is 27.8°C.

iii) Management

- Prompt and complete removal and destruction of infected plants and fruits from the orchard is very important.
- Spraying of mancozeb or Foltaf or Ridomil or Aliette or chlorothalanil all at 0.2% at fortnightly interval helps to control the aerial infections of the disease.

2.4.6. Watery Fruit Rot, *Rhizopus stolonifer*

i) Symptoms: The fruit rot is characterized by the production of irregular water-soaked lesions which gradually enlarge and get covered with white and dark brown fungus growth and sporangiospores (Fig. 2.11). The fruit become watery and emits a foul odour. Papaya orchard and storage room are the sources of inoculum. Infection occurs through wounds. The rot rapidly destroys the entire fruit and quickly spreads to the other fruits.

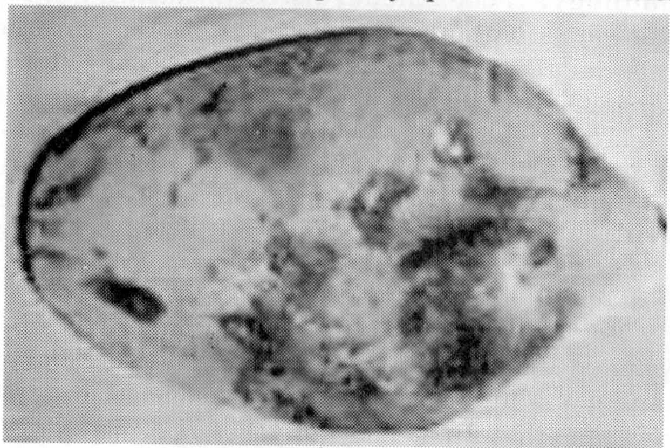

Fig. 2.11. Watery rot on papaya fruit (Courtesy: K. Srinivasan, TAFE, Chennai).

ii) Management: The disease can be checked by storing the fruits at 10°C or lower temperature. Hot water treatment of fruits at 49°C for 20 min can check the rot.

2.4.7. *Ascochyta* Fruit Rot, *Ascochyta caricae*

i) Symptoms: The pathogen causes both trunk and fruit rot (Fig. 2.12). The disease generally occurs on half grown or mature fruits which show small circular water-soaked spots. The lesions may become sunken and brownish black. The rot usually spreads irregularly outwards. It becomes most severe at 30°C and 100% humidity.

Fig. 2.12. *Ascochyta* blight on papaya (Courtesy: K. Srinivasan, TAFE, Chennai).

ii) Management: The disease can be controlled by spraying the fruits after set with 2: 2: 50 Bordeaux mixture at an interval of 21 to 30 days. The disease incidence can also be reduced by fruit dip in 2-aminobutane carbonated solution (1%) or by fumigation with 2- aminobutane.

2.4.8. *Phomopsis* Fruit Rot, *Phomopsis caricae*

i) Symptoms: The affected fruits develop a water-soaked spot which increases in size. The rotten area is depressed and cracks at a later stage. It turns dark brown to black. Fruits at all the stages of development can be infected by the pathogen.

ii) Management: Post-harvest treatment of fruits with Borax, benomyl, Captan, Difolatan and mancozeb was effective against the disease.

2.4.9. *Macrophomina* Fruit Rot, *Macrophomina phaseolina*

i) Symptoms: The fruit rot of papaya is responsible for 5-20% losses during summer. The disease is characterized by small, water-soaked spots which appear as circular specs on the fruit surface. The spots rapidly enlarge and become sunken. The pathogen advances deep into the fruit and causes rotting and disintegration (Fig. 2.13). The affected tissue hardens and becomes dotted with sclerotia of the pathogen. The affected fruits turn brown to black and the pulp becomes brownish black. Susceptibility increases with increase in size and maturity of fruit. Maximum decay of fruit occurs at 30°C and 100% relative humidity (Kapoor and Chohan, 1974b).

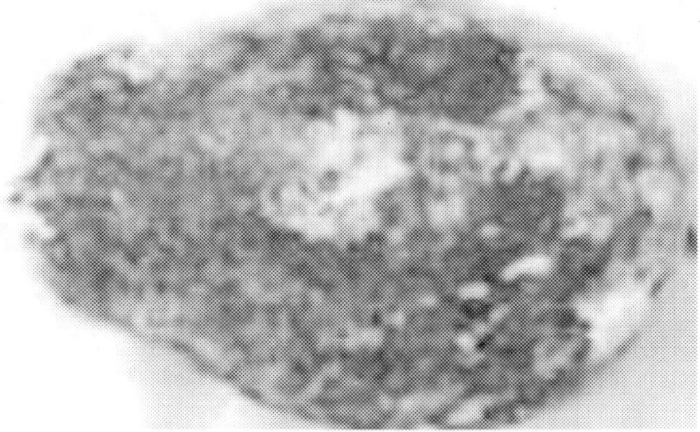

Fig. 2.13. *Macrophomina* fruit rot in papaya (Courtesy: K. Srinivasan, TAFE, Chennai).

ii) Management: Kapoor and Chohan (1974a) found that ziram, ferbam and captan were effective for the control of the disease.

2.4.10. *Fusarium* Fruit Rot, *Fusarium solani*

i) Symptoms: The pathogen causes post-harvest dry fruit rot of papaya in West Bengal. Affected fruits show small, stem end or surface lesions. Profuse, white hyphae are seen associated with the rot.

ii) Management: Hot water treatment of fruits at 49°C for 20 min gave good control of the disease (Pathak *et al.*, 1972).

2.4.11. *Botryodiplodia* Fruit Rot, *Botryodiplodia theobromae*

i) Symptoms: The fungus incite stem end rot and surface fruit rot (Fig. 2.14). This may be confused with *Ascochyta caricae*, but water-soaked margins are wider and softer with internal discolouration.

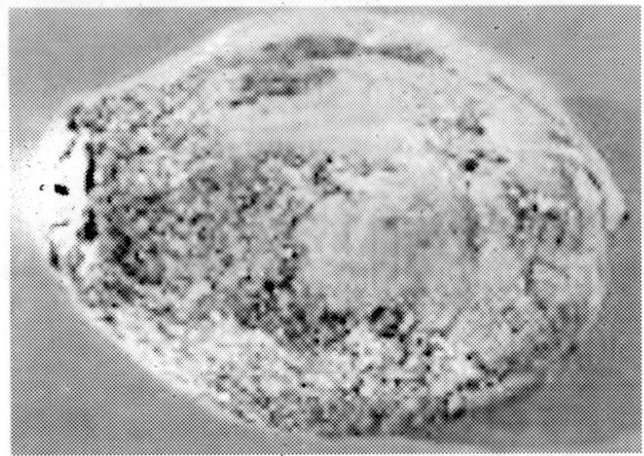

Fig. 2.14. *Botryodiplodia* fruit rot on papaya (Courtesy: K. Srinivasan, TAFE, Chennai).

ii) Management : Hot water treatment of fruits at 49°C for 20 min helps to keep the disease under check.

2.5. PINEAPPLE

Pineapple (*Ananas comosus*) is grown in 85,400 ha with annual production of 1.353 million tonnes (Tables 2.13 and 2.14). There has been phenomenal increase in production and productivity (15.8 MT/ha) due to adoption of improved production technologies especially in Karnataka, West Bengal and Kerala. However, productivity is still low in North Eastern States (8-10 MT/ha) owing to traditional methods of cultivation compared to West Bengal (26.38 MT/ha).

Table 2.13. Major pineapple producing countries of the world (2005-05) (Bijay Kumar, 2007)

Country	Area (in '000 ha)	Production (in '000 MT)	Productivity (in MT/ha)	% share of world production
Brazil	62	2292	37.1	12.7
Thailand	98	2183	22.2	12.1
Philippines	49	1788	36.3	9.9
Coasta Rica	27	1605	59.9	8.9
China	76	1442	19.0	8.0
India	85	1353	15.8	7.5
Indonesia	80	925	11.6	5.1

Nigeria	121	917	7.6	5.1
Kenya	14	600	42.0	3.3
Mexico	15	552	36.2	3.1
Others	268	4386	16.4	24.3
Total	896	18043		

Table 2.14. Area, production and productivity of leading pineapple growing states (2005-06) (Bijay Kumar, 2007)

States	Area (in '000 ha)	Production (in '000 MT)	Productivity (in MT/ha)
West Bengal	13.4	379.2	26.3
Assam	14.6	241.0	16.5
Manipur	2.6	134.5	52.1
Kerala	12.8	109.3	8.6
Bihar	4.2	108.0	25.5
Tripura	5.2	106.4	20.5
Karnataka	10.6	88.6	8.4
Meghalaya	9.5	83.7	8.8
Arunachal Pradesh	7.9	37.0	4.7
Nagaland	2.0	24.8	12.4
Tamil Nadu	0.7	21.7	33.2
Others	2.0	18.9	9.7
Total	85.4	1353.1	15.8

2.5.1. Base Rot, Fruit Rot and Leaf Spot, *Ceratocystis paradoxa*

The disease, also called as stem end rot, core rot or black rot, is important and quite common in Assam State. Market surveys have shown that the loss of fruits due to rottage may be as high as 10-15%.

i) Symptoms: The leaves show grey spots with dark margins which turn olive brown or white. As the disease advances, tissues dry up and leaves become distorted. Water-soaked lesions are observed on fruits which are yellowish but finally turn dark. The affected tissues rot (Fig. 2.15). The fungus invades through wounds or cut ends and spreads inter- or intra-cellularly. The fungus survives in the soil for long periods.

ii) Management

- For fresh planting, suckers should be selected from the disease-free fields.
- Disinfection of nursery soil with Brassicol or Vapam to kill the pathogen.
- Dipping of suckers in 1% Bordeaux mixture before planting has been advocated.

- Destruction of affected plants is foremost to check the disease.
- The stem end rot is checked by exposure of the harvested fruits to the sun for 2 hours.
- The cut end of the fruit stalk should be dipped in 10% benzoic acid in alcohol.
- Dipping of fruits in 1000 ppm thiobendazole or 2000 ppm benomyl for 5 min gave good protection against the disease (Sridhar, 1975).
- Dusting of stem ends with 0.1% benzoic acid in kaolin, packed in bamboo baskets effectively checked the rot during 4-5 days transit (Mallikarjunaradhya *et al.*, 1979). The treatment was found to be commercially feasible on account of low cost.
- The packing cases and baskets should be sprayed with 3% formalin solution.

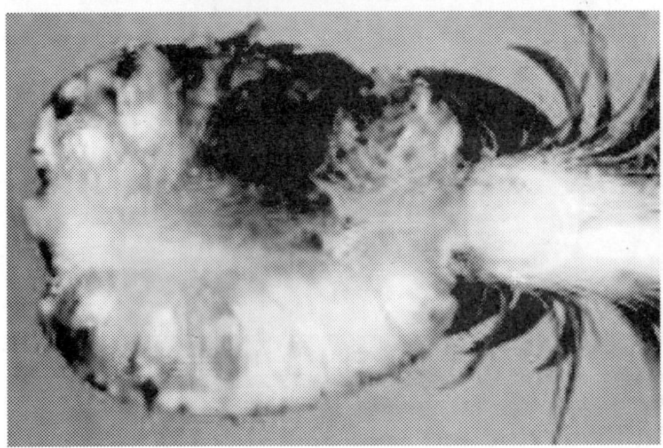

Fig. 2.15. Fruit rot of pineapple.

2.5.2. Heart or Stem Rot, *Phytophthora nicotianae* var. *parasitica*

i) Symptoms: The initial infection may appear at the apical end of the stem, at the base of the stem or at the roots. This results in loss of turgidity and a slight twisting of central leaves and in later stages by the withering and discolouration to yellowish pink and brown (Fig. 2.16). The inner whorl of leaves can be easily detached from the stem by slight pull. The stem may show brownish discolouration of the margins between healthy and diseased tissue. The fungus survives in the soil. The disease incidence is more in areas with poor drainage.

ii) Management

- Dipping of suckers in 1% Bordeaux mixture helps in reducing the disease incidence.
- Planting of suckers in well drained soil can minimize the disease incidence.

Fig. 2.16. Heart rot of pineapple (Courtesy: K. Srinivasan, TAFE, Chennai).

2.5.3. Anthracnose, *Colletotrichum ananas*

Anthracnose has been reported from Ratnagiri area of Maharashtra.

i) Symptoms: The infection appears in the form of irregular blotches with minute fruiting bodies that originate mostly from apex of the leaves. Sometimes marginal spotting is also seen. It results in withering and dropping of leaves and ultimately the growth of the infected culms and development of fruits is seriously affected. The disease is favoured by high humidity and takes about a fortnight to develop on young leaves. During June-July, black specks are observed on leaves which lead to withering.

ii) Management

- The disease can easily be checked by spraying carbendazim or Topsin-M both at 0.1% or 0.2% chlorothalanil.

2.5.4. Leaf Blotch, *Pestalotia microspora*

The disease was first reported by Rao and Mhaskar (1973) from Kodagu district of Karnataka State.

i) Symptoms: The disease manifests itself in the form of pale brown circular to irregular small necrotic lesions on the leaves mostly originating from the tip or central part which soon enlarge. These lesions later coalesce, turn dark brown, have rough surface and irregular outlines with slightly depressed centre. At this stage, the leaves turn slightly yellowish and exhibit blotch look. The infected areas soon develop numerous black sub-erumpent fructifications of the pathogen. The disease is favoured by prevailing cool humid weather. Oil palm, apple and guava are the alternate hosts for the pathogen. The temperature range of 25-28°C and RH of 75-100% are optimum conditions for sporulation of the fungus (Rao and Mhaskar, 1981).

ii) Management: The disease can be checked by spraying zineb or chlorothalanil both at 0.2% or 0.4% copper oxychloride.

2.6. JACKFRUIT

2.6.1. Leaf Spot, *Colletotrichum gloeosporoides*

i) Symptoms: The disease is characterized by dark brick red spots on both surfaces of leaves. In mature spots, the centres become grayish with erupted dark acervuli. The margins of such spots turn dark brown.

ii) Management: The disease can be effectively managed by spraying 0.1% carbendazim or Topsin M or 0.2% Foltaf.

2.6.2. Dieback, *Botryodiplodia theobromae*

i) Symptoms: This is the most destructive disease of the jackfruit. The onset of die-back symptoms becomes evident by discolouration and darkening of the bark at some distance from the tip. The dark area spreads and young green twigs start withering first at the base and then extending outwards along the veins of leaf edges. The affected leaves turn brown and their margins roll upwards. At this stage, the twig or branch dies, shrivels and drops. There may be exudation of gum from affected branches. Such branches have also been found to be affected by shoot borers. Infected twigs show internal discolouration. In early stages, epidermal and sub-epidermal cells of twigs appear slightly shriveled.

ii) Management: Pruning of infected twigs followed by spraying of 0.1% carbendazim or Topsin M or 0.2% chlorothalanil have been recommended for the management of the disease. Management of shoot borers and shot hole borers by suitable insecticides is also important.

2.6.3. Pink Disease, *Botryobasidium salmonicolor*

The disease is widespread in tropical and sub-tropical regions.

i) Symptoms: It appears as a pinkish powdery coating on the stem representing profuse conidial production. Young woody branches of the affected trees lose their leaves and show die-back.

ii) Management: The affected branches should be pruned and burnt and the cut ends should be pasted with Bordeaux paste or copper oxychloride.

2.7. STRAWBERRY

Strawberry (*Fragaria vesca*) is an attractive, luscious, tasty and nutritious fruit with a distinct and pleasant aroma and delicate flavour. It is rich in vitamin C and iron and is mainly consumed as fresh. Jam and syrup are also prepared from strawberry. It is cultivated in tropical and sub-tropical areas round the year. It is commercially cultivated in Himachal Pradesh, Uttar Pradesh, Maharashtra, West Bengal, Nilgiri Hills, Delhi, Haryana, Punjab and Rajasthan.

2.7.1. Leaf Spot, *Mycosphaerella fragariae*

i) Symptoms: The disease is common in hills. Spots are seen on the leaflets and petioles, fruits and stalks. Lesions are small, deep purple, the centre being brown and later turning white, resembling birds' eye. On the lower surface, the veins are reddish purple. The leaflet may be killed. The disease appears when days are warm and nights cool.

ii) Management: Plant in well drained soil, keeping down weeds. Wider spacing can reduce the leaf spot incidence. The older leaves of the runner plants should be removed before fruit setting. Spraying of carbendazim at 0.1% also gives effective control of the disease.

2.8. MULBERRY

Sericulture industry is dependent on mulberry (*Morus* spp.) cultivation as the leaves form the main food of silkworms. Under the existing conditions, the yield is 3 to 3.5 tonnes/ha which can be stepped up to 30 tonnes by adopting improved cultivation aspects. Mulberry varieties are also grown for fruits. They bear small white, black or red fruits which are sweet and can be eaten. The fruits are a delicacy in Kashmir and Afghanistan. The fruits can be dried and powdered for feeding poultry and hogs.

Excellent quality of jam can be prepared from its juice. The vegetable prepared from green fruits is beneficial for stomach ailments. The fruits are good for curing mental diseases, madness, hysteria, constipation and headache. The stem and root bark is good against worms. It is also good for throat infections.

2.8.1. Powdery Mildew, *Uncinula clintoni*

i) Symptoms: Enormous number of spores are borne on superficial growth of fungus giving the host surface a dusty or powdery appearance. Later, black minute fruiting bodies may develop on the powdery mass.

ii) Management: Two sprays of 0.1% karathane at 15 days interval gives effective management of the disease.

2.9. KARONDA

Karonda, *Carissa carandus* belonging to the Family Apocynaceae is an evergreen spiny shrub or small tree. It is widely grown in India, South Africa and Malaya. Fruits are sub acidic and sweet when ripe and rich in vitamin C and minerals. Fruits are eaten fresh but more useful for pickling, chutney and for culinary purposes. It also makes excellent jelly.

2.9.1. Anthracnose, *Gloeosporium carissae*

The disease is widespread in Uttar Pradesh.

i) Symptoms: The characteristic appearance of black grayish lesions or spots on leaves are noticed.

ii) Management: Spraying of 0.3% copper oxychloride is effective against the disease.

2.10. AVOCADO

2.10.1. Leaf and Fruit Spot, *Colletotrichum gloeosporoides*

i) Symptoms: Circular to irregular dark brown spots appear on leaves which later become bigger with brick red margins. On fruits, the raised spots are observed. The infected fruits as well as leaves with numerous spots drop prematurely. The infected fruits remaining on the tree do not attain proper size and become deformed. The fruits having latent infections develop the symptoms of anthracnose during storage and transit.

ii) Management: Spraying of 0.1% carbendazim or Topsin M or 0.2% mancozeb or chlorothalanil at fortnightly interval is recommended for management of the disease.

2.10.2. Root Rot, *Phytophthora cinnamoni*

i) Symptoms: The disease leads to decline through rotting of rootlets, girdling of the trunk and dropping of blighted leaves. The disease becomes noticeable by gumming on the surface of the affected bark which turns brown and develops longitudinal cracks. The bark rots and the tree dry because of the girdling effect. The feeder roots and rootlets also show black rotting which makes the plants devoid of nutrients. The leaves of affected trees wilt which ultimately drop. Ill drained soil and excessive irrigation are the predisposing factors for the disease development. Piling up of soil around the collar region and injury to the roots and base of the stem also favours infection.

ii) Management: The disease can be effectively controlled by soil drenching with 0.2% aliette around the base followed by spraying on the leaves. Trunk injection with aliette is also effective. Obtaining planting material from disease prone areas should be avoided. Avocado cv. Duke is reported to be resistant to the disease.

2.11. MANGO

India is the largest producer of mango in the world with an annual production of 12.538 million tonnes from an area of 2.021 million ha (Tables 2.15 and 2.16). Mango occupies 36.7 per cent of the area under fruits and it contributes 21.3 per cent of the total fruit production in the country.

Table. 2.15. Major mango producing countries of the world (2005-06), Bijay Kumar, 2007)

Country	Area (in '000 ha)	Production (in '000 MT)	Productivity (in MT/ha)	% share of world production
India	2021	12538	6.2	40.1
China	434	3676	8.5	11.8
Thailand	285	1800	6.3	5.8
Mexico	183	1679	9.2	5.4
Pakistan	165	1606	9.7	5.1
Indonesia	273	1413	5.2	4.5
Brazil	84	1348	16.0	4.3
Philippines	172	1003	5.8	3.2
Nigeria	138	812	5.9	2.6
Egypt	35	380	10.9	1.2
Others	578	4996	8.6	16.0
Total	4369	31251		

Table. 2.16. Area, production and productivity of leading mango growing states (2005-06) (Bijay Kumar, 2007)

States	Area (in '000 ha)	Production (in '000 MT)	Productivity (in MT/ha)
Andhra Pradesh	399.3	3194.3	8.0
Uttar Pradesh	251.5	2673.3	10.6
Karnataka	124.5	1236.8	9.9
Bihar	140.2	1222.7	8.7
Gujarat	96.0	772.1	8.0
Maharashtra	444.5	638.6	1.4
Tamil Nadu	125.1	537.8	4.3
West Bengal	70.1	513.3	7.3
Kerala	88.0	511.1	5.8
Orissa	125.3	428.8	3.4
Others	156.2	809.1	5.2
Total	2020.6	12537.9	6.2

2.11.1. Seedling Wilt, *Sclerotium rolfsii*

i) Symptoms: Blighting of the infected seedlings is the common symptom which is visible when the seedlings get dry. Mustard seed size fungal sclerotia appear in the vicinity of dead seedlings.

ii) Management

- Select healthy stones, make them pulp-free and dress them.
- Collection and destruction of diseased seedlings.
- Avoid contaminated soil in future plantings.
- Spray 2% Bordeaux mixture.

2.11.2. Die-back or Stem End Rot, *Botryodiplodia theobromae*

Die-back is prevalent in mango growing states of India.

i) Symptoms: The disease is characterized by dying back of twigs from top downwards, particularly in the older trees, followed by complete defoliation giving an appearance of scorching by fire. Dark lesions on bark appear which increase in size resulting in death of young twigs. The upper leaves lose their healthy green colour and gradually turn brown accompanied by upward rolling of leaf margin. In advanced stages, such leaves shrivel, fall off in a month or more, leaving the shriveled twigs almost bare. Cracks appear on branches which exude gum before they die. When the graft union of young plant is affected, it usually dies.

ii) Epidemiology: High summer temperatures predispose the mango plants to the attack of the disease by reducing the vitality of the plant. RH of about 80%; maximum and minimum temperatures of 31.5°C and 25.9°C, respectively, and rains favour disease development. The optimum temperature for spore germination was 30°C.

iii) Management

- Harvesting of mangoes on clear dry day, preventing snapping of the pedicel and avoiding injury to the fruits at all stages of handling helps to reduce the disease incidence.
- Preventive measures such as selection of scion from healthy trees, sterilization of the budding knife, keeping the grafted plants in a relatively dry environment and gradual exposure to full sunlight are effective.
- Pruning of diseased twigs 7-8 cm below the infection site followed by spraying of Bordeaux mixture (5-5-50) or 0.3% COC before the onset of monsoon was effective in controlling the disease (Prakash and Raoof, 1989).
- Mango cv. Siroli was found resistant.

Application of fresh cow dung to the cut surfaces of twigs and pruned branches kept 93.33% mango plants completely free from the incidence of gummosis and bark splitting up to 6 months (Table 2.17). The positive response of fresh cow dung pasting on trees may be due to the presence of high population of Actinomycetes which are reported effective against anthracnose, stem end rot diseases and gummosis of mango.

Table. 2.17. Effect of pasting fresh cow dung and copper oxychloride on cut surfaces of pruned mango trees.

Treatment	No. of shoot emergence/pruned branch	% plants free from gummosis and bark splitting	Cost (Rs./tree)
Fresh cow dung	39	93.33	3.5
Copper oxychloride	36	88.88	34.0

2.11.3. Powdery Mildew, *Oidium mangiferae*

Powdery mildew is one of the most devastating diseases of mango affecting almost all varieties. It is most catastrophic disease of subcontinent and Egypt. In India, the disease is wide spread and a serious threat to successful production of mango. The disease is most prevalent during the dry period of December to March or extended up to May depending on the diurnal variations in different geographical areas. The disease causes severe yield losses ranging from 30 to 90% (Prakash and Srivastava, 1987).

i) Symptoms: Powdery mildew on inflorescence is the most common stage of this disease occurring throughout the peninsular India (Fig. 2.17). Symptoms of powdery growth on leaves and shoots are more common in hilly areas of North India. Initially white or grey growth of the fungus appears on the panicles which later turn black resulting in flower shedding. There will be reduction in number and size of flowers which results in reduced yields. The fruits drop at peanut stage which are often malformed and off coloured.

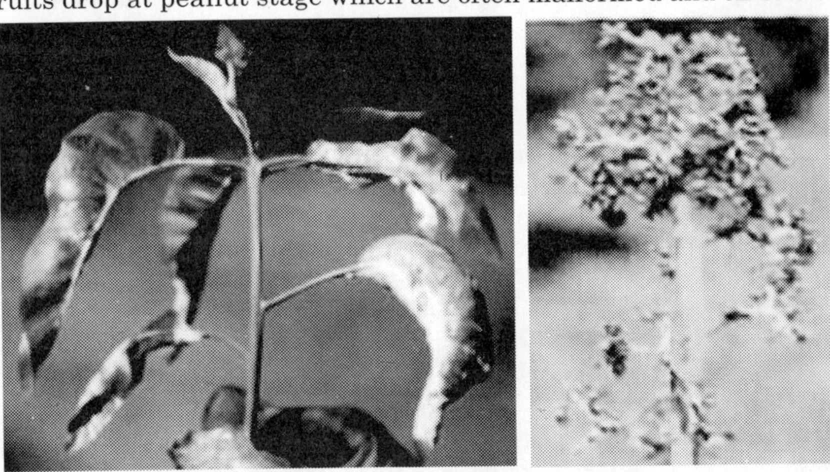

Fig. 2.17. Powdery mildew on mango leaves and inflorescence (Courtesy: K. Srinivasan, TAFE, Chennai).

ii) Epidemiology: The fungus survives as dormant mycelium and conidia on host in resting phase. It is also known to survive on some of the annual or perennial hosts. The dry weather favours the disease initiation, development and spread. Cooler nights, light intermittent rain, cloudy weather, heavy morning mist and increased humidity favour the disease development and spread at the time of flowering. In North India, the disease is favoured by minimum temperature of 15-17°C with RH of 65-85% coupled with high wind velocity.

iii) Management

- The infected gardens should be kept free from weeds and other alternate hosts. Removal and burning of infected leaves, shoots and flowers helps to reduce the spread of the disease.
- Spraying with 0.1% carbendazim or 0.05% tridemefon or 0.3% wettable sulphur gave effective control of powdery mildew. The first spray may be given soon after flowering, followed by 2 more sprays at fortnightly intervals.
- Mango cvs. Totapuri, Neelum, Zardalu, Totapuri-Khurd and Janardan Pasand were found resistant to powdery mildew (Gupta, 1976).

2.11.4. Anthracnose, *Colletotrichum gloeosporoides (Glomerella cingulata)*

Anthracnose is widespread and destructive disease of mango both in field and storage. It is responsible for blossom blight, peduncle blight, leaf spot, twig blight, wither tip, fruit russeting and fruit rot. Bose *et al.* (1973) reported that young plantations of Bombay Green cv. were completely wiped out from Terai region of Uttar Pradesh as a result of severe wither tip.

i) Symptoms: The pathogen infects leaves, twigs, inflorescence and fruits (Fig. 2.18). On leaves symptoms appear as brownish black sunken spots which later increase in size, coalesce giving blighted appearance leading to shedding of leaves. On the twigs, brownish black necrotic lesions appear which later coalesce, show twig blight and die-back symptoms. The fruit infection at early stage appears as black necrotic spots. These spots later coalesce resulting in reduced fruit growth and size. On mature fruits, the spots become more conspicuous resulting in reduced market value of the produce.

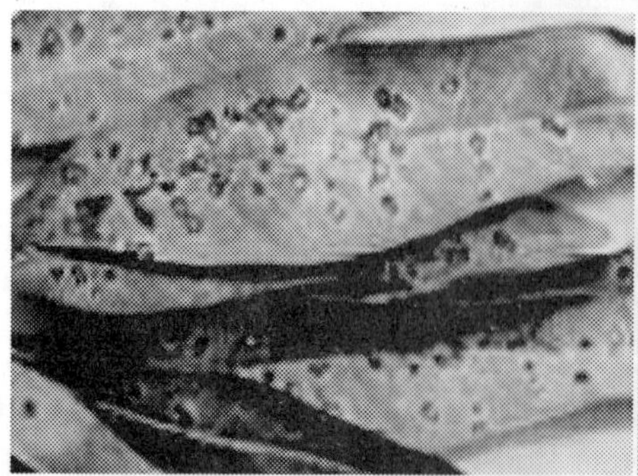

Fig. 2.18. Anthracnose on mango leaves and fruit (Courtesy: K. Srinivasan, TAFE, Chennai).

ii) Epidemiology: The infected leaves, dried twigs and fallen debris serve as primary source of inoculum in the spread of the disease. Conidial spore mass also serve as secondary source in the spread of the disease. Prolonged period of rainfall during flowering helps in severe outbreak of the disease. The temperature range of 24-32°C at shoot, flower and fruit development stage are considered as most ideal for infection. The disease is generally observed during July-August period of rainy season.

iii) Management

- Pruning affected branches and application of Bordeaux paste on cut ends.
- Spraying 1% Bordeaux mixture or 0.2% captan or 0.1% thiophanate or 0.25% mancozeb twice at 15 days interval on leaves.
- Pre-harvest spraying of 0.1% benomyl or carbendazim, Topsin M or Baycor at 15 days interval four times before harvest resulted in the lowest percentage of fruit rotting (Ullasa and Rawal, 1982).
- Dipping of fruits in 500 ppm benomyl and 900 ppm thiobendazole gave good control of the disease.
- Hot water treatment of fruits at 50-55°C for 15 min. gave good control.
- *Bacillus subtilis* was found relatively effective by controlling 50% of the disease on fruits at post-harvest.

2.11.5. Sooty Mold

The sooty mold is caused by fungi such as *Meliola mangiferae, Capnodium mangiferae, C. ramosum, Trichospermum acerinium, Microxyphium columnatum, Lewptoxyphium fumago* and *Triopospermum myrti.* The disease is very common whenever honey dew or sugary substance secreting insects, such as hoppers, scales, coccids and mealy bugs are found (Prakash and Srivastava, 1987).

i) Symptoms: The disease is characterized by the presence of a black velvety thin membranous covering on the leaf lamina (Fig. 2.19). In severe cases, the tree completely turns black with mold on the entire surface of twigs and leaves. The affected leaves curl and shrivel under dry conditions. The fungus multiplies on the honey dew secreted by the insects and spreads on the plant surface making it black and ugly owing to the masses of black spores on the leaf surface. During flowering time, its attack results in reduced fruit set and sometimes causes fruit fall. It is also noticed on fruits of late mango varieties (Fig. 2.19) (Prakash, 1988).

The fungus is non-pathogenic because its mycelium does not enter the host tissue to absorb nutrients but draws nutrients from the sweet honey dew. Thus, the damage by the fungus is not direct as it thrives saprophytically and interferes with the normal functioning of plant by cutting off the effective photosynthesizing leaf area.

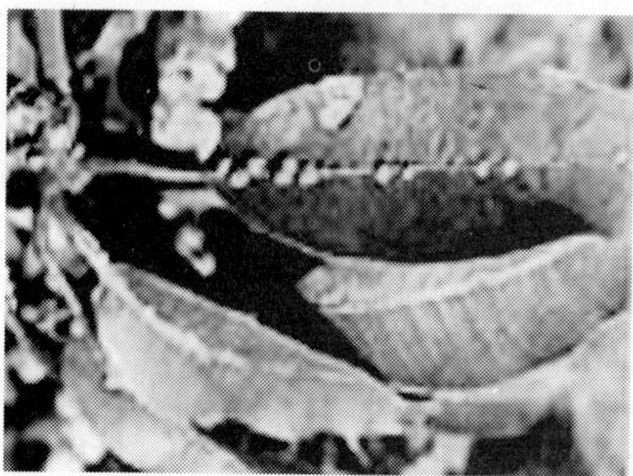
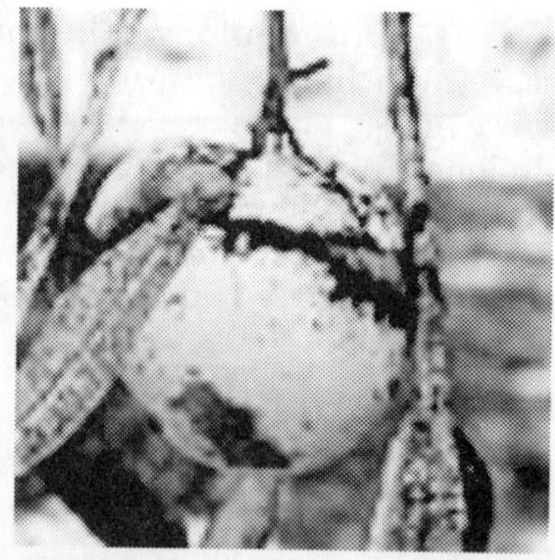

Fig. 2.19. Sooty mold on mango leaves and fruit (Courtesy: K. Srinivasan, TAFE, Chennai).

ii) Epidemiology: The disease is severe in old and dense orchards where light intensity is low. Trees exposed to eastern side (sun light) had less incidence while the trees in centre of the orchard had 95% incidence. High incidence of insects and the sugary substance secreted by them favour development of sooty mold. Continuous and heavy rainfall washes down this substance but high humidity proved congenial for growth of the fungus.

iii) Management: If honey dew secreting insects are killed by suitable insecticides (covering both the sides of leaves), the mold dies out for want of a suitable growth medium. Spraying of Elosal (900 g/450 litres of water) at 10-15 days interval proved quite effective. Spraying Wettasul + Metacid + gum Arabica (0.2 + 0.1 + 3.0%) at 15 days interval could control sooty mold (Prakash, 1988). Spraying insecticides and then removing the sooty mold growth in dry flakes by subsequent application of soluble starch was very effective method of control.

The incidence of sooty mold on fruits can be controlled by dipping them in a bleaching solution (¼ lb each of chloride of lime and boric acid /gallon of water) and subsequently washing them.

Mango cv. Alphanso has been reported to be resistant to sooty mold (Peethambharam and Aravindakshan, 1975).

2.11.6. Pink Disease, *Botryobasidium salmonicolor (Pellicularia salmonicolor)*

The pink disease, also called as thread blight, rubellosis and cobweb, is widespread in tropical and subtropical regions of India, and has been reported from south India.

i) Symptoms: The disease is noticed as a pinkish powdery coating on the twigs and branches. By then, the fungus would have invaded the bark to get established in the internal tissues and interfere with the transport of nutrients. The fungal growth often girdles the stem. Severely infected bark shreds and the wood gets exposed. Leaves turn yellow and dry, shoot and branches of the affected plants wilt and dry. The pink colour on the tissues is owing to the profuse conidial production by the fungus. In advanced cases, the fungus may produce pustular or nectar stage. These pustules are orange red and arranged systematically in rows along the stem.

The disease is seen during or just after rainy season and persists from one season to another through dormant mycelium inside the bark. The cankerous tissues serve as a potential source of infection in wet season.

ii) Management: The disease can be kept under control by cutting and removing the affected branches and painting Bordeaux paste on cut ends. Spraying with 1% Bordeaux mixture during June followed by 2 more sprays at 15 days interval helps in controlling the disease. The disease can also be controlled by use of lime sulphur and oil-based copper fungicides.

2.11.7. Twig Blight, *Phoma glomerata*

The disease is prevalent in West Bengal, Bihar, Maharashtra, Karnataka, Madhya Pradesh, Rajasthan, Gujarat, Uttar Pradesh and Goa (Prakash and Srivastava, 1987).

i) Symptoms: The symptoms of this disease are noticeable on old leaves only. Initially, the lesions are minute, irregular, yellow to light brown and scattered over the leaf lamina. As the lesions enlarge, their colour changes from brown to cinnamon and these become irregular. Fully developed spots are characterized by dark margin and dull grey necrotic centres. When infection is severe, the spots coalesce to form 3.5-13.0 cm patches which result in withering and defoliation of infected twigs (Fig. 2.20).

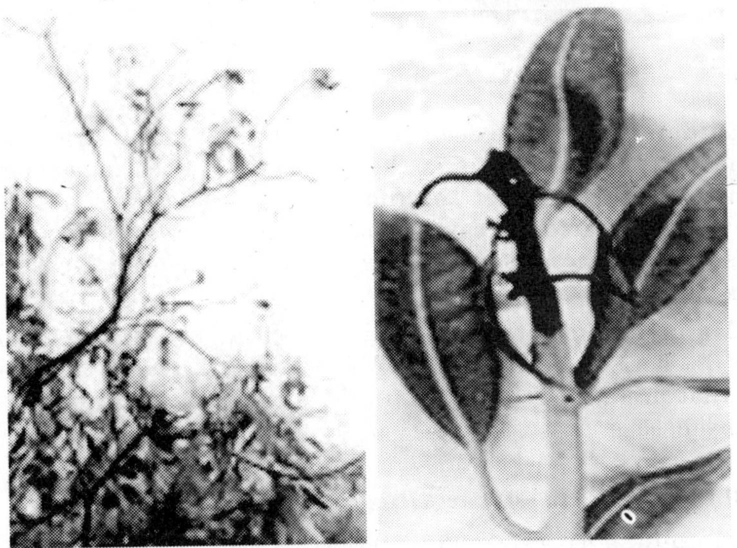

Fig. 2.20. Twig blight on mango (Courtesy: K. Srinivasan, TAFE, Chennai).

ii) Management: Spraying of 0.2% benomyl followed by 0.3% copper oxychloride were found effective.

2.11.8. Stem End Rot, *Diplodia natalensis*

i) Symptoms: This is the main post-harvest disease of ripe mango fruits. The rot not only spoils 4-6% fruits, but also adversely affects their export value (Fig. 2.21). In the initial stage, the epicarp darkens around the base of the pedicel. In next few hours, the affected area enlarges to form a circular black path which under humid conditions, extends rapidly and turns the whole fruit completely black within 2-3 days. The pulp of the diseased fruit becomes brown and soft. The pathogen incites soft rot with the aid of pectinolytic and cellulolytic enzymes.

Fig. 2.21. Stem end rot on mango.

ii) Epidemiology: The dead twigs and bark of mango trees harbour the pathogen. With the onset of rains, the atmosphere of the mango orchard gets contaminated with spores of the fungus. The pathogen can infect rapidly either when inoculated in cavities made on pedicels of the fruits or on exposed surface of the attached pedicel or in the stem end scar from where the pedicel is removed.

iii) Management: Stem end rot infection can be reduced by dipping the fruits in 6% borax solution at 43°C for 3 min. Rawal and Ullasa (1985) reported that pre-harvest sprays with 0.1% carbendazim was quite effective against the disease.

2.11.9. Black Mold Rot, *Aspergillus niger*

i) Symptoms: The affected fruits show yellowing of base and development of irregular, hazy, grayish spots which coalesce into dark brown or black lesions (Fig. 2.22). The mesocarp of the rotten area becomes depressed and soft. The diseased fruits show rapid decrease in ascorbic acid content. The pathogen infects injured fruits. Mango cv. Totapuri is highly susceptible.

ii) Management: A fruit dip treatment with 1500 ppm benomyl can control the fruit rot (Bhargava and Singh, 1975).

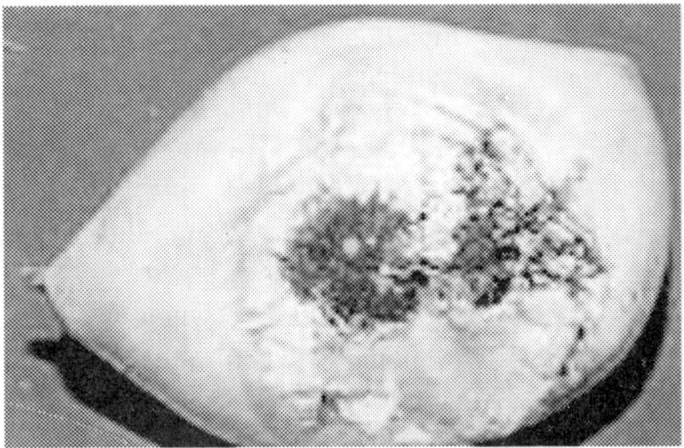

Fig. 2.22. Black mold rot on mango fruit (Courtesy: K. Srinivasan, TAFE, Chennai).

2.11.10. Soft Rot, *Rhizopus arhizus*

i) Symptoms: The soft rot was responsible for the decay of 6.3% fruits in Delhi markets. The pathogen incites a typical soft rot (Fig. 2.23). Development of the disease is slow and of low intensity at low temperature (7-8°C), but the disease develops rapidly between 20° and 40°C (Thakur, 1972).

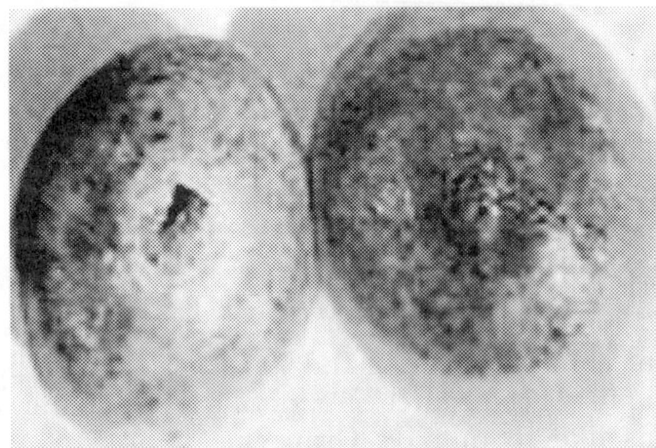

Fig. 2.23. Soft rot on mango fruit (Courtesy: K. Srinivasan, TAFE, Chennai).

ii) Management: Thakur and Chenulu (1970) found that 2-aminothiozole with 2-aminopyridine each at 5% concentration protected the mango fruits from soft rot for 20 days.

IDM

- Spraying 0.3% copper oxychloride for the management of die-back, anthracnose and red rust diseases during September - October.
- Spraying 0.2% wettable sulphur for the control of powdery mildew disease.
- Spraying of 2% starch to clear sooty mold.

Table. 2.18. Calendar of IPM activities in Mango

Month/s	Target pest	IPM activities
October	Die-back, Anthracnose	Flooding of orchards. Pruning of infected and dried branches, 10 cm below the dried portion and pasting of copper oxychloride. Spraying of 0.3% copper oxychloride after pruning. Removal of diseased foliage/ twigs infected with anthracnose (twig blight stage).
November	Die-back, Anthracnose	Removal of weeds in orchards which harbour diseases. Spraying of 0.3% copper oxychloride for die-back.
January	Powdery mildew	Removal of weeds and powdery mildew infected young leaves.
March	Powdery mildew	Spraying 0.2% wettable sulphur.
April	Powdery mildew	Second spraying 0.2% wettable sulphur after fruit setting. Removal of powdery mildew infected leaves and malformed panicles.
May	Sooty mold	Spraying of starch solution to clear sooty mold.
June	Anthracnose	Removal of diseased foliage/twigs infected with anthracnose.
July	Red rust	Spraying of 0.3% copper oxychloride for red rust and anthracnose.
August	Red rust, Anthracnose, Damping-off (Nursery)	Spraying of 0.3% copper oxychloride for red rust and anthracnose. Providing proper drainage in nursery. Sanitation and removal of weeds from orchard.

2.12. GRAPEVINE

Grapevine (*Vitis vinifera*) is one of the important commercial crops grown in different climatic zones of the world for table delicacies and wine preparation. It is fairly good source of minerals like calcium, phosphorus and iron and vitamins like B_1 and B_2. Grapevine is grown in 64,300 ha with annual production of 1.630 million tonnes (Tables 2.19 and 2.20). In grapes, India has recorded the highest productivity (25.4 t/ha) in the world.

Table. 2.19. Major grape producing countries of the world (2005-06), Bijay Kumar, 2007)

Country	Area (in '000 ha)	Production (in '000 MT)	Productivity (in MT/ha)	% share of world production
Italy	755	8554	11.3	12.6
USA	378	7088	18.7	10.5
France	855	6790	7.9	10.0
China	453	6592	14.5	9.7

Spain	1161	6054	5.2	8.9
Turkey	530	3850	7.3	5.7
Iran	315	2964	9.4	4.4
Argentina	212	2830	13.4	4.2
Chile	178	2250	12.6	3.3
Australia	153	2027	13.2	3.0
South Africa	113	1683	14.9	2.5
India	64	1631	25.4	2.4
Others	2222	15468	7.0	23.0
Total	7385	67696		

Table. 2.20. Area, production and productivity of leading grape growing states (2005-06) (Bijay Kumar, 2007)

States	Area (in '000 ha)	Production (in '000 MT)	Productivity (in MT/ha)
Maharashtra	45.1	1275.0	28.3
Karnataka	10.4	193.2	18.5
Tamil Nadu	2.6	84.8	32.6
Andhra Pradesh	1.9	40.3	21.2
Punjab	1.1	30.2	27.0
Haryana	0.1	3.3	33.0
Madhya Pradesh	0.1	2.6	25.0
Mizoram	0.5	0.6	1.2
Jammu & Kashmir	0.2	0.3	1.5
Others	2.2	0.3	0.1
Total	64.3	1630.7	25.4

Currently, about 20,000 tonnes of fresh grapes are exported, accounting for just 2% of the total production. Out of this, about 12,000 tonnes are exported to the Middle East markets and the rest to U.K, Europe and few South-East Asian markets. Among these markets, U.K. is the most paying besides being reliable. Middle East markets are not dependable and are also less paying.

2.12.1. Downy Mildew, *Plasmopara viticola*

Downy mildew is prevalent in USA, Europe, North and South America, India, Australia and New Zealand. It is the most serious disease of grapevine throughout south India, especially Andhra Pradesh and Karnataka (Sohi, 1983). The disease has become a major limiting factor for production of grape berries in Northern Karnataka. Substantial crop loss occurs when flower or young berries are killed.

i) Symptoms: The disease is first observed as small translucent (oil spots) pale yellow spots with indefinite borders on the upper surface of the leaves. On the lower surface of leaves directly below the spots, a downy growth of the fungus appears (Fig. 2.24). Severely affected vines are readily defoliated, preventing the ripening of fruit and maturation of canes and exposing fruits to damage by sun burn. Under favourable conditions (20-25°C temperature and 80% RH), the pathogen attacks young shoots, tendrils, inflorescences and berries resulting in flower/fruit shedding. Complete crop failure is encountered in South India whenever there is continuous rainfall during the early part of December (Sohi, 1983).

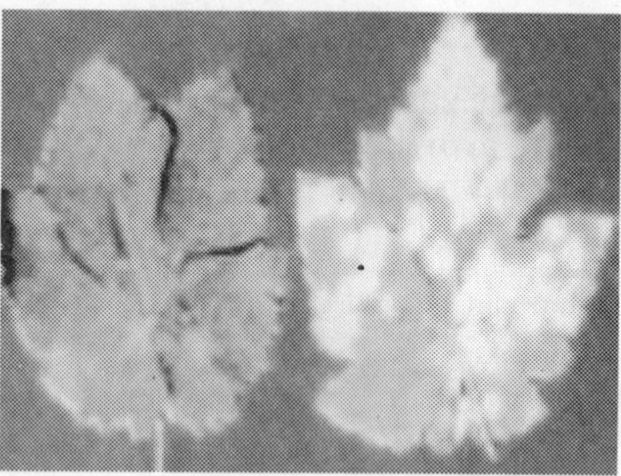

Fig. 2.24. Downey mildew on grapevine leaves.

ii) Epidemiology: The most favourable temperature for germination of sporangia is between 10° and 23°C. Disease development is favoured during rainy season when there is a heavy dew, RH is above 90% and temperature is between 23° and 25°C. The pathogen survives on the infected leaves and vines as oospores and also dormant mycelium. The secondary spread is through wind-borne sporangia and zoospores, which are found on the new flush.

iii) Management

- Fallen leaves, twigs, flowers and berries that may contain hibernating oospores should be collected and burnt.
- Canes should be kept above the ground level and free circulation of air should be allowed by providing proper spacing and pruning.
- After pruning, spraying 1% Bordeaux mixture mixed with groundnut oil, coconut oil and neem oil reduces infection.
- When the flushes are formed, spraying thrice with difolatan or chlorothalanil or metalaxyl or dimethomorph all at 0.2% at weekly intervals is effective.
- The biocontrol agent, *Fusarium proliferatum* is effective against downy mildew under field conditions.

- Cvs. Amber Queen, Cardinal Champa, Champion, Excelsior, Red Sultana are reported to be resistant.

2.12.2. Powdery Mildew, *Uncinula necator*

Powdery mildew is prevalent in almost all grape growing regions of the country. The disease is more problematic than downy mildew in relatively dry regions. It is prevalent in Australia, Greece, Hungary, India, Syria and the USA.

i) Symptoms: The disease infects the plants at any stage of growth. It affects all the aerial parts. Powdery patches appear on leaves, canes, tendrils, flowers and young fruit bunches (Fig. 2.25). On leaves, the powdery patches enlarge and upper leaf surface becomes dusty. The infected leaves may curl upwards in dry weather. The powdery growth turns grey and finally becomes dark in colour. Diseased vines give a sickly look and show restricted growth. Infected berries do not develop, their skin becomes hard and ultimately cracks. In case of early infection, the fruits do not develop normally, become deformed and crack.

Fig. 2.25. Powdery mildew infection on grapevine leaf and fruits (Courtesy: K. Srinivasan, TAFE, Chennai).

ii) Epidemiology: The disease is more prominent before the onset of rains (February-June) in South India, while it is more prominent during October-November under North Indian conditions. It is favoured by warm dry weather and retarded by sunshine. Disease development is adversely affected by rainfall. The fungus survives as dormant mycelium and as cleistothesia on the shoots and buds from season to season. The disease spreads by air-borne conidia.

iii) Management

- The diseased plant parts should be removed and destroyed.
- Overcrowding of vine should be avoided by thinning of leaves to improve ventilation.
- Avoid pruning during sensitive periods of first and second fortnights of November.
- Spraying of wettable sulphur or dinocap both at 0.2% or 0.1% carbendazim or 0.05% folicur at first appearance of the disease followed by 2 to 3 sprays at 10-12 days interval gives effective control.
- Biocontrol agent, *Ampelomyces quisqualis* grown on nutrient-treated cotton twine suspended over grapevines suppressed powdery mildew throughout the rainy season (pycnidiospores of mycoparasite suppressed powdery mildew).
- Grapevine cvs. Red Sultana, St. George and 1613 are highly resistant (Sohi and Sridhar, 1970).

2.12.3. Anthracnose, *Elsinoe ampelina*

Anthracnose, also called as birds' eyespot, has been reported to be prevalent in Europe, South Africa, India, Eastern parts of the USA and West Australia. It is a serious disease prevalent in almost throughout India and is a limiting factor in grape production in Andhra Pradesh, Karnataka, Tamil Nadu, Rajasthan, Uttar Pradesh, Punjab and Haryana. The disease is responsible for 10-15% loss in yield in Punjab and Haryana (Bedi *et al.*, 1969).

i) Symptoms: The characteristic symptoms of the disease include grey-black spots with red brown margins which develop rapidly on actively growing leaves. Later, the affected leaves become shot-holed and tattered. On stem, brown to black spots develop into sunken cankers with slightly raised edges (Fig. 2.26). As cankers coalesce, stems are girdled, weakened and killed. Slightly sunken brown spots, sometimes with grey centres, develop on immature berries (Fig. 2.26). Cankers also weaken canes and reduce their future productivity.

ii) Epidemiology: The disease is severe during July-August and November-December months. Infection in new sprouts takes place during rainy season. Heavy rains after pruning leads to higher disease incidence. The pathogen survives as dormant mycelium in the cankers on the stem and also on infected twigs. Secondary spread is through conidia, which are carried by wind and rain water.

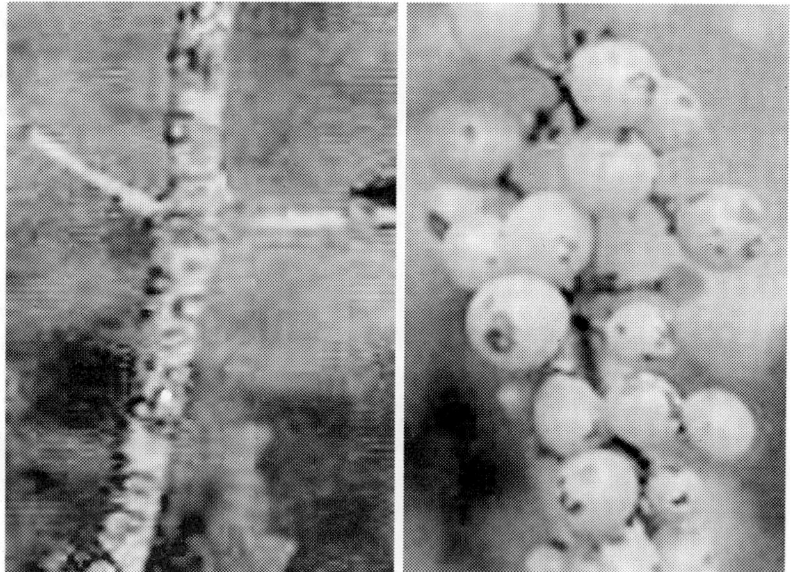

Fig. 2.26. Anthracnose on stem and berries of grapevine (Courtesy: K. Srinivasan, TAFE, Chennai).

iii) Management

- Removal and destruction of all the diseased leaves and twigs after pruning is the foremost requirement.
- Spraying pruned canes and leaves with a mixture of ferrous sulphate at 2.5 kg + 0.5 pint sulphuric acid in 4-5 litres of water helps to manage the disease.
- Spraying 1% Bordeaux mixture or 0.25% copper oxychloride or 0.1% carbendazim or 0.2% mancozeb or 0.2% difolatan or 0.2% chlorothalanil or 0.1% bitertenol (Baycor) or 0.1% difenconazole at bud swell, fruitlet stages and 21 days after the second spray effectively controls the disease.
- Cvs. Angur Kalan, Bharat Early, Delight, Camphor Ash, Naibel, Concord, White Beauty Seedless, Muscat Hamburg (Gulabi), Himrod, Hussaini, Karachi Niagra, Khalili, Niabel, Schuyler White, St. Valior, White Muscat, Large White, Golden Queen, Bangalore Blue, Isabella and Golden Muscat are resistant.

2.12.4. Black Rot, *Guignardia bidwelli*

After its introduction in France by about 1880, the black rot had spread to all grape growing regions of Europe. In India, the disease has been observed in Madurai (Tamil Nadu) and Bangalore (Karnataka) areas.

i) Symptoms: On fruits, the disease begins to show as light, brownish, soft, circular spots which increase in size and the entire berry is discoloured (Fig. 2.27). The decayed berries begin to shrivel within a week and are transformed into hard, black mummies. Circular red spots appear on the leaves (Fig. 2.27) and later the margins become black. Minute black dots representing fruiting bodies of the fungus are arranged in a ring near the outer edge.

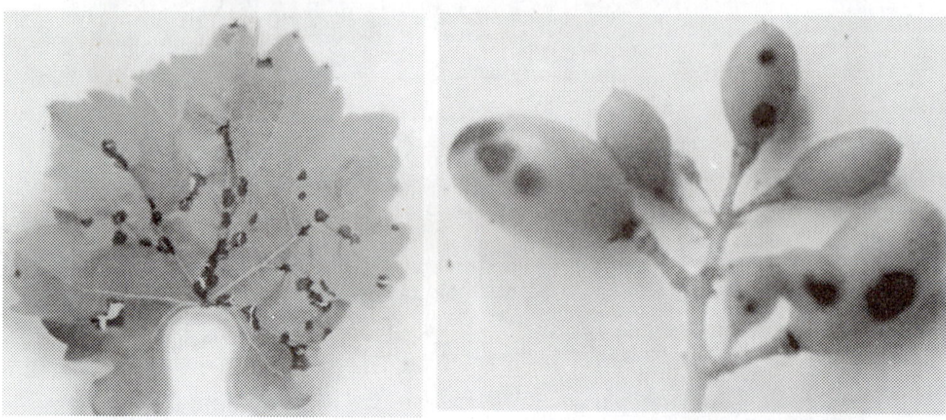

Fig. 2.27. Black spots on leaves and berries grapevine.

ii) Epidemiology: Black rot is more destructive in warm and humid regions than in cooler and drier areas. Frequent rains and humid climate are conducive for the disease development. Perithecia develop on mummified grape berries and the ascospores are discharged when mummies are wet. Ascospores produce germ tube and penetrate directly through the cuticle. Primary infection occurs on young leaves and fruit pedicels. Pycnidia are rapidly produced and pycnidiospores spread through meteoric water. They survive the winter and germinate in the following season. Leaf wetness and temperature are considered as important factors in disease forecasting.

iii) Management

- Collection and destruction of diseased leaves and berries helps to reduce the inoculum.
- Spraying with 1% Bordeaux mixture or ferbam or captan or chlorothalanil all at 0.2% when the new shoots are 15 to 25 cm long and repeating the sprays before and after bloom helps in reducing the disease incidence.

2.12.5. Grey Mold Rot, *Botrytis cinereria*

The disease occurs in the entire grape growing regions of the world. It is one of the principal causes of post-harvest spoilage in storage. The pathogen is capable of growing at low temperatures.

i) Symptoms: In early stage, tissue just beneath the surface of fruit is infected loosening the skin from the flesh. The affected area turns light brown. The fungal infection advances into the inner flesh resulting in a soft watery mass of decayed tissue. Under moist atmosphere, the fungus sporulates on the surface of the fruit and the typical powdery grey mold stage becomes evident. Infected fruits shrivel and turn dark brown. The disease starts in midseason and continues to develop until harvest in the absence of rain. The fungus infects stigma and style and becomes latent in the necrotic stigma and style tissues at the stylar end of the berry. In a compact infected bunch, fruits inside may split during growth. Infection of such fruits results in bunch rot. Even a single field infected berry may cause 'nest rot' in transit and storage.

ii) Epidemiology: The fungus spreads through contact. Susceptibility to disease increases with maturity of fruits. Bunches picked after rainfall shows more disease incidence.

iii) Management

- Pruning and thinning should be done to lower the humidity around the clusters and thinning of berries in compact clusters reduced the severity of grey rot.
- Pre-harvest infection can be controlled by three applications of 0.2% captan at monthly intervals before rainfall.
- Fruits for transit should be precooled at 4.44°C and then for storage should be cooled to -0.55°C to 0°C.
- Sodium bisulphate, which releases SO_2 when in contact with moist air, can be used with packing material to control this disease.
- *Trichoderma harzianum* was found very effective against grey rot.

2.12.6. Blue Mold Rot, *Pencillium digitatum*

i) Symptoms: The disease is characterized by a scanty growth of the pathogen which is white at first and turns bluish green later. The pathogen is associated with the wounds or cracks in the skin. The affected tissue becomes soft and watery and develops moldy flavour.

ii) Management: Grapes fumigated with SO_2 in storage do not usually suffer from this disease. Tandon *et al.* (1975) recommended Captan for reducing decay of grapes by *Pencillium* and *Aspergillus*.

2.12.7. *Cladosporium* Rot, *Cladosporium herbarum*

i) Symptoms: The disease causes a black firm, shallow decay which is usually localized on one side of blossom end. The fruits become flat or wrinkled on the affected side. A sparse surface growth of grey-green fungus may be seen on the fruits kept at room temperature. The infection occurs through uninjured skin or through wounds or cracks at the blossom end.

ii) Management: The infection can be reduced by sulphur fumigation, cooling and storage at 0.55°C to 0°C.

2.12.8. Bitter Rot, *Greenaria uvicola*

i) Symptoms: The pathogen produces brown, hard depressed patches on the berries which subsequently bear numerous black, dot-like fructifications. Grapes harvested with small lesions develop typical rot symptoms and deteriorate in transit and storage. Twenty five per cent of Anab-e-Shahi bunches were found infected at Hyderabad (Reddy and Reddy, 1983). The number of infected berries in a bunch varied from 1 to 54 and almost 30% were infected in the vineyard.

ii) Management: The pruned canes should be collected and destroyed as the fungus survives on pruned canes which serve as primary source of inoculum.

Post-harvest dip treatment of fruits in captan and difolatan each at 0.2% were effective in reducing rotten berries to 12.8 and 10.1%, respectively as compared to 78.8% in untreated control (Ullasa and Rawal, 1982).

2.12.9. Dead Arm, *Phomopsis viticola*

i) Symptoms: The infection takes place through wounds at pruning time. It develops slowly. When it becomes severe, the long arms fully laden with bunches, suddenly fade and bunches dry up (Fig. 2.28).

Fig. 2.28. *Phomopsis* on stem and leaves of grapevine (Courtesy: K. Srinivasan, TAFE, Chennai).

ii) Management: The pruned canes should be collected and destroyed. The dead canes or arms should be pruned well beyond the region of healthy tissue. Cut ends should be applied with Bordeaux paste. Spraying of 1% Bordeaux mixture at fortnightly intervals reduces the disease incidence.

2.12.10. Rust, *Phakopsora vitis*

i) Symptoms: The disease has been reported from Salem, Dharmapuri, Nilgiris and Coimbatore districts of Tamil Nadu. Generally the variety Black Prince is severely affected. The disease is generally observed during the winter season. The fungus produces numerous orange coloured sori on the lower surface of the leaves (Fig. 2.29). In severe cases of infection, the entire leaf surface is covered by the sori leading to defoliation. The uredospores are found in several layers, binucleate with hyaline or coloured walls. Teliospores are sessile, non-echinulated, single celled with one germ pore. The wall is pigmented. The fungus attacks generally blue coloured grapes than green coloured grapes.

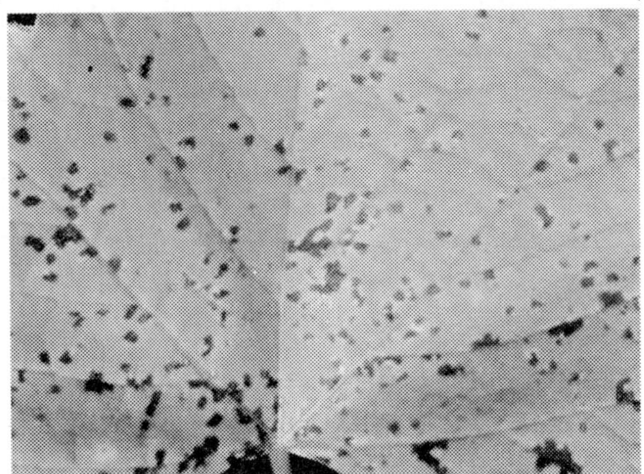

Fig. 2.29. Grape leaf infected with rust disease.

ii) Management: The disease can be managed by dusting sulphur at 25 kg/ha or spraying 0.2% wettable sulphur or zineb.

2.13. GAUVA

Guava rank fifth in area and production with an estimated production of 1.823 million tonnes from an area of 0.182 million hectares (Table 2.21). Guava occupies 3.3 per cent of the area under fruits and contributes 3.1 per cent of the total fruit production in the country.

Table. 2.21. Area, production and productivity of leading guava growing states (2005-06) (Bijay Kumar, 2007)

States	Area (in '000 ha)	Production (in '000 MT)	Productivity (in MT/ha)
Maharashtra	31.3	223.6	7.1
Bihar	27.7	199.0	7.2
Uttar Pradesh	15.5	157.8	10.2
Karnataka	7.3	154.1	21.2
West Bengal	9.9	153.0	15.5

Punjab	7.8	135.7	17.5
Andhra Pradesh	10.1	131.6	13.1
Uttarakhand	14.3	99.6	7.0
Gujarat	7.0	98.7	14.1
Tamil Nadu	8.5	92.2	10.9
Orissa	13.8	89.4	6.5
Others	28.7	288.5	10.0
Total	181.7	1823.3	10.0

2.13.1. Guava Wilt, *Fusarium oxysporum* f. sp. *psidii*, *Helicotylenchus dihystera*

It is a very destructive disease which is widely distributed in almost all guava growing regions of India. It is caused by the fungus, *Fusarium* along with spiral nematode. Guava wilt is prevalent in Uttar Pradesh, Delhi, West Bengal, Bihar, Madhya Pradesh, Andhra Pradesh, Haryana, Punjab and Rajasthan. Singh and Lal (1953) recorded 5 -10% loss annually in 12 districts of Uttar Pradesh which was estimated around Rs. 0.8 to 1.0 million. In West Bengal, guava wilt reduced the yields by 80%, from 11.35 tonnes/ha in healthy plantations to about 1.82-2.27 tonnes/ha in affected plantations (Chattopadhyay and Sen Gupta, 1955).

i) Symptoms: Symptoms appear during rainy season. The leaves turn to brown colour, dry and drop. The infected leaves show deficiency of nitrogen and zinc. The infected branches and twigs show typical die-back symptom. As the disease advances, the entire plant dies in three to four weeks. The cross section of root shows typical blackening of xylem vessels.

ii) Epidemiology: The fungus survives in dead roots of trees and as saprophyte in the form of chlamydospores in roots. The pathogen infects through wounds, natural openings and root hairs. The fungus blocks the vascular bundles disrupting the uptake of water and nutrients. This disease is more common in alkaline soils with moderate moisture.

iii) Management

- Avoid water stagnation. Remove affected plants and burn them. Treat the soil with permalon beg.
- Soil drenching with 0.2% carbendazim around tree in March, June and September coupled with spraying 0.3% zinc sulphate twice in March and September is advocated.
- Use Lucknow-49, Banarsi (Andhra strain), Dholak Sind, Nasik, White Guava No. 6299, Supreme, Clone 32-12 cvs. which are resistant to wilt.
- Use of nitrogen fertilizer is also helpful to manage the disease.

- Application of gypsum or lime to the soil (1 kg/tree) or neem cake at 10 kg/tree after root exposure at a depth of 15 cm was found effective in checking the disease incidence.
- Biological control with *Aspergillus niger* AN 17 (Pusa Mrida), *Pencillium citrinum, Trichoderma harzianum, T. virens* and intercropping with marigold and turmeric gave good control of wilt. *A. niger* was found most effective which can be mass multiplied on FYM and applied every year as it controls wilt and also provide nutrition to guava plants (Tables 2.22 and 2.23).

Table 2.22. Evaluation of bioagents against guava wilt.

Bio-agent	Average wilting (%)	% Control
Pencillium citrinum	47.33	52.67
Aspergillus niger AN 17 (Pusa Mrida)	21.66	78.34
Trichoderma harzianum	24.66	75.34
Control	100.00	0.00

Table 2.23. Effect of intercropping on the incidence of guava wilt.

Intercrop	Total plants	Healthy	Wilted	Partially wilted	% Wilting
Control	20	16	3	1	17.5
Turmeric	20	19	0	1	2.5
Marigold	20	20	0	0	0.0
Garlic	20	18	1	1	7.5

2.13.2. Anthracnose, *Gloeosporium psidii*

The disease is a serious problem in Uttar Pradesh, Punjab and Karnataka. It causes die-back, twig blight, wither tip and fruit spots. This is also a serious post-harvest disease which causes fruit rot in guava.

i) Symptoms: The plant shows die-back symptoms. Pin-head spots are first seen on the unripe fruits which gradually enlarge measuring 5-6 mm in diameter (Fig. 2.30). They are dark brown to black in colour, sunken, circular and have minute, black stromata in the centre of the lesions which produce creamy spore masses in moist weather. Several spots coalesce to form bigger lesions. The infected area in unripe fruits become corky and hard, and often develops cracks in case of severe infection. On ripe fruit, the infection causes softening of tissue and lesions attain a diameter of 10-20 mm. Unopened buds and flowers are also attacked and cause their shedding. Spread of infection is very rapid on fully mature green fruits, while young fruits do not normally take infection, perhaps due to the differences in the concentration of K-ions in the tissues.

ii) Epidemiology: The infected fruits, twigs and branches serve as primary source of inoculum. Wounds caused by insects also facilitate in pathogen entry. Optimum temperature for fruit infection is 30°C. The temperature range of 10°-35°C and relative humidity of 95% are most ideal conditions for disease development. Deficiency of potassium further hastens the disease spread.

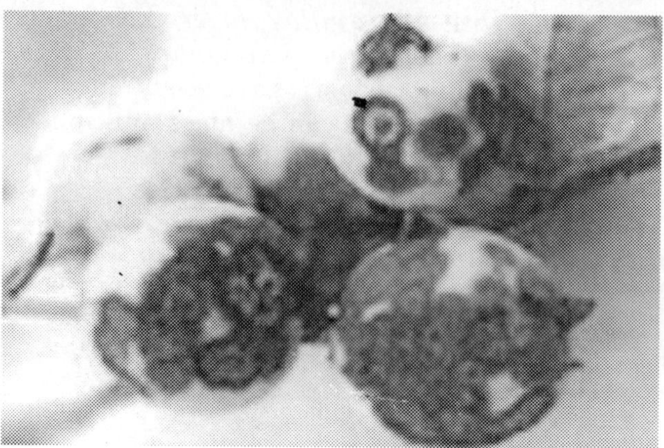

Fig. 2.30. Anthracnose on guava fruits (Courtesy: K. Srinivasan, TAFE, Chennai).

iii) Management

- Removal and burning of the infected twigs, branches and fruits immediately after rainy season.
- Spraying of 1% Bordeaux mixture or 0.2% difolatan or 0.2% zineb or 0.1% topsin M as soon as the disease appears in the orchard has been recommended.
- Harvested fruits can be fumigated with sodium metabisulphate or bleaching powder to avoid post-harvest anthracnose/fruit rot.
- Apple guava (light fleshed) has moderate resistance to anthracnose disease.
- Gupta and Chatrath (1973a) reported that Thiram, Kirti-Copper and Cuman were good protectants as pre-inoculation sprays. Post-inoculation dip treatment with tetracycline proved effective against the disease. Irradiation of fruits with 100 Kr gamma rays for 5, 18 and 24 hours after inoculation gave good control of fruit rot (Gupta and Chatrath, 1973b).

2.13.3. Fruit Canker, *Pestalotia psidii, Colletotrichum gloeosporoides*

The disease is distributed in Maharashtra, Karnataka, Himachal Pradesh and Uttar Pradesh.

i) Symptoms: The disease generally appears on green fruits and rarely on leaves. The first evidence of infection on fruit is the appearance of minute, brown or rust coloured, unbroken, circular, necrotic areas which in advanced stages, tears open the epidermis in acircinate manner. The margin of lesion is elevated and a depressed area is noticeable inside. The crater-like appearance is more noticeable on fruits than on leaves (Fig. 2.31). The canker is confined to a very shallow area and does not penetrate deep into the flesh of the fruit. In older cankers, white mycelium with numerous spores is noticeable. In severe cases, raised, cankerous spots develop in great numbers and the fruits break open to expose seeds. The infected fruits remain under developed, become hard, malformed and mummified and drop in great numbers.

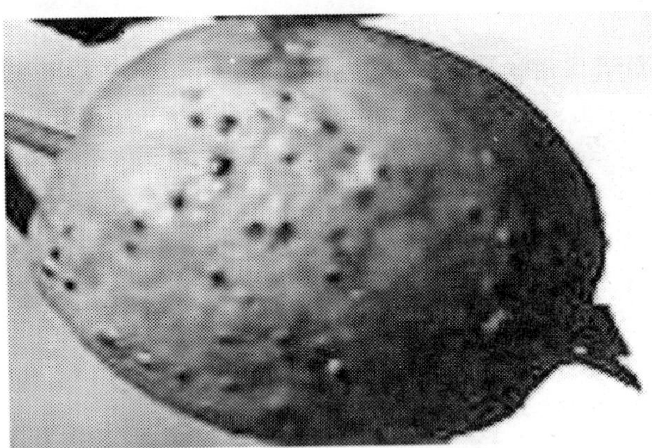

Fig. 2.31. Canker on guava fruit (Courtesy: K. Srinivasan, TAFE, Chennai).

ii) Epidemiology: Kaushik *et al.* (1972) reported that the maximum disease occurs at 25-30°C and at high relative humidity. The pathogen is primarily a wound parasite.

iii) Management

- The spread of disease (in early stages of infection) can be controlled by spraying of 1% Bordeaux mixture at 2-3 litres/tree at 15 days interval.
- Post-harvest washing of fruits with 200 ppm aureofungin can protect the fruits for five days (Kaushik *et al.*, 1972). The fruit rot can also be checked by fruit dip treatment with Aretan. This treatment also checks fruit rot incited by *Colletotrichum gloeosporoides* (Khanna and Chandra, 1976).
- Guava cv. Nasik was almost immune; Safeda and Apple Colour were highly resistant; Sind was resistant and Dholka was moderately resistant to canker.

2.13.4. Scab, Grey Blight and Leaf Spot, *Pestalotia psidii*

i) Symptoms: Reddish brown spots appear on leaves and fruits (Fig. 2.32). As the disease advances, these spots turn grey in colour. The symptoms appear as deep scabby spots on fruits.

ii) Management : Collection and destruction of diseased plant parts and spraying with 0.2% mancozeb as preventive measure helps to manage the disease.

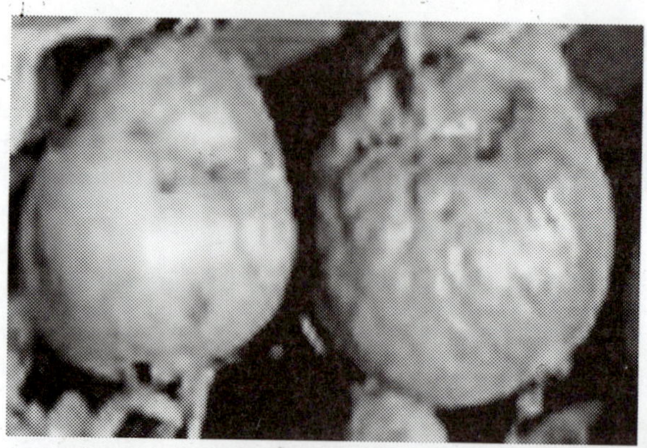

Fig. 2.32. Scab on fruits (Courtesy: K. Srinivasan, TAFE, Chennai).

2.13.5. *Phytophthora* Fruit Rot, *Phytophthora nicotianae* var. *parasitica*

The disease was responsible for considerable losses in Mysore, Karnataka during rainy season on Allahabad Safeda and is now known to occur in Tamil Nadu, Andhra Pradesh, Punjab, Maharashtra and Rajasthan (Sohi and Sridhar, 1971). The incidence of the disease varied from 5 to 20%.

i) Symptoms: The pathogen attacks unripe fruits at the styler end. The whitish cottony growth develops very rapidly as the fruit ripens and covers almost entire surface within 3-4 days in humid weather (Fig. 2.33). The fruits near soil level, covered with dense foliage, under high relative humidity are most severely affected. The skin of the fruit below the whitish cottony growth becomes a little soft, turns light brown to dark brown and emits a characteristic unpleasant smell. These fruits either remain intact or drop off. When the disease appears on young and immature fruits, they shrink, turn dirty brown to dark brown, hard in texture and either remain intact as mummified fruit or drop off.

Fig. 2.33. *Phytophthora* rot of guava fruits (Courtesy: K. Srinivasan, TAFE, Chennai).

ii) Management

Kothari (1968) recommended weekly sprays with 2: 2: 50 Bordeaux mixture or 0.2% copper oxychloride for the control of fruit rot. Monthly sprays with 0.2% zineb or 10 ppm aureofungin during fruiting gave effective control of the disease (Sohi and Sridhar, 1971).

2.13.6. *Botryodiplodia* Fruit Rot, *Botryodiplodia theobromae*

i) Symptoms: The soft and watery rot incited by the pathogen causes considerable damage during storage and transit. The affected fruits show brownish discolouration mostly at the stem end which generally proceed downwards in irregular wavy manner (Fig. 2.34). In advanced stages, numerous small pycnidia are produced over the entire surface of the fruit. The maximum decay occurs at 30°C.

Fig. 2.34. Fruit rot on guava (Courtesy: K. Srinivasan, TAFE, Chennai).

ii) Management: Careful handling and storage of fruits at 15°C and quick disposal can minimize losses.

2.13.7. *Phoma* Fruit Rot, *Phoma psidii*

i) Symptoms: The disease is characterized by brown spots over fruit surface which ooze out cream pycnidiospores.

ii) Management: The rot can be checked by treating the fruits with TBZ. Storage of fruits at 15°C also reduces the disease incidence.

2.14. LOQUAT

2.14.1. Wither Tip

i) Symptoms: The disease has been reported from Delhi and Western Uttar Pradesh. It causes shedding of leaves and withering of tips of branches. On leaves, light green spots appear first which later turn brown. Usually, the spots develop at the tip or margin of leaves. Acervuli develop as black dots on twigs and leaves from which pinkish masses of spores ooze out.

ii) Management: Pruning of infected branches and treatment with Chevastelon solution (6% cold solution of potassium dichromate mixed with

6% cold solution of copper sulphate) is recommended for the control of the disease (Lele and Butani, 1975).

2.14.2. Collar Rot, *Sclerotium* sp., *Diplodia natalensis*

i) Symptoms: The disease has been reported from Uttar Pradesh which results in complete girdling. *D. natalensis* infects invariably as secondary pathogen.

ii) Management: Scraping of infected bark and application of Bordeaux paste or copper oxychloride helps to manage the disease.

2.14.3. Twig Blight and Canker, *Cytospora chrysosperma*

i) Symptoms: The disease has been reported from Delhi and Western Uttar Pradesh. The affected plants show sick look, less foliage and blighted appearance. On diseased twigs, embossed, globose and cankerous pimples appear. The bark becomes loose and gives amber-yellow exudates which become horny on drying.

ii) Management: Pruning of infected branches and treatment with Chevastelon solution is recommended for the control of the disease (Lele and Butani, 1975).

2.15. APPLE

Apple (*Malus pumila*) is the most important temperate fruit of the North-Western Himalayan region. It is predominantly grown in Jammu and Kashmir, Himachal Pradesh and hills of Uttarakhand accounting for about 90% of the total production. Its cultivation has also been extended to Arunachal Pradesh, Sikkim, Nagaland and Meghalaya in North-Eastern region and Nilgiri hills in Tamil Nadu. Apple is grown in an area of 2, 34,700 ha with annual production of 1.755 million tonnes (Tables 2.24 and 2.25).

Table 2.24. Major apple producing countries of the world (2005-06) (Bijay Kumar, 2007)

Country	Area (in '000 ha)	Production (in '000 MT)	Productivity (in MT/ha)	% share of world production
China	1851	24020	13.0	38.7
USA	153	4409	18.8	7.1
Iran	201	2662	13.2	4.3
Turkey	121	2570	21.2	4.1
Italy	57	2192	38.4	3.5
Poland	170	2075	12.2	3.3
France	47	1857	39.1	3.0
Russian Federation	384	1773	4.6	2.9
India	235	1756	7.5	2.8

Chile	37	1350	37.0	2.2
Argentina	45	1272	28.3	2.0
Others	1531	16109	10.5	26.0
Total	4832	62044		

Table 2.25. Area, production and productivity of leading apple growing states (2005-06) (Bijay Kumar, 2007)

States	Area (in '000 ha)	Production (in '000 MT)	Productivity (in MT/ha)
Jammu & Kashmir	107.9	1093.3	10.1
Himachal Pradesh	88.6	540.4	6.1
Uttarakhand	28.8	112.3	3.9
Arunachal Pradesh	9.1	9.6	1.1
Others	0.3	0.0	0.0
Total	234.7	1755.6	7.5

2.15.1. Scab, *Venturia inequalis*

Scab is the most destructive disease of apple and present in all the countries of the world where apples are grown. The disease is particularly severe in high rainfall and humid areas. In India, it is present in all the apple growing areas of Jammu and Kashmir, Himachal Pradesh, Uttar Pradesh, Sikkim, Arunachal Pradesh and Nilgiris. Losses from scab result from mid-season defoliation of trees, reduction in fruit production, devaluation in fruit grade, weakening of the trees, failure of fruit bud formation and increased expenses on spray operations. The first epidemic of scab in Kashmir valley in 1973 completely ruined the apple crop worth US$ 540,000. During 1983, it rendered 10% apple crop (30,000 MT) unfit for consumption in Himachal Pradesh which has to be destroyed and as a result the state suffered a loss of Rs. 1.5 crores (Gupta, 1985).

i) Symptoms: Scab may appear on leaves, fruits, petioles and green twigs. The most striking symptoms of scab are commonly observed on leaves and fruits. Fruits may show small, rough, black, circular lesions on their skin (Fig. 2.35) while on the tree or after keeping in cold storage. Fruits picked from infected trees appear apparently healthy but are too sticky to keep even in cold storage. Such fruits soon develop scab symptoms even at low temperature and may not last long in storage. The affected fruits rot due to secondary infection of the lesions. Secondary infections on leaves are so numerous that the entire leaf surface appears covered with scab, commonly referred sheet scab. Lesions on young fruits resemble those on leaves but turn dark brown to black and become corky or scab-like with time. Infections are often limited to one or two spots per fruit. Secondary infections are clumped together.

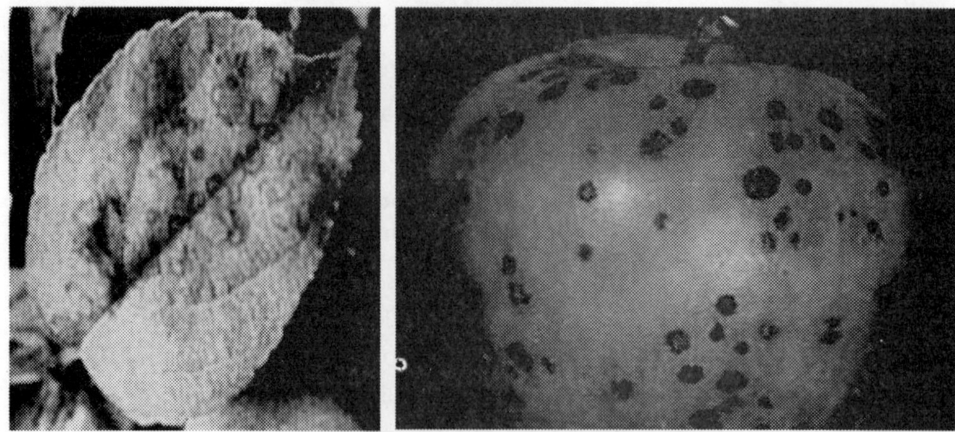

Fig. 2.35. Apple fruit infected with Scab.

ii) Epidemiology: High relative humidity, high rainfall and temperatures between 16° and 20°C favour the fast development of the disease. Generally wet conditions prevailing in early autumn (October-November) of the previous year followed by similar conditions during spring are attributed to scab epidemics. Prolonged late summer during monsoon is also the critical period for epidemics (Gupta, 1985). Cool and moist climatic conditions due to rain or snow at higher elevation and in shady portions of the orchards are congenial for the development of pseudothecia in the over wintering leaves. When moisture is not a limiting factor and the fallen leaves remain pliable, fungus starts producing pseudothecia even at low temperatures of 4° to 8°C. But such pseudothecia develop slowly throughout the winter, attain full size and mature at an optimum temperature of 15°C. Prolonged wet weather during winter and spring does not allow normal development of pseudothecia and even the ascospore maturity gets delayed by a month or so. Pseudothecia formed in autumn and winter mature in spring to produce ascospores, the chief inoculum for primary infection. The secondary spread is through conidia.

iii) Management

- Clean cultivation, ploughing to remove plant debris harbouring the fungus, raking and destruction of fallen leaves and pruned material can reduce the primary infection.
- Spraying of 0.2% captan or 0.25% dodine at short intervals after petal fall or a single application of 0.3% difolatan at green bud stage followed by 0.2% captan at petal fall controls the disease. Dodine reduces spore formation and scab spots and inhibits germination of spores produced on such spots.
- Spraying of 5% urea suppresses pseudothecial production and accelerates decomposition of leaves. Additional spray of 2% urea once in autumn prior to leaf fall and again just before bud break followed by

two sprays with 0.2% captan or 0.25% mancozeb at 10-15 days interval after petal fall is also recommended.

- Cvs. Emra, Red Free, Prima, Priscilla, Macfree, Liberty, and hybrids Ambstarking, Ambroyal, Ambrich and Ambred are resistant.
- The biocontrol agent, *Athalia bombacina* was found effective.

2.15.2. Powdery Mildew, *Podosphaera leucotricha*

i) Symptoms: It is found in all the apple growing countries of the world. The disease infects leaves, blossoms, green shoots and the fruits (Fig. 2.36). On leaves, the fungus appears as felt-like patches or as a solid mat on the surface, particularly on their under surfaces. Initial infections on the underside of the leaf may cause chlorotic patches or spots on the upper surface. Infected leaves tend to crinkle, curl or roll upwards along the edges, giving them a narrow appearance. During severe infection, leaves will usually drop prematurely during the summer. Infected shoots are stunted. Severe infection leads to defoliation of the mildewed shoots with a subsequent curtailment of its growth and reduction in number of its potential buds. Small grayish or white patches are seen on the underside of leaves. Diseased twigs are crinkled. Fruits show stunted growth along with typical rosetting.

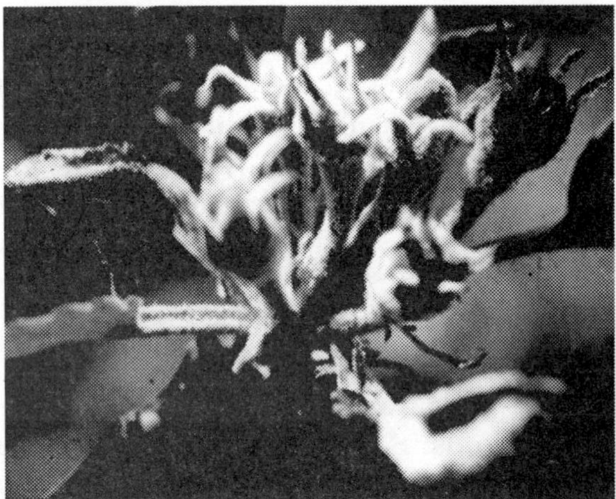

Fig. 2.36. Powdery mildew on apple.

ii) Epidemiology: Prevalence of more than 90% relative humidity and temperature ranging between 19° to 22°C favoured the conidial germination. However, dry weather favoured further disease development. Increased light intensity also restricts disease development. The fungus over winters in the form of mycelium in diseased vegetative buds and fruits. Secondary spread is through wind-borne conidia.

iii) Management

- Pruning after shedding of leaves, thinning out and cutting back of laterals, removal and destruction of all diseased parts reduces the disease incidence.

- Pre-bloom spray of lime sulphur (1: 60 dilution) starting at the dormant stage and continuing throughout the bloom period at fortnightly interval controls the disease.
- Spraying of wettable sulphur at 0.3% is recommended during pre-bloom and bloom period.
- Spraying of binapacryl at 450 g a.i./ha in 225 litres of water gives effective control.
- Application of 0.1% carbendazim is also effective against powdery mildew.
- Apple cvs. Maharaja Chunth and Golden Chinese have been found resistant to powdery mildew disease.

2.15.3. Seedling Blight, *Sclerotium rolfsii*

Seedling blight is of common occurrence in apple nurseries and affects 2-3 year old seedlings. Heavy losses have been reported by this disease wherever apple is grown. In India, the disease is distributed in Karnataka and Himachal Pradesh. In Himachal Pradesh, seedling mortality up to 40% has been reported in some of the registered nurseries (Bhardwaj and Agarwala, 1986a).

i) Symptoms: Blighting of the infected seedlings is the common symptom which is visible when the seedlings get dry. Mustard seed size fungal sclerotia appear in the vicinity of dead seedlings.

ii) Disease Cycle: A soil temperature of 25°-30°C, pH 6.0 and 30% or above soil moisture have been found to be most suitable for disease development (Bhardwaj and Agarwala, 1986b). Seedling mortality is directly proportional to soil temperature. The seedling mortality is greater in sandy loam soil than in loamy sand and clay loam soil.

iii) Management

a) Chemical Methods: Soil drenching with 3000 ppm of thiram, 300 ppm of brassicol and 400 ppm of aureofungin have been found effective against the pathogen.

b) Host Resistance: Rootstocks M 2, M 4 and MM 106 are found to be resistant to seedling blight.

2.15.4. Black Rot/Black Canker, *Physalopora obtusa*

The disease is also known as die-back or smoky blight canker.

i) Symptoms: The disease appears in India in three phases, i.e, stem canker, leaf spot (frog eye leaf spot) and fruit rot (black rot) of which the canker phase is the most destructive (Fig. 2.37). In this, reddish brown sunken lesions develop on trunk and branches which turn smoky with a series of alternate rings. The wood below is stained reddish brown. Spurs, branches and twigs above the canker are killed. Numerous pimple-like protuberances may appear over the bark of the blighted twigs or along the margins of the canker.

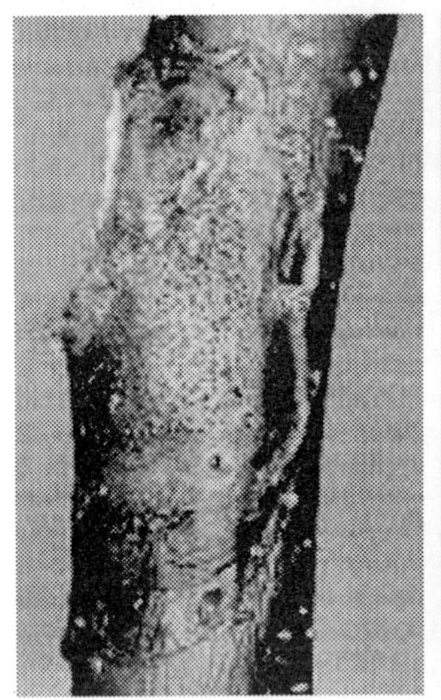

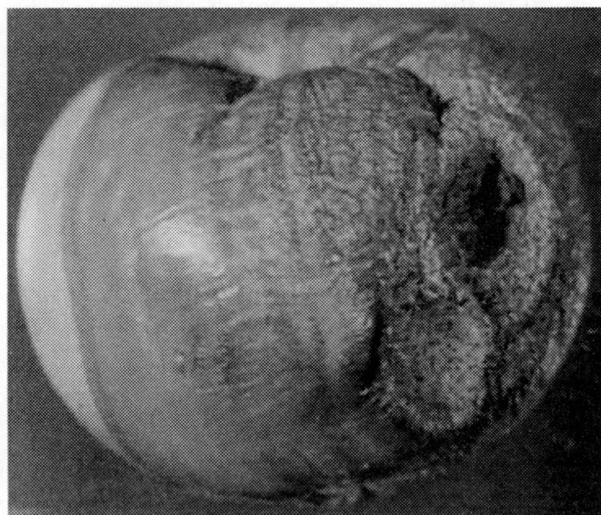

Fig. 2.37. Black rot on stem and fruit of apple.

ii) Epidemiology: The pathogen survives winter in the form of dormant mycelium and fruiting structure in cankered mummified fruits and on dead wood. High humidity (>75%) and a temperature of 20-22°C favour disease development (Singh and Agarwal, 1988).

iii) Management: Avoid mechanical injury. Apply wound dressers on cut areas. Remove girdled limbs. Cankers must be scraped with clean knife and covered with Bordeaux (Cu So_4 – 1 part, lime and linseed oil – 2 parts each) or Chaubatia paste (Lead oxide – 1 part, Copper carbonate – 1 part, Linseed oil – 11 parts). Remove all the mummified fruits, dead and pruned twigs from the orchard and burn.

2.15.5. Collar Rot, *Phytophthora cactorum*

The disease is also known as crown rot and trunk canker is universally present in all the apple growing regions of the world including USA, UK, Canada, New Zealand, The Netherlands, Germany, Australia and India. On susceptible varieties, it causes extensive losses even resulting in death of apple trees within a few years.

i) Symptoms: The above-ground symptoms are often confused with white root rot. The infection starts from the collar region and spreads to the under-ground parts. Bark at the soil level becomes slimy and rots resulting in cankered areas. The attacked trees are recognized by chlorotic foliage with red colouration of veins and margins.

ii) Epidemiology: The fungus is known to survive in most of the orchard soils but a soil temperature of 12°-20°C associated with pH 5-6 is found to be the best for its survival. Majority of infections in apple occur through mycelial penetration of stems near the ground line but zoospore infection may occur since these are liberated in the soil. Pieces of bark containing the oospores may also serve as the source of infection, which usually occurs during damp and cool weather in the spring. Temperature of 20°-25°C along with high soil moisture is favourable for sporangial production but oospores are produced at low moisture level at the same temperature.

iii) Management

- Clean the infected collar area with a sharp knife and apply Bordeaux or Chaubatia paste. Grafting of rootstocks 30 cm above the ground level helps in reducing the disease incidence.
- Drenching the area around tree trunk with fungicides like Dithane M-45 or metalaxyl was found effective.
- Use of antagonists like *T. viride*, *T. harzianum*, *Enterobacter aerogenes* and *Bacillus subtilis* along with soil solarization have also been found to protect the plants from root rot infection.
- Rootstocks M2, M4, MM104, MM113 and MM114 have been reported to be resistant.

2.15.6. White Root Rot, *Dematophora necatrix*

i) Symptoms: This is a soil-borne disease which is most important as the fungus can cause death of plants. The disease appears on underground plant parts and causes complete rotting of the roots. The fine roots are attacked first which are completely devoured and infection spreads to main root through secondary root system. Lateral roots turn dark brown and become infected with white flocculent fungus during monsoon months. Bronzing of leaves, stunted growth and size are important above-ground symptoms. Root rot affected trees are usually associated with a heavy blossom and fruiting next year. However, in succeeding years, few leaves emerge and much of the immature fruits fail to reach maturity. Infected trees often persist for 2-3 years depending upon infestation of fungus.

ii) Epidemiology: Small black sclerotia may be formed over the dead bark of wood. Infection of new roots is brought about by the mycelial growth in the soil or contact of plant with old dead roots left in the soil from previous infected plants. The disease incidence is severe when soil moisture is excessive during June to September and 100-140 cm rainfall. Soil type and soil pH between 6.0-6.5 are more favourable for survival of the pathogen. High textured soil is more favourable for development of disease.

iii) Management

- Hot water treatment of infected seedlings at 45°C for 1 hr, digging up of isolation trenches, amending the structure of clayey soils by adding more organic matter, following central drainage system have proven quite effective in preventing the spread of disease and increasing tree life.

- Rotten roots must be cut and the cut ends painted with Bordeaux or Chaubatia paste.
- Soil solarization by polyethylene tarping further improved the control.
- Soil drenching with carbendazim, benomyl, thiram, thiabendazole, pyrazophos, aureofungin has been recommended.
- Soil treatment with neem cake + *T. harzianum* was effective.
- The affected trees should be approach grafted with seedlings.
- Use of antagonists like *T. viride*, *T. harzianum* and *Enterobacter aerogenes* along with soil solarization have also been found to protect the plants from root rot infection.

2.15.7. Blotch or Premature Leaf Fall Complex, *Marssonina coronaria*

Since 1995, this disease has spread in Himachal Pradesh and caused ravage in apple resulting in premature leaf fall in the mid-season.

i) Symptoms: Disease appears both on leaves and fruits. Dark brown patches develop on upper surface of leaves varying from 5-10 mm in diameter as typical blotch symptoms (Fig. 2.38). Under favourable climatic conditions, these spots coalesce to form large dark brown blotches and the surrounding areas turn yellow. This yellowing advances towards petiole and induces mid-season defoliation. Irregular necrotic blotches on mature leaves characterize the disease. In affected orchards, apple fruits nearing maturity are commonly seen hanging on the defoliated trees. Repeated premature leaf shedding results in failure of the crop in the following season. Golden Delicious varieties are most severely affected.

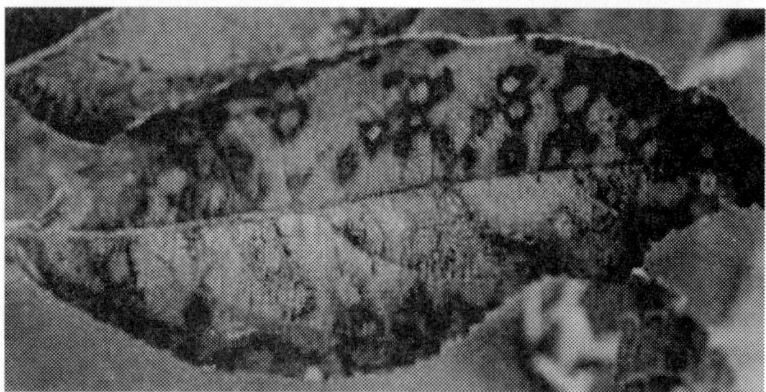

Fig. 2.38. Leaf blotch on apple.

ii) Epidemiology: Temperature above 20°C in June-July and relative humidity above 70% continuously for 5-6 days induce defoliation. Prolonged hot and dry weather affect the primary infection and disease development adversely.

2.15.8. Soft Rot, *Pencillium expansum*

The disease is responsible for major losses in India.

i) Symptoms: In the early stages, the rotted area is light in colour and soft and watery in texture. The rotted area does not become sunken although it is very soft in texture. The rot advances rapidly. In humid atmosphere, grey blue cushions of fruiting bodies of the pathogen appear on the diseased skin.

ii) Management : The skin injury should be avoided by stringent precautions in picking, grading and packing. The diseased and injured fruits should be discarded. Growth and spread of the rot is very slow at low temperature and may be almost entirely inhibited at 0 to 1.6°C. Therefore, the fruits should be placed in cold storage after packing. Sulphur dioxide fumigation by sodium metabisulphite was significantly effective in checking the disease (Kaul and Munjal, 1982). Apple coated with paraffin oil and pungent mustard oil withstood rotting. Fruits wrapped with SOPP-impregnated wrappers provided some protection against *P. expansum*. Benomyl at 1000 ppm was the best treatment for the management of soft rot at 25°C (Roy, 1975). Dipping fruits in 1000 ppm diphenylamine for 5 min or in 500 ppm of aureofungin for 20 min provided the effective control of the disease (Naik and Joshi, 1973).

2.15.9. Whisker's Rot, *Rhizopus arrhizus*

i) Symptoms: The name "Whisker's" is derived from the appearance of fruits covered with fungal mycelium and black fruiting bodies extended into thin cavities. The flesh becomes sour in taste without any undesirable odour. Occasionally, the infected fruits appear to be free from any fungal growth, but on cutting, sparse mycelial growth with black sporangial heads are seen in cavities. Rotting advances rapidly at 32 to 38°C. Humidity appears to have slight influence on disease development.

ii) Management: The fruits should be stored in cold storage maintained at 10°C. Overripe fruits should not be stored. Maximum protection to fruits from rot is provided by 5% 2-aminothiozole with 5% thiourea. Fumigation with sodium metabisulphite, wrapping with dichlonitroaniline wraps and coating with pungent mustard oil proved useful in controlling the disease (Kaul and Munjal, 1982).

2.15.10. Bitter Rot, *Glomerella cingulata*

i) Symptoms: The disease appears as a faint, light brown discolouration beneath the skin. The discolouration increases in diameter and extends in a cone shape. The rough circular lesion becomes depressed. Tiny, raised, black dots appear beneath the cuticle and give rise to acervuli which break the cuticle. In moist weather, sticky pink masses of spores are seen which may be arranged in clearly defined rings. With the increase of lesions, the entire fruit rots.

The perennial cankers, dead wood and mummified fruits serve as source of inoculum. Infection occurs directly through cuticle after formation of the appresorium. Lenticulate infection is more common and mature fruits are more susceptible.

ii) Management: The disease can be prevented by one spray of copper fungicide. Treatment with 2% trisodium phosphate solution can check the disease in storage.

2.15.11. Pink Mold Rot, *Trichothecium roseum*

i) Symptoms: The rot affects all delicious varieties of apple. The pathogen incites core rot of apple and is quite prevalent and destructive. The attack is both external and internal but is sometimes only internal. The internal infection becomes evident by a pinkish fluffy growth of the fungus. The rot is usually found associated with scab following which the disease has two well marked stages. In the first, rot appears as sunken, about 5 cm wide brown bands encircling the scab spots with fungal threads on the rotted area producing pink spore masses under moist warm conditions. In the second or final stage of rot is characterized by chocolate brown sunken area with an irregular outline which remains rather firm and dry. The affected tissue has a bitter taste. In the absence of scab, rotting may begin from any point of injury on the fruit surface starting as light brown lesions radiating from the point of infection, producing a firm dark brown rot. In advanced stages, light pink conidia appear, the rotted flesh becomes musty and the seed cavity contains a pink spore mass. The pathogen enters through mechanical injuries, such as bruises, cuts, stem punctures and open calyx tubes. Direct entry is possible through the lenticels.

ii) Management: Fruits should be placed in cold storage below 10°C to reduce the incidence of the disease.

2.15.12. Core Rot, *Alternaria alternata*

i) Symptoms: The disease is commonly found in apples having open pore at the calyx end. The spores of *A. alternata* splashed by rain drops fall at the calyx end. When such fruits are stored, the fungus grows within the core and penetrates inside the flesh. The maximum incidence of the disease (16-62%) in Jaipur (Rajasthan) markets was observed in April (Durrani, 1976). Typical rot symptoms of the disease are nearly round, brown to black lesions, often centered around a skin break or weakened tissue. The spots are firm, dry and shallow. The surface of spots becomes dark brown to black in the advanced stages. The rotted tissue becomes spongy and the affected flesh turns black.

ii) Management: Treatment of fruits with thioacetamide suppressed development of symptoms up to 35 days (Vir and Gaur, 1971). Post-harvest dip treatment of fruits with 500 to 1000 ppm of aureofungin could prevent the rot up to six days after inoculation (Durrani, 1976).

Table. 2.26. Calendar of IPM activities in Apple

Crop stage	Target pests	IPM activities
Dormant season module (October-November)		
Dormant (after natural leaf fall)	Sunburn, lichen & green moss	Apply Bordeaux paste + linseed oil in 1: 1 ratio (1 kg of each in 9 litres of water) to stem portions only.
Spring season module (April-June)		
Pink bud, petal fall	Powdery mildew	Spray 0.05% carbendazim for powdery mildew and scab. Remove suckers and water sprouts at the base of tree trunk and on major scaffold limbs and paint all the cut ends with Bordeaux paste.
Summer season module (July-September)		
Fruit set (pea-sized fruits)	Marssonina blotch, Scab	Spray 0.3% mancozeb for premature leaf fall and scab.
Fruit development	Marssonina blotch, Scab	Spray 0.3% mancozeb for premature leaf fall and scab.
Fruit development (20 days after earlier spray)	Marssonina blotch, Scab	Spray 0.3% mancozeb for premature leaf fall and scab.
Fruit development (20 days after earlier spray), pre-harvest & harvesting stage	Marssonina blotch, Scab, Sooty blotch, Fly speck	Spray 0.25% mancozeb for diseases and good quality fruits.
Post-harvest stage	Premature leaf fall, Scab, Skin injury	Collection and destruction of fallen leaves. Spray 0.05% carbendazim or 0.3% copper oxychloride to delay leaf fall till late autumn and to prevent skin injury and sun burn.
Fruit development (rainy season)	White root rot	Sow cabbage and garlic (non-host crops) and solarize soil. Apply 0.05% carbendazim or Bordeaux mixture in 30-60 cm deep holes around the tree basin. Apply bacterial antagonists – *Enterobacter aerogenes* and *Bacillus subtilis*. Add deodar needles and neem cake in soil around the stem.
Early rainy season	Collar rot	Drench soil around the tree trunk with ridomil or Bordeaux mixture. Apply bacterial antagonists – *Enterobacter aerogenes* and *Bacillus subtilis*. Sow cabbage and garlic (non-host crops).
Leaf yellowing stage	Premature leaf fall, Scab	Spray urea at 10 kg/ha for early decomposition of diseased leaves in tree basins.

Recipe for Various Pastes Used During Dormant Season

Bordeaux Paste: With the start of natural defoliation process indicating dormancy, apple plant stem and main branches have to be covered with Bordeaux paste so that direct sunlight do not cause sunburn damage to stem (otherwise it may lead to canker development in the following season).

Copper sulphate	-	1 kg
Lime	-	3 kg
Water	-	9 litres
Linseed oil	-	1 litre

Chaubbatia Paste: This is applied for early recovery of wounds in apple by natural healing at a faster rate so as to prevent development of canker. The paste can be prepared by slow heating of copper carbonate in linseed oil and adding red lead oxide to formulate the paste.

Copper carbonate	-	500 g
Red lead oxide	-	500 g
Raw linseed oil	-	1 litre

Cow Dung Paste: This is a conventional technique for the control of canker as wound sealant. It was used initially as stem dressing paste to avoid cold injury in winter. Later, it proved as a good alternative to fill larger wounds.

Fresh cow dung	-	1 kg
Clay soil	-	1 kg
Copper sulphate powder	-	10 g

2.16. PEACH AND PLUM

Peach (*Prunus persica*) is a temperate fruit rich in proteins, sugar, minerals and vitamins. In India, peach is grown in Himachal Pradesh, Punjab, Haryana, Delhi, west Uttar Pradesh, Jammu & Kashmir and Assam.

2.16.1. Leaf Curl, *Taphrina* spp.

The disease has been known in Europe since the early part of nineteenth century. In India, it is prevalent in Himachal Pradesh, Jammu and Kashmir and Kumaon hills of Uttar Pradesh where cool and moist weather prevails.

i) Symptoms: Symptoms of leaf curl are confined to current year's growth of leaves, twigs, blossoms and fruits. The leaf blade along the midrib becomes thickened and puckered. Infected twigs become swollen and are killed. Infected flowers and fruits drop rapidly. Peach is the main host but symptoms also appears on other stone fruits, i.e. plum, apricot, almond and cherry.

ii) Epidemiology: The causal fungus is carried over the seasons in twigs and with the onset of spring, infection takes place.

iii) Management

a) Cultural Methods: Removal and burning of infected shoots reduce the spread of the disease. Removal of weed hosts which acts as secondary hosts is essential.

b) Chemical Methods: A dormant spray of 6:10:100 Bordeaux mixture with an adhesive and a winter spray with 1.2% Bordeaux mixture before bud burst stage give effective control of the disease. A single spray of carbendazim or copper oxychloride before the break of dormancy (swelling of buds) is highly effective in controlling the disease (Sharma *et al.*, 1987, 1988).

2.16.2. Scab, *Venturia carpophila*

Scab has been recorded only on peach in India (Baruah *et al.*, 1980). Elsewhere, it has been reported on plum, cherry and apricot.

i) Symptoms: Symptoms on fruits appear as circular, dark olivaceous spots 2-4 mm in diameter which later coalesce forming a uniform dark olivaceous velvety patch. They are superficial, i.e. not below the epidermal layer of the cells. The fungus thus reduces the fruit quality. On twigs, fungus produces dark brown, oval lesions of 3-6 mm in diameter.

ii) Management

a) Cultural Methods: Cultural practices are very effective in preventing infection. Low lying planting sites should be avoided and the trees should be thinned to allow free air circulation.

b) Chemical Methods: Spraying plants with thiophanate methyl before blooming and benomyl one month before picking of fruits gave effective management of peach scab.

2.16.3. Brown Rot of Fruits, *Monilinia fructicola, M. fructigena, M. laxa, M. laxa* f.sp. *mali*

Among the fruit rots, brown rot is the most important in stone fruits including peach.

i) Symptoms: Symptoms of the disease are blighting of blossom and leaves, canker production on woody tissues and rotting of fruits. Blossom blight is the first symptom during spring and attacked parts turn grey to dark brown. The fungus spreads through peduncle and reach branches causing twig blight. Stem cankers usually develop from blighted twigs or fruit spurs. Fruit rot is the most destructive phase of the disease. Rotted fruits either fall down or hang as firm mummies (Sharma and Kaul, 1988).

Conidia and ascospores produced on the mummified fruits and canker spots serve as a source of primary infection.

ii) Management

a) Chemical Methods: Spraying plants with thiophanate methyl before blooming and benomyl one month before picking of fruits gave effective management of brown rot disease. Storage of fruits under controlled environment having a mixture of 11% carbon monoxide, 4% oxygen and 5% carbon dioxide gave complete control of *M. fructicola* on peaches (Kader *et al.*, 1982). Post-harvest dipping of fruits in benomyl, carbendazim, iprodione, triforine has also been found effective in controlling the disease in storage (Sharma and Kaul, 1989). Boron at 750 μg/ml gave effective control of whisker's rot of peaches (Sharma *et al.*, 1989).

2.17. PEAR

2.17.1. Leaf Spot, *Mycosphaerella chaubattiensis*

i) Symptoms: The disease is confined to the leaves. The mature spots are easily recognized by their grayish white centres and well defined margins.

ii) Management: Spraying 0.1% carbendazim in third week of June, 0.3% zineb 20 days later and again 0.1% carbendazim after 20 days gave effective control of the disease (Shah *et al.*, 1985).

2.17.2. Leaf Blight, *Fabraea maculata*

i) Symptoms: The pathogen produces symptoms on leaves, shoots and fruits. On leaves, spots appear first as purple dots which later turn to circular deep purple or dark brown lesions of 6 mm diameter. Small black pimple-like fruiting bodies also appear in the centre of these spots. On the fruits, black sunken spots appear. Infected fallen leaves serve as primary inoculum.

ii) Management: Early application of ferbam or Bordeaux mixture has been successfully used for its control.

2.17.3. Rust, *Gymnosporangium sabinae*

i) Symptoms: Rust is a serious disease of pear in Europe when grown in the neighbourhood of the alternate host *Juniper sabianae*. The rust on *Juniper* is perennial and resembles that on apple and quince. The leaves are extensively infected and defoliation results.

ii) Management: The disease can be controlled by removal of *Juniper* trees from within 150 m of pear trees.

2.17.4. Fruit Decay

i) Symptoms: The most important fungi responsible for pear decay are *Botrytis cinerea, Cytospora ambiens, Glomerella cingulata, Mycosphaerella fructigena, Pencillium expansum* and *Tricothecium roseum*. The cumulative incidence of different pathogens varies from 2.1 to 19.7% (Sharma, 1989a).

ii) Management: Pre-harvest application of 0.1% carbendazim and post-harvest dipping of fruits in vincozoline or cold storage of fruits at 6°C helps to reduce the disease incidence (Sharma, 1989b).

2.18. POMEGRANATE

Pomegranate (*Punica granatum*) is an ancient and favourable table fruit in tropical and sub-tropical regions of the world. It is commercially grown for its sweet-acidic fruits which provide a cool refreshing juice and is valued for its medicinal properties. Its popularity is also due to the ornamental nature of the plant especially when bearing bright red flowers throughout the year.

In India, it is considered as a crop of the arid and semi-arid regions because it can be grown even under conditions of severe drought and frost. Pomegranate is grown in an area of 1,16,400 ha with annual production of 0.849 million tonnes (Table 2.27). The fruit rind, juice, leaf and roots are used in the preparation of various Ayurvedic medicines. Fruit is good for stomach inflammations and heart pain, diarrhea and dysentery. Juice is useful for leprosy patients. The stem, root and bark are useful in destroying parasitic worms and cure of fever. The dry rind is useful in teeth and gum disorders. The fruit juice contains glucose and fructose. The juice and seed contain large quantities of tannin and agolic acid which are essential in curing several diseases.

Table 2.27. Area, production and productivity of leading pomegranate growing states (2005-06) (Bijay Kumar, 2007)

States	Area (in '000 ha)	Production (in '000 MT)	Productivity (in MT/ha)
Maharashtra	91.0	593.6	6.5
Karnataka	12.4	138.6	11.2
Andhra Pradesh	7.7	65.0	8.4
Gujarat	3.8	38.2	10.1
Tamil Nadu	0.4	11.1	25.0
Rajasthan	0.3	1.8	6.0
Chhattisgarh	0.1	0.6	6.7
Himachal Pradesh	0.6	0.2	0.4
Total	116.4	849.1	7.3

2.18.1. Cercospora Leaf Spot, *Cercospora punicae*

i) Symptoms: Light brown zonate spots appear on the leaves and fruits (Fig. 2.39). Black and elliptical spots appear on the twigs. The affected areas on the twigs become flattened and depressed with raised edge. Such twigs dry up. In severe cases, the whole plant dies.

Fig. 2.39. *Cercospora* leaf spot on leaves and fruit of pomegranate (Courtesy: K. Srinivasan, TAFE, Chennai).

ii) Epidemiology: The pathogen spreads through wind-borne conidia. The disease is serious during September-November.

iii) Management

- Pruning and destruction of diseased twigs helps to reduce the inoculum.
- Spraying with 0.1% thiophanate methyl or 0.2% chlorothalanil or 0.25% mancozeb is recommended.

2.18.2. Anthracnose, *Colletotrichum gloeosporoides*

i) Symptoms: The disease appears as small, regular to irregular dull violet or black spots on the leaves. These spots are surrounded by yellow margins. Later, the spots coalesce and become bigger. The infected leaves turn yellow and drop off. The fungus also infects fruits also (Fig. 2.40).

ii) Epidemiology: The pathogen spreads through wind-borne conidia. The disease is severe during August-September when there is high humidity and the temperature between 20° and 27°C.

iii) Management: The disease is effectively controlled by spraying with 0.1% carbendazim + 0.2% copper oxychloride or 0.1% thiophanate methyl or 0.2% mancozeb or 0.15% kitazin at fortnightly interval.

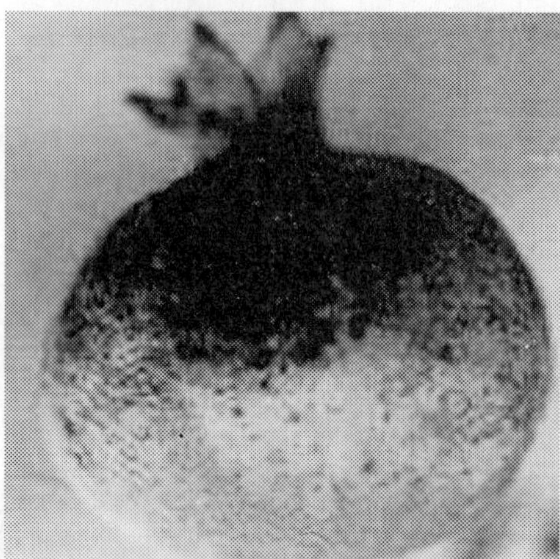

Fig. 2.40. Anthracnose on pomegranate fruit (Courtesy: K. Srinivasan, TAFE, Chennai).

2.18.3. Sphaceloma Leaf Spot, *Sphaceloma punicae*

i) Symptoms: The disease attacks leaves, shoots, calyx and fruits. Rusty spots appear on leaves. Infected leaves turn yellow and die. Rust coloured pustules appear on fruits.

ii) Epidemiology: Drizzling rains and abundant dew favour disease development and spread.

iii) Management: Spraying with 0.1% carbendazim or 0.1% thiophanate methyl helps to manage the disease.

2.18.4. *Phomopsis* Fruit Rot, *Phomopsis* sp.

i) Symptoms: The disease starts from calyx and gradually spreads over the entire fruit. Pycnidia of the pathogen appear on the affected area. The dry rot starts from leaves and move to the fruits.

ii) Management: One spray with Perenox can check the secondary infection of the disease. The damage can be greatly reduced by three sprays at 10 days interval with 0.5% perenox with Albolium.

2.18.5. Soft Rot, *Rhizopus arhizus*

i) Symptoms: The disease starts as small spots which increases in size and coalesce. The disease appears to be restricted to the rind, but internally the entire fruit decays into a pulpy mass. Under dry conditions, cracking originates from point of infection. Maximum disease intensity has been observed at 80% relative humidity.

ii) Management: Storage of fruits at 0°C does not show any symptoms of the soft rot. Packing straw with SO_2 producing chemicals can protect the fruits from soft rot (Kanwar and Thakur, 1972). Satisfactory control of the disease has been achieved by treatment of fruits with linseed, mustard and castor oils (Kanwar and Thakur, 1974).

2.19. BER

Ber, *Ziziphus mauritiana* (Syn. *Z. zuzuba*) is a native of Indian Subcontinent. It grows in tropical and sub-tropical regions, found all over India. Fruits are used as an ingredient in the preparation of 'Joshanda" (an Ayurvedic medicine used in chest trouble). The fruits are laxative and invigorating. The leaves are used to treat conjunctivitis. They are antihelminthic and good in stomatitis, gum bleeding, asthma and liver complaints. Bark is used for the treatment of diarrhea. The root is used as decoction in fever and its powder is applied to ulcers and old wounds. The kernels are sedative and as soporific which are used to stop vomiting. It is also used for abdominal pain in pregnancy.

2.19.1. Powdery Mildew, *Oidium erysiphoides* var. *ziziphi*

The disease is distributed in almost all ber growing regions of India.

i) Symptoms: The initial symptoms are seen on flower buds as powdery growth of the fungus. As the disease advances, the powdery growth will be seen on small fruits of pinhead stage. In severe cases, the entire fruit is covered with white powdery growth of the fungus resulting in reduction of yield and market quality of fruits. The spots change into light brown to dark brown discolouration, infected area becomes slightly raised having cankers and fruits crack. Finally, shedding of young fruits occurs.

ii) Epidemiology: The pathogen survives on the branches, twigs and bud wood during the absence of flowers and fruits which may serve as the primary source of infection. The pathogen may also survive on alternate hosts including weeds. The secondary spread of the disease takes place by the air-borne conidia. The disease is more prevalent during dry period. The rainfall at four weeks lead time has significant positive reaction with disease development. The decrease in night temperature and increase in day temperature further favour the disease spread.

iii) Management

- Pruning in between trees during second fortnight of April to first fortnight of May reduces the incidence of powdery mildew.
- Spraying with 0.3% wettable sulphur alternated with 0.1% tridemifon is recommended for the management of ber powdery mildew in northern Karnataka (Jamadar and Desai, 1998). Benomyl, carbendazim, karathane and sulfex were also found effective fungicides for the management of the disease. The spraying should be done during third week of November when the fruits attain peanut size. Generally, 2-3 sprays are required for effective control.

- Ber cvs. Safeda Rohtak, Sua, Noki, Chonchal, Sanaur 5, Katha Phal, Sanaur 1, Illachi Jhajjar, Kakrola Gola, Kala Gora, Pathani and Mirchia were found resistant (Kapur *et al.*, 1975; Jeyarajan and Cheema, 1972). The ber cv. Umran was moderately resistant to powdery mildew.

2.19.2. *Alternaria* Leaf Spot, *Alternaria* spp.

i) Symptoms: The disease is characterized by the formation of small, irregular brown spots on the upper surface of the leaves. On the lower surface, dark brown to black spots are formed. The spots coalesce to form big patches. The leaves drop later.

ii) Epidemiology: The disease development is favoured at 20° to 30°C with an optimum at 25°C. Plant debris serve as potential source of primary infection. High humidity and frequent rainfall seem to be more important than temperature for the disease development. Frequent rainfall possibly helps in dissemination of spores which cause secondary infection (Madaan and Chand, 1985).

iii) Management

- The disease can be controlled effectively by spraying 0.2% difolatan or zineb (Gupta and Madaan, 1978).
- Ber cvs. Bahadurgarhia, Govindagarh Special, Gola Gurgaon, Popular Gola, Seo Bahadurgarhia, Safeda Rohtak, Jhajjar Special and Mirchia were found resistant to leaf spot (Gupta and Madaan, 1980).

2.19.3. Black Leaf Spot, *Isariopsis indica* var. *ziziphi*

Black leaf spot, also called as *Isariopsis* moldy leaf, is reported from Haryana state, India.

i) Symptoms: Sooty tuft-like circular to irregular black spots develop on leaf surface. When infection advances, it covers a large area on the lower surface of the leaves and upper surface shows brownish discolouration.

ii) Epidemiology: The disease starts during October-November. The fungus survives in plant debris and soil which serve as primary source of infection. The secondary infection occurs through spores present in air.

iii) Management

- The disease can be controlled by spraying 0.2% carbendazim or difolatan (Gupta and Madaan, 1985).
- A mycoparasite, *Hansfordia pulvinata* was found growing on the diseased spots which can keep the disease under check (Gupta and Madaan, 1979a).

- Ber cvs. Safeda Rohtak, Mudia-Murhera, Sua, Sanaur 1, Pathani, Jhajjar Selection, Seo Bahadurgarhia and Jhajjar Special were found resistant to black leaf spot (Gupta *et al.*, 1980).

2.19.4. Sooty Mold

i) Management: Spraying 1% Bordeaux mixture starting at the appearance of the disease and there after 3 more sprays at 10-15 days interval is recommended.

2.19.5. *Cercospora* Leaf Spot, *Cercospora ziziphi, C. jujubae*

The disease caused by *C. jujubae* was reported from Punjab, Maharashtra, Delhi and Madhya Pradesh; while *C. ziziphi* was reported from Karnataka, Andhra Pradesh and Haryana.

i) Symptoms: Circular to oval spots appear on leaves measuring up to 4 mm in diameter, epiphyllous, yellow at first and turning to brown surrounded by dark brown margin.

ii) Management

- Spraying with 0.2% zineb gave satisfactory control of the disease.
- Ber cvs. Safeda Rohtak, Kakrola Gola, Bahadurgarhia, Reshmi, Jhajjar Selection, Popular Gola and Seo Bahadurgarhia were found resistant (Gupta and Madaan, 1985).

2.19.6. *Cladosporium* Leaf Spot, *Cladosporium zizyphi, C. herbarum*

The disease is reported to be caused by *C. zizyphi* in Maharashtra, Bihar and Punjab; while the disease is caused by *C. herbarum* in Haryana.

i) Symptoms: The disease appears on lower surface of leaves as small, light brown to brown irregular spots.

ii) Management

- Spraying with 0.2% copper oxychloride twice at two weeks interval at the appearance of the symptoms is recommended.
- Ber cvs. Banarasi, Villaity, Govindgarh Selection 3, Jhajjar Selection and Jogia are reported to be resistant (Gupta and Madaan, 1985).

2.19.7. Rust, *Phakospora zizyphivulgaris*

The disease has been reported from Bihar, Maharashtra, Haryana and West Bengal.

i) Symptoms: On the lower surface of leaves, small, irregular, reddish brown uredo pustules appear which later spread to cover the whole leaf. The infected leaves start drying and ultimately defoliation takes place. The disease is observed during January-February on both wild and cultivated varieties.

ii) Management

- Spraying with 0.2% zineb or mancozeb or wettable sulphur is recommended for the management of the disease.
- Cvs. Banarasi, Seo, Katha-Gurgaon, Laddu, Dandan, Sanar-1, Gola-Gurgaon-2, Safeda Selection, Sanar-3, Kishmish, Narma and Safeda are reported to be resistant (Gupta and Madaan, 1985).

2.19.8. *Glomerella* Fruit Rot, *Glomerella cingulata*

i) Symptoms: The disease is characterized by the production of brownish discolouration at the stem end which gradually spreads around the fruit. Under humid conditions, dot-like bodies appear profusely over the surface of the fruit, oozing salmon coloured spore masses. Perithecia were observed in advanced stages. Severely infected fruits become mummified. Ber cvs. Umran was found to be most susceptible followed by Dandan and Kaithli.

ii) Management: Ber cv. Wild Jhar Ber was the least susceptible to fruit rot.

2.19.9. *Phoma* Fruit Rot, *Phoma hissaernsis*

i) Symptoms: The infected fruits are characterized by small slightly depressed dark brown spots near the stem end. The erumpent pycnidia develop on these spots. The lesions become irregular in shape measuring 15-25 mm in diameter. The fungus survives in diseased debris which become the primary source of infection.

ii) Management: The disease can be controlled by spraying 0.2% zineb.

2.19.10. *Alternaria* Fruit Rot

i) Symptoms: The disease is caused by *Alternaria* state of *Pleospora infectoria*. It is characterized by slightly depressed brown to dark brown, circular lesions on the fruits. Sometimes concentric rings are also present on these spots. The smaller spots coalesce to form the bigger spots. The fungus survives in debris and soil. The fruits touching the soil surface gets initially infected and later on the disease spreads through spores present in air.

ii) Management: Spraying of 0.2% zineb or 0.2% difolatan has been reported to be effective against the disease.

2.19.11. *Colletotrichum* Fruit Rot, *Colletotrichum gloeosporoides*

i) Symptoms: The disease appears during February-March when the fruits start ripening. The symptoms start with small slightly depressed, light brown, water-soaked lesions on fruits. Later they enlarge and coalesce to form bigger spots. Under humid conditions, the acervuli of the pathogen are formed on the spots in masses. The fungus survives in diseased debris in soil and becomes the primary source of infection on the fruits touching the soil surface. The secondary infection occurs through the spores present in air or through the splash of rain.

ii) Management: The disease can be controlled by spraying 0.2% copper oxychloride 2-3 times at 21 days interval.

2.20. CUSTARD APPLE

The custard apple, also known as *Sitaphal*, Sugar apple, Sweet sop and *Sharifa*, is hardy and tolerant to drought. It grows very well even in shallow soils and its flowering coincides with maximum moisture availability period. It also sheds off leaves during stress periods to evade moisture loss from plant tissues through transpiration and thus a most appropriate fruit crop for arid region. Custard apple fruits are very popular in Deccan plateau and are also grown commercially in smaller scale in Assam, Andhra Pradesh, Gujarat, Bihar, Karnataka, Madhya Pradesh, Maharashtra, Orissa, Uttar Pradesh and Rajasthan. Large area is reported from Andhra Pradesh.

The fruit is used for table purpose. The edible portion or pulp is creamy or custard-like, granular with good blend of sweetness and acidity. The pleasant flavour and wild aroma have a universal liking. The ripe fruits are rich in carbohydrates and a good source of Vitamins A and C. Glucose is the predominant sugar. Fruit pulp can be used in ice creams, confectionary and certain milk products. The seeds contain 27-30% oil which can be used in soap and paint industry. The seed cake containing 4% nitrogen is beneficial as manure. Powdered seed when mixed with mung seed at 100 ppm is effective against storage pests for 100 days. Seed extract is effective for the control of houseflies. Crushed leaves have insecticidal properties. Roots are reported to have purgative properties.

2.20.1. Leaf Blight, *Alternaria* sp.

i) Symptoms: The disease is present in almost all the custard apple growing areas. Affected leaves show light brown spots which later develop into large blotches. These infections have an irregular outline and sometimes involve a major part of the lamina. The greenish black fruiting bodies are observed on the surface of the affected parts. Such leaves ultimately drop off. The infections have been observed on the twigs of affected leaves.

ii) Management: The disease can be managed by spraying the plants with 0.2% zineb or 0.3% copper oxychloride at fortnightly interval.

2.20.2. Dry Rot, *Colletotrichum gloeosporoides*

i) Symptoms: The initiation of the disease takes place at the stem end of fruit. The infection is characterized by blackening of fruits at the stem attachment which spreads towards the periphery. The black spots with grayish centres are also observed on the individual compartments. The affected parts become hard and mummified. Sometimes, the entire fruit turns black, hard and remain attached to the twig. Small circular to irregular spots with brick red margins are observed on the leaves. The centre of these spots turns white with dark acervuli.

ii) Management: The disease can be effectively managed by removing affected fruits, pruning and destroying the twigs bearing such fruits. Spraying with 0.1% carbendazim or Topsin M or 0.2% Dithane M-45 at biweekly interval are helpful in controlling the secondary spread of the disease.

2.20.3. Fruit Rot, *Glomerella cingulata*

i) Symptoms: The disease has been reported from Assam. The infection starts from blossom-end in the form of a dry, blackish brown spot which spreads very slowly in all directions and covers the entire fruit. Later, the infected fruits get shriveled and drop or remain on the tree.

ii) Management: Spraying 1% Bordeaux mixture at 2-3 litres/tree when infection starts at blossom end helps in reducing the disease incidence. Spraying with 0.1% carbendazim or Topsin M or 0.2% Dithane M-45 at biweekly interval are also helpful in controlling the secondary spread of the disease.

2.21. PHALSA

Phalsa, *Grewia subinaequalis* (Syn. *Grewia asiatica*) belonging to Family Tiliaceae is of Indian origin. It can be grown in all parts of north India except at higher elevations. Its plantation is mainly confined to the states of Haryana, Punjab, Gujarat, Rajasthan, Uttar Pradesh and Maharashtra. It can come up very well as filler crop. The tiny fruits have to be picked from bush several times during the fruiting season. It comes to the market during April-May.

Phalsa fruits are used as table fruit and in preparation of squash and juice. Jam and pickle can also be prepared from phalsa fruits. Ripe fruits are sub acidic and good source of Vitamins A and C.

2.21.1. Leaf Spot, *Cercospora grewiae*

i) Symptoms: On affected leaves, tiny brown lesions appear on both sides of leaves and cause premature leaf fall particularly during rainy season.

ii) Management: Spraying 0.3% maneb or 0.2% copper oxychloride keeps the disease under control.

2.22. AMLA

Amla (*Emblica officinalis*) is an important fruit crop owing to its hardy nature, suitability to various waste lands, high productivity/unit area (15-20 tonnes/ha), nutritive and therapeutic value. It can be cultivated in salt-affected and ravenous areas and semi-arid regions of Karnataka. There has been a steady increase in area and production of amla as a result of identification and development of suitable varieties [Kanchan (NA 4), NA 6 and NA 7] and development of improved production technologies.

A budded/grafted amla tree starts bearing third year onwards after planting, attain full bearing within 10-12 years and continue to bear for 60-75 years under well managed conditions. An amla tree may bear 100-300 kg fruits/tree, giving 15-20 MT/ha.

Amla being a deep-rooted, deciduous tree with sparse foliage, is an ideal plant amicable for 2- and 3-tier cropping system. Fruits, vegetables, flowers and a few medicinal and aromatic plants are well suited for intercropping in amla orchards. Some models are: amla + ber, amla + guava, amla + ber + phalsa, amla + dhaincha + wheat/barley, amla + dhaincha + onion/garlic/ fenugreek/ brinjal and amla + dhaincha + German chamomile.

Amla fruits are rich in Vitamin C (ascorbic acid content-500-625 mg/100g pulp). Amla fruits are highly valued among indigenous medicines. It is acrid, refrigerant, diuretic and laxative. Dried fruits have been reported to be useful in hemorrhages, diarrhea, dysentery, anemia, jaundice, dyspepsia and cough. 'Trifla' and 'Chavanprash' are well known indigenous medicines in Ayurvedic system prepared from amla.

2.22.1. Rust, *Ravenelia emblicae*

Rust is an important disease of amla in Rajasthan. The disease has also been reported from Rehmankhera (Lucknow) and Rajgarh (Pratapgarh) areas of Uttar Pradesh.

i) Symptoms: On fruits, the pustules are initially black which later develop into a ring (Fig. 2.41). The pustules coalesce and cover a big area on the fruit. The black spores get exposed by rupturing a papery covering. The fruit gives a dirty appearance. On leaves, pinkish brown pustules develop which may appear singly or in groups (Fig. 2.41).

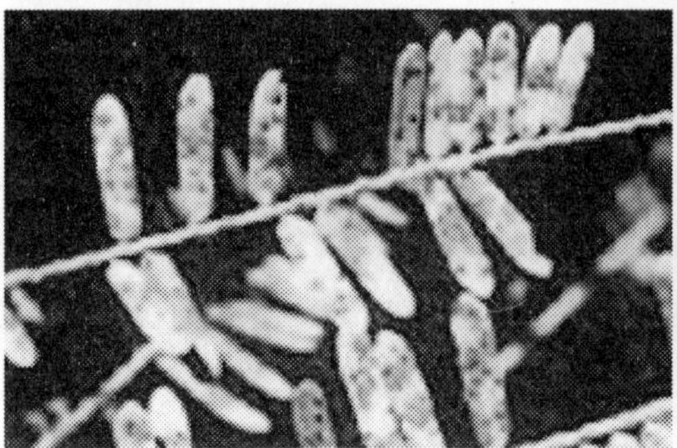 

Fig. 2.41. Rust on leaves and fruits of amla (Courtesy: K. Srinivasan, TAFE, Chennai).

ii) Management: Three sprays of 0.5% wettable sulphur at an interval of one month from July was found effective against rust disease. Three sprays of 0.1% Bayleton at 15 days interval controlled the disease up to 95%. Sprays of 0.2% zineb are also effective.

2.22.2. Anthracnose, *Colletotrichum* sp.

i) Symptoms: The disease occurs on leaves and fruits during August-September (Fig. 2.42). Initial symptom of the disease appears in the form of minute, circular, brown to grey spots with yellowish margin on leaflets. The central area of spots remains grayish with dot-like fruiting bodies. On fruits, the depressed lesions, which later turn dark in the centre forming acervuli which are often arranged in rings. The lesions may vary in size and shape with spore masses appearing on fruiting bodies at high humidity. Consequently, the fruits become shriveled and rot (Misra and Shivpuri, 1983).

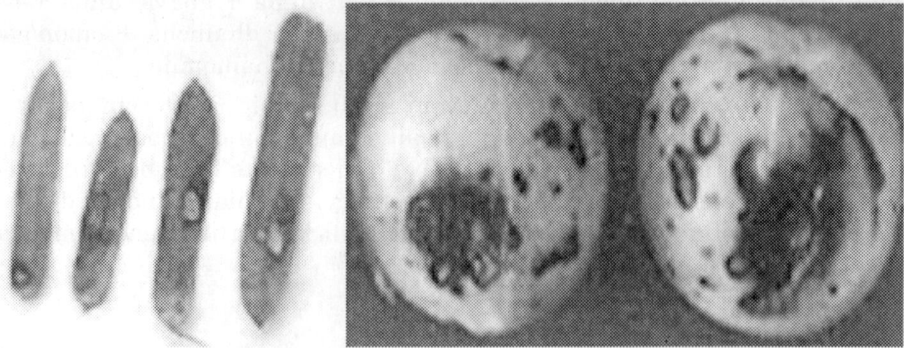

Fig. 2.42. Anthracnose on leaves and fruits of amla (Courtesy: K. Srinivasan, TAFE, Chennai).

2.22.3. Soft Rot, *Phomopsis phyllanthi*

i) Symptoms: Smoke brown to black, round lesions develop on fruits within 2-3 days after inoculation. The infected part later turns to olive brown with water-soaked area which covers the entire fruit within 8 days resulting in deformed fruits. The disease is more prevalent during December to February. Although the fungus causes infection in young and mature fruits, the latter are preferred. Injury is essential for the development of the disease (Lal *et al.*, 1982).

The pathogen also infects *Pyrus malus, Psidium guajava, Citrus sinensis, C. medica, Musa paradisiaca, Luffa cylindrica, Abelmoschus esculentus* and *Citrullus vulgaris.*

ii) Management: Spraying of 1500 ppm difolatan or 100 ppm Dithane M-45 or 100 ppm carbendazim just after harvest was found effective for the management of the disease.

2.22.4. Bark Splitting, *Rhizoctonia solani*

i) Management: Soil treatment with *Aspergillus niger* is effective for the management of bark splitting disease.

2.23. FIG

The fruits are valued for their laxative property and are rich in protein digestive enzymes. It is also effective in the treatment of piles, constipation, asthma, sexual weakness, corns and chronic cough. It also removes physical and mental exertion and endows the body with renewed vigour and strength. It is an excellent tonic for weak people who suffers cracks in lips, tongue and mouth.

2.23.1. Rust, *Cerotelium fici*

i) Symptoms: Small round, brown to black eruptive lesions develop generally on lower surface of leaves (Fig. 2.43). As the disease advances, defoliation and premature fruit drop takes place.

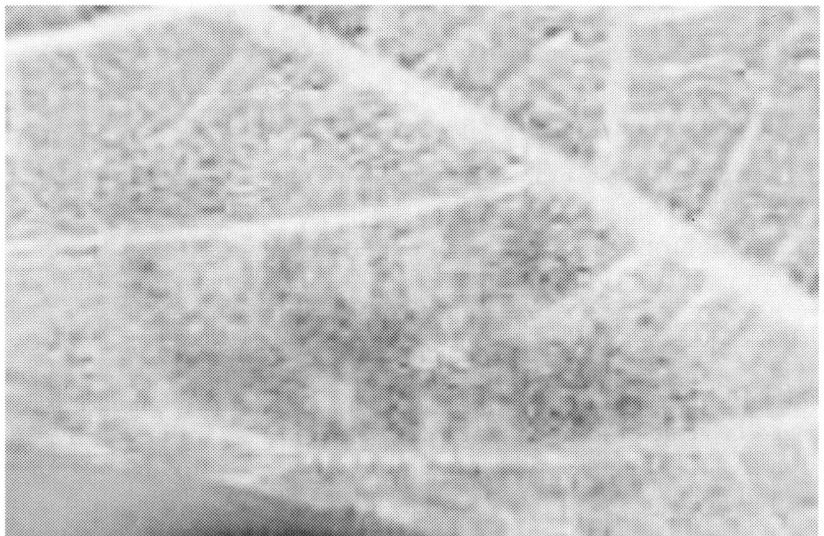

Fig. 2.43. Rust on leaf of fig (Courtesy: K. Srinivasan, TAFE, Chennai).

ii) Epidemiology: The pathogen spreads through wind-borne urediospores.

iii) Management: Spraying with 0.25% mancozeb or 0.2% zineb or 0.1% propiconazole is recommended for the control of the disease.

2.23.2. Leaf Spot, *Cylidrocladium scoparium*

The disease has been reported from Uttar Pradesh.

i) Symptoms: In the beginning, minute brown spots appear on the leaves (Fig. 2.44). They enlarge to uniform size which are zonate, prominent, reddish lesions with dark brown margins. Later the lesions coalesce to form irregular patches. The tissues in the lesions disintegrate and fall off leaving a hole. Infected leaves are shed 20-30 days earlier than healthy leaves.

ii) Management: Spraying with 0.25% mancozeb + 0.1% copper oxychloride is recommended.

Fig. 2.44. Leaf spot on fig (Courtesy: K. Srinivasan, TAFE, Chennai).

2.23.3. *Ascochyta* Leaf Spot, *Ascochyta caricae*

i) Symptoms: Symptoms appear on the upper surface of leaves after rains. Spots are irregular, grey measuring 0.5 to 2.0 cm in diameter.

ii) Management: Singh and Singh (1986) reported that the disease can be effectively controlled by spraying 0.1% benomyl or carbendazim.

2.24. JAMUN

The jamun, *Syzygium cuminii* (*Eugenia jambolana*) belonging to Family Myrtaceae is considered to be native to India or East Indies. Fully ripe fruits are eaten fresh and can be processed into beverages, jelly, jam, squash, wine, vinegar and pickles. It is a good source of iron. The powdered seed is useful in the treatment of diabetes.

It is widely grown in the larger part of India from Indo-Gangetic plains in the north to Tamil Nadu in the south. It is one of the most hardy fruit crop and can be easily grown in neglected and marshy areas where other fruit plants can not be grown successfully.

2.24.1. Anthracnose, *Gloeosporium euginiae*

i) Symptoms: Characteristic appearance of black dead lesions or spots on leaves and sometimes on stem is noticed.

ii) Management: One or two sprays of 0.3% copper oxychloride gives effective management of the disease.

2.24.2. Leaf Spot and Fruit Rot, *Glomerella cingulata*

i) Symptoms: Affected leaves show small scattered spots which are light brown or reddish brown in colour.

ii) Management: Affected fruits rot and shrivel. The disease can be controlled by spraying 0.2% zineb.

2.25. TAMARIND

The major breakthrough in recent years has been the identification of tamarind types with less acidic pulp, commonly referred as sweet tamarinds. In Thailand, more than 50 sweet tamarind cultivars are grown, while in Philippines, 8 selected cultivars are popular among the farmers. Presently, sweet tamarind plantations are interplanted with annual crops in Thailand. Small scale plantations of sweet tamarind can be raised for the fresh fruit market.

India, the largest producers and exporters of tamarind, continue to grow sour types. Tamarind is most suitable for growing in regions that have extended spells of dry weather. Those that farm under adverse climatic conditions will benefit most due to the draught resistant nature of this species. Due to ease of cultivation, particularly in marginal areas which are often unsuitable for other crops and the potential for domestic use and export, tamarind could be considered a "Tree for the future". Tamarind attains maximum production during tenth year which yields 4 tonnes/ha (40 kg/tree) with a net income of Rs. 38,000 annually.

The leaves are an important source of food and herbal medicine and the edible pulp of ripe fruits is used as flavouring agent in cooking, soups, jams, chutneys, sauces, juices, etc. The fruit pulp of tamarind is the richest natural source of tartaric acid (8 to 18%) and is the chief acidulant used in the preparation of foods. The major industrial product of tamarind is the seed kernel powder which is an important material used in sizing of textiles, paper and jute. Tamarind seed protein contains highly favourable amino acids and could complement methionine and cystine in the diet, to alleviate protein malnutrition which is widely prevalent in developing nations.

Table 2.28. The major tamarind growing countries and its production (Gunasena & Hughes, 2000)

Countries	Production (in MT)
India	2,50,000 (1964)
Thailand	1,40,000 (1995)
Coata Rica	221 (1997)
Mexico	37
Puerto Rico	23 (1977)

During 1993, exports reached 11,145 tonnes valued at Rs. 10.8 million (US$ 2,40,300), and were in the form of fresh fruit, pulp, seed powder and paste. The importing countries are mainly Pakistan followed by UAE, Japan and Yemen. Over 75% of the imports are of sweet type, having a premium demand over sour types.

2.25.1. Powdery Mildew, *Erysiphe polygoni* and *Oidium* sp.

i) Symptoms: The powdery mildew disease is of common occurrence in nursery seedlings. The disease causes defoliation and early growth is severely retarded.

ii) Management: In order to have effective control of powdery mildew, 3-4 sprays of 0.2% wettable sulphur at 15 days interval is the most economical method (Siddaramaiah and Kulkarni, 1982).

2.25.2. Leaf Spot, *Cercospora tamarindi*

i) Symptoms: In Uttar Pradesh, India, tamarind leaves have been attacked by *Cercospora tamarindi* resulting in severe defoliation (Khan *et al.*, 1988).

ii) Management: The leaf spot can be easily controlled by spraying with maneb, benomyl or captan.

Table 2.29. Management of fungal diseases in fruit crops

Crop	Disease/Causal agent	Symptoms	Pesticide Dose/litre	Time of Application
Banana	Panama wilt, *Fusarium oxysporum* f. sp. *cubense*	Yellowing of leaves acropitically, hanging of leaves around pseudo stem, cracking and discolouration of vascular bundles	Carbendazim-1g	Dipping before planting followed by drenching at bimonthly intervals starting from 5th month of planting
	Sigatoka leaf spot, *Cercospora musicola*	Black streaks on the leaves in a serial manner surrounded by yellow halo. Affected leaves dry off.	Carbendazim-1 g Bitertanol/ Thiophanate Methyl-0.7ml Tridemorph-2 g Chlorothalonil/ Iprodione-2 g	During rainy season/ November
	Leaf spot/ Anthracnose, *Gloeosporium musae*/ *Colletotrichium gloeosporioides*	Regular to irregular spots on leaves with dark margins and grayish centre. On fruits typical dark brown depressed regions	Carbendazim-1 g Bitertanol/ Thiophanate Methyl/ IPUB/ Chlorothalonil/ Iprodione-2 g	In humid weather July-December
	Leaf blight, *Macrophomina phaseolina*	Dark brown patches on the leaves specially on the leaf lamina	Carbendazim-1 g Bitertanol/ Thiophanate Methyl-0.7 ml Tridemorph Chlorothalonil/ Iprodione-2 g	February/ March

	Stem end rot, *Botryodiplodia theobromae*	Blackening of fingers from stem end. The infected tissue becomes soft and fruits emit foul smell	Carbendazim-1 g Bitertanol/Thiophanate Methyl-2 g/ IPUB/ Chlorothalonil/ Iprodione-2 g	Pre-harvest spray and post harvest dip
Papaya	Powdery mildew, *Oidium caricae / Laveilulla taurica*	White mealy growth on leaves, petiole and young shoots	Wettable sulphur-2 g Triadimefon-1 g	September-January
	Damping off (Complex of organisms)	Causes drooping of seedlings	Captan/Carbendazim-2 g	Seed Treatment
	Anthracnose, *Colletotrichum gloeosporioides*	Dark brown spots of chocolate colour	Carbendazim-1 g, Thiophanate Methyl-2 g, Chlorothalonil/Mancozeb-2 g	15 days before harvesting
	Stem rot/Foot rot (Caused by group of fungi)	Rotting of stem near collar region	Chlorothalonil/Iprodione-2 ml, Metalaxyl/ Mancozeb/Al-Phosetyl-2 g	As and when required
Citrus	Powdery mildew, *Acrisporium tingitanium*	As in other crops	Tridemorph-0.5 ml Triadimefon/ Myclobutanil-0.6 g Wettable sulphur-2 g	September-March
	Stem rot/Fruit rot/ Leaf fold, *Phytophthora nicotianae* var. *parasitica*	Light water soaked spots on leaves starting from lamina. On fruits dark brown discolouration with soft tissue	Bordeaux mixture-1% Al-Phosetyl/ Metalaxyl/ Mancozeb-2 g	Soil drenching/ spray
	Wither tip/ Anthracnose, *Colletotrichum gloeosporioides*	Browning of tissue from tip to inside of leaf. Such leaf drop off giving barren look of the plant. Small twigs with dark black acervuli. On fruits chocolate colour	Carbendazim/ Bitertanol/ Thiophanate Methyl/ IPUB-1 g Chlorothalonil/ Iprodione-2 g	In humid weather and rainy season
	Leaf blight, *Alternaria citri*	Dark brown zonate spots on leaves. On fruits light brown to dark brown spots are observed	Iprodione-2 g Copper oxychloride-3 g	Winter months
Sapota	Leaf spot, *Pheophleospora indica*	Reddish brown circular spots	Ziram /Zineb/ Iprodione-2 g	
Grape-vine	Downy mildew, *Plasmopara viticola*	Pale olive spots with milky white mycelium on the under surface of leaves	Copper oxychloride-3 g Metalaxyl + Mancozeb-2 g Al-phosetyl-2 g Mancozeb-2 g	3rd leaf stage (Contact fungicides should be sprayod at 3-7

				days interval, systemics should be sprayed at 10-15 days interval)
	Powdery mildew, *Uncinula necator*	Grayish white growth on leaves and berries	Wettable sulphur-2 g Triademafon-1 g Myclobutanil-0.5 g Cyperconzole-0.5 g Fenarimol-0.5 gm	1st week of December/ cloudy nights
	Anthracnose, *Sphaceloma ampelina*	Dark brown sunken spots on the aerial parts	Carbendazim-1 g Thiophanate methyl-1 g Bitertanol-1 g Mancozeb-2 g	Rainy season
	Greenaria berry rot, *Greenaria uvicola*	Water-soaked yellowish spots	Captan-2 g Thiophanate methyl-1 g Bitertanol-1 g	Rainy season
	Alternaria leaf blight and stem rot	Dark brown zonate spots	Ziram-4 ml Iprodione/Zineb-2 gm Bitertanol-1 gm	October-November
Mango	Powdery mildew, *Oidium mangiferae*	Mealy white growth on leaves & inflorescence	Wettable sulphur-2 g Tridemorph/Dinocap-2 g Triadimefon/ Myclobutanil-1 g; Hexaconazole-1 g	December 2nd week
	Anthracnose, *Colletotrichum gloeosporioides*	Leaf spot and fruit rot with characteristic spots	Carbendazim/ Thiophanate methyl-1 g Benomyl/IPUB-1 g Iprodione/ Chlorothalonil-2 g	Rainy season spray/fruit dip before storage
	Stem end rot, *Botryodiplodia theobromae*	Light brown to black spots on the stem end side of the fruits	Carbendazim/ Thiophanate Methyl-1g Benomyl/IPUB-1 g Iprodione/ Chlorothalonil-2 g	Pre harvest spray and post harvest dip
	Fruit rot, *Alternaria alternata*	Light brown to irregular spots	Iprodione/ Chlorothalonil-2 g	Pre harvest spray and post harvest dip
	Blossom Blight	Drying of inflorescence which turns light brown to black. Infected flower fail to set fruits	Ziram-4 ml Mancozeb+ Dinocap-2g+1ml, Iprodione-2 g	Flowering stage
Guava	Fruit Canker, *Pestalotiopsis psidi*	Cankerous growth on fruit leading to cracking	Ziram-4 ml Zineb/ Iprodione-2 gm	Rainy season

	Fruit rot, *Phomopsis psidi*	Chocolate coloured spots from the styler end.	Chlorothalonil/ Iprodione-2 g Carbendazim/Bitertanol/ Thiophanate Methyl-1 g	Rainy season
	Anthracnose, *Colletotrichum gloeosporioides*	Dark brown to chocolate coloured depressed spots on fruits. On leaves red coloured pustules.	Carbendazim/ Thiophanate Methyl-1 g	June-Sept
Ber	Powdery mildew, *Oidium zizyphi* var. *indicae*	White mealy growth on leaves and fruits	Wettable sulphur-2 g Dinocap-1 g, Triadimefon/ Myclobutanil/ Fenarimol-0.5 ml	July-Sept
Pomegranate	Leaf spot, *Cercospora punicae / Colletotrichum gloeosporiodes*	Light brown to dark brown spots on leaves.	Mancozeb-2 g	Rainy season
	Fruit rot, *Cercospora punicae / Colletotrichum gloeosporioides*	On fruits dark brown depressed spots	Chlorothalonil-2 g Carbendazim-1 g Thiophanate Methyl/ Bitertanol/Benomyl-1 g	

Table 2.30. Biological control of fruit crop diseases

Fruit Crop	Disease/Pathogen/s	Potential biocontrol agent/s	Mode of application
Banana	Panama wilt, *Fusarium oxysporum* f. sp. *cubense*	*T. viride, Aspergillus niger, Pseudomonas fluorescens, T. viride + P. fluorescens*	Sucker treatment Pseudo stem injection Soil treatment
Citrus	*Phytophthora* spp.	*Trichoderma viride / T. harzianum* at 100 kg/ha.	Soil treatment
Citrus (Mandarin)	Root rot, *Phytophthora nicotianae* pv. *parasitica*	*T. viride, T. harzianum, T. virens*	Soil treatment
Citrus (Orange)	Blue mold, *Pencillium italicum*	*T. harzianum*	Fruit treatment
	Root rot, *Macrophomina phaseolina*	*Glomus fasciculatum, Gigaspora gigantia*	Soil treatment
	Root rot (*Phytophthora* sp.)	*Trichoderma* spp.	Soil treatment
Strawberry	Grey mold, *Botrytis cineria*	*T. harzianum*	Through honey bees
Mulberry	Leaf spot, *Cercospora moricola*	*T. viride, T. harzianum*	Foliar spray
	Stem canker & Die-back, *Botryodiplodia* spp.	*T. virens, T. harzianum, T. pseudokoningii*	Cutting & soil treatment
	Cutting rot, *F. solani*	*T. virens, T. harzianum, T. pseudokoningii*	Cutting & soil treatment

	Collar rot, *Phoma sorghina*	*T. virens, T. harzianum, T. pseudokoningii*	Cutting & soil treatment
Grapevine	Powdery mildew	*Ampelomyces quisqualis*	Foliar spray
	Grey rot, *Botrytis cinerea*	*T. viride*	Aerial spray
Guava	Anthracnose, *Pestalotia psidii, Colletotrichum gloeosporoides*	*T. harzianum*	Spray
	Fruit rot, *Lasiodiplodia theobromae, Phomopsis psidi, C. gloeosporoides, Pestalotiopsis versicolor, Rhizopus arrhizus*	*Trichoderma* sp.	Fruit treatment
	Wilt, *Gliocladium roseum* & *F. solani*	*Pencillium citrinum*	Soil treatment
		Aspergillus niger (AN 17)	Soil treatment
		T. harzianum	Soil treatment
Mango	Fruit rot, *Rhizopus arrhizus, Lasiodiplodia theobromae*	*Trichoderma* sp.	Fruit treatment
	Anthracnose, *Colletotrichum gloeosporoides*	*T. harzianum*	Foliar spray
		Streptosporangium pseudovulgare	Foliar spray
	Powdery mildew, *Oidium mangiferae*	*S. pseudovulgare*	Foliar spray
Passion fruit	Collar rot, *R. solani*	*T. harzianum, Trichoderma* sp.	Seed treatment
Amla	Bark splitting, *Rhizoctonia solani*	*Aspergillus niger*	Soil treatment
Apple	White root rot, *Dematophora necatrix*	*T. viride, T. harzianum, T. virens*	Soil treatment
	Scab, *Venturia inequalis*	*Chaetomium globosum, Cladosporium* spp., *Trichothecium roseum*	Foliar spray

Chapter 3

VEGETABLE CROPS

China is the largest producer of vegetables contributing 47.3% of the total world production (Table 1.1). Vegetables occupy an area of 22.518 million ha in China with a production of 434.332 million MT. Productivity of vegetable crops is highest (33.8 MT/ha) in Spain.

India is the second largest producer of vegetables next only to China contributing 11.9% of the total world production. Vegetables occupy an area of 7.164 million ha with a production of 109.050 million tonnes (Bijay Kumar, 2007) (Tables 3.1 and 3.2). Vegetable crops contribute 58.93% share to the total production of horticultural crops. India produces 14.7% of the world onions. In vegetables, India occupies the first position in peas; second in cabbage, cauliflower, onion, brinjal; third position in potato and fourth in tomato. The per capita availability of vegetables has increased from 76 g in 1951 to 256 g in 2001 as against 280 g/day recommended by ICMR for a balanced diet. Increase in production is mainly due to increase in productivity. Productivity is still lower in India (15 t/ha) than the world average productivity (15.7 t/ha). There is scope for increasing the productivity of vegetables in India further. The vegetable production target for 2020 has been fixed at 150 million tonnes.

Table 3.1. Major vegetable producing countries of the world

Country	Area (in '000 ha)	Production (in '000 MT)	Productivity (in MT/ha)	% share of world production
China	22518	434332	19.3	47.3
India	7164	109050	15.2	11.9
USA	1307	36845	28.2	4.0
Turkey	1060	26290	24.8	2.9
Egypt	600	16165	26.9	1.8
Italy	593	15994	27.0	1.7
Iran	631	15760	25.0	1.7

Russia	928	15404	16.6	1.7
Spain	395	13356	33.8	1.5
Japan	424	11745	27.7	1.3
Mexico	687	11241	16.4	1.2
Korea	329	10965	33.3	1.2
Others	15984	200416	12.5	22.0
Total	52620	917563	15.7	---

Source: FAO (except Indian data) (Indian data source: Indian Horticulture Database 2006)

Table 3.2. Area, production and productivity of vegetable crops in India (2005-06)

Vegetable crops	Area ('000 ha)	Production ('000 MT)	Productivity (MT/ha)	% share of India production
Brinjal	553.9	9136.3	16.5	8.4
Cabbage	267.4	6921.6	22.1	5.4
Cauliflower	291.8	5260.1	18.0	4.8
Okra	376.1	3684.0	9.8	3.4
Onion	695.1	9248.4	13.1	8.5
Peas	314.0	2298.7	7.3	2.1
Potato	1550.2	29093.7	18.8	26.7
Sweet potato	129.4	1148.6	8.9	1.1
Tapioca	252.7	7924.9	31.4	7.3
Tomato	534.5	9361.8	17.5	8.6
Others	2199.5	25971.4	11.8	23.8
Total	7163.9	109049.5	15.2	

(Source: Indian Horticulture Database 2006)

Potato, tomato, brinjal, chilli, cabbage, cauliflower, onion, peas, gourds, okra and melons are now grown on commercial scale with increasing demand, better output, change in production system and development of suitable cvs. These crops have high potential for foreign exchange earning. Today the emphasis on vegetable breeding is on value addition, disease resistance and better quality.

The important vegetables exported include onions-7.1%, cabbage-4%, cauliflower-4%, okra-3%, peas-3%. Traditional vegetables like onion, potato, okra, bottle gourd and chilli are exported to Malaysia, Singapore, Gulf, Sri Lanka, Bangladesh, Pakistan and Nepal. Non-traditional vegetables like asparagus, cabbage, capsicum, sweet corn, baby corn, green peas, French bean, cucumber, gherkin, cherry tomato, Brussels sprout, broccoli, processed and frozen French bean and lima bean are exported to European countries, Australia, Gulf and S.E. Asian countries. French bean, lima bean, sweet corn and capsicum are exported to Gulf. Some of the new products introduced in export market are tomato paste in bulk aseptic packs, processed button mushrooms, freeze dried and instant frozen vegetables. Dehydrated onion and tomato products have great potential as export commodities.

Vegetable crops are more susceptible to diseases caused by fungi that result in reduction of yield and quality. Seed-borne diseases and their inoculum in vegetables have increased with adoption of OGL and GATT agreement. Diseases affecting from seedling stage to harvest and spoilage caused by pathogens during transit, storage and marketing are the main constraints in vegetable production. A total of 65 million tonnes as pre-harvest losses in vegetables due to diseases alone in the world have been estimated that accounts for 10.5%. Introduction of hybrids in vegetable crops has resulted in new disease problems like *Pseudocercospora* leaf blight in tomato and brinjal in India. Disease intensity of enation leaf curl in okra, chilli leaf curl, gummy stem blight, *Didymella* leaf blight, *Alternaria* leaf blight in cole crops and viral diseases in cucurbits have increased throughout the country. Intensive and monoculture of vegetables also increased the disease pressure of soil-borne pathogens such as *Sclerotium*, *Sclerotinia* and *Rhizoctonia*. The off-season cultivation of vegetables has prolonged the survival period of pathogens in the soil. The severity and incidence of major fungal diseases of various vegetable crops are presented in Table 3.3.

Table 3.3. Prevalence of fungal diseases of vegetable crops.

Crop	Disease	Pathogen	Severity (%)	Incidence (%)
Brinjal	Phomopsis blight	*Phomopsis vexans*	10-70	32.5
	Collar rot	*Sclerotium rolfsii*	10-30	20.0
	Leaf blight	*Alternaria alternata*	5-35	25.2
Tomato	Alternaria blight	*Alternaria solani*	15-90	40.0
	Septoria leaf spot	*Septoria lycopersici*	5-20	10.0
	Collar rot	*Sclerotium rolfsii*	10-40	20.0
	Fruit rot	*Rhizoctonia, Cladosporium, Pythium*	5-20	15.0
Chilli/Capsicum	Die-back	*Colletotrichum capsici*	15-60	35.0
	Anthracnose	*Colletotrichum capsici*	10-30	15.0
	Leaf blight & Fruit rot	*Alternaria alternata* *Alternaria solani*	5-20	13.7
	Leaf spot	*Cercospora capsici*	10-35	20.0
Cucurbits	Downy mildew	*Pseudoperonospora cubensis*	10-80	30.0
	Anthracnose	*Colletotrichum lagenarium*	30-90	55.0
	Cercospora leaf spot	*Cercospora citrulina*	20-50	30.0
	Vine & Fruit rot	*Sclerotium, Rhizoctonia, Pythium, Sclerotinia*	10-20	15.0

	Phytophthora blight	*Phytophthora cinnamoni*	25-60	35.0
Cowpea/Pea/ Beans	Leaf blight	*Cercospora cruenta, Pseudocercospora cruenta*	20-90	60.0
	Stem rot	*Macrophomina bataticola*	5-20	10.0
	Phoma leaf blight	*Phoma phaseolina*	3-15	12.0
	Rust	*Uromyces phaseoli*	5-30	10.0
	Powdery mildew	*Erysiphe pisi*	10-90	50.0
	Downy mildew	*Peronospora pisi*	5-60	15.0
	White rot	*Sclerotinia sclerotiorum*	2-35	25.0
Cole crops	Leaf spot	*Alternaria brassicae, A. brassicola*	10-40	26.0
	Downy mildew	*Peronospora parasitica*	5-25	10.0
Okra	Leaf blight	*Cercospora abelomoschi, C. hibisina*	25-95	70.0
Onion/Garlic	Purple blotch	*Alternaria porri*	15-80	40.0
	Leaf blight	*Stemphylium vesicarium*	20-40	25.0
	Downy mildew	*Peronospora destructor*	10-25	10.0
Vegetable nursery	Damping off	*Rhizoctonia, Alternaria, Phytophthora, Fusarium, Colletotrichum, Pythium, Sclerotium, Phomopsis*	30-85	50.0

3.1. POTATO

Potato (*Solanum tuberosum*) is one of the important vegetable crops grown in India. It is being cultivated in 1.550 million hectares producing 29.094 million tonnes of potatoes with an average yield of 18.8 tonnes per hectare (Bijay Kumar, 2007) (Tables 3.4 and 3.5). It thrives best in cool climate. Therefore it is a summer crop in the hills and a winter crop in the plains.

Table 3.4. Major potato producing countries of the world

Country	Area (in '000 ha)	Production (in '000 MT)	Productivity (in MT/ha)	% Share of world production
China	4884	73460	15.0	22.4
Russia	3071	37280	12.1	11.4
India	1550	29094	18.8	8.9
Ukraine	1516	19462	12.8	5.9
USA	439	19091	43.5	5.8
Germany	277	11624	42.0	3.6
Poland	588	10369	17.6	3.2
Belarus	462	8185	17.7	2.5

Netherlands	156	6777	43.4	2.1
France	156	6605	42.2	2.0
Others	6102	105368	17.3	32.2
Total	19200	327314	---	---

Source: FAO (except Indian data)
(Indian data source: Indian Horticulture Database 2006)

Table 3.5. Area, production and productivity of leading potato growing states in India (2005-06)

State	Area (in '000 ha)	Production (in '000 MT)	Productivity (MT/ha)
Uttar Pradesh	443.2	9986.8	22.5
West Bengal	320.6	7106.6	22.2
Bihar	308.9	5702.5	18.5
Punjab	75.0	1223.5	16.3
Gujarat	41.0	1088.5	26.6
Madhya Pradesh	53.3	799.7	15.0
Assam	90.1	629.1	7.0
Uttarakhand	22.2	450.6	20.3
Haryana	19.0	374.4	19.7
Karnataka	57.5	368.4	6.4
Others	119.5	1363.6	11.4
Total	1550.2	29093.7	18.8

(Source: Indian Horticulture Database 2006)

3.1.1. Early Blight, *Alternaria solani*

Early blight of potato, caused by the fungus *Alternaria solani,* poses a significant risk to crop productivity in the field and to tuber quality in storage. This disease can occur over a wide range of climatic conditions and can be very destructive, sometimes resulting in complete defoliation of plants. Early blight is found in all potato growing regions and is a major disease in hot climates, especially where crops are irrigated. This disease is most severe on maturing or nutritionally deficient potato vines.

i) Symptoms: Initial infection is most frequent on older leaflets on lower portions of plants. Lesions first appear as small, (1-2 mm) dark brown to black spots. The spots enlarge and develop concentric rings of raised and depressed necrotic tissue which gives the lesion a characteristic "target spot" appearance. Advanced lesions often have angular margins because of limitation by leaf veins and a narrow chlorotic halo frequently surrounds the spot (Fig. 3.1). As the disease progresses, affected leaves turn yellow and senescent and either dry up or fall off.

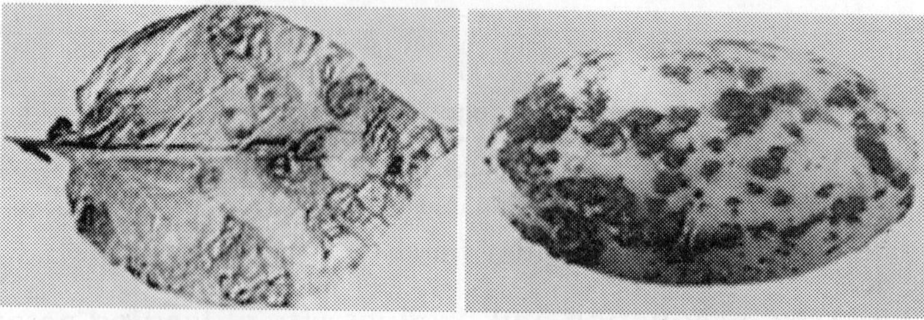

Fig 3.1. Early blight on potato leaf and tuber.

Tuber infection occurs less frequently than leaf infection and are usually evident only after several months of storage. Tuber lesions are dark, sunken, irregularly shaped and often surrounded by a raised violet border (Fig. 3.1). The underlying flesh is dry, leathery, and usually brown. In advanced decay, the tissue is often water-soaked and yellow to greenish yellow. Lesions can enlarge during storage, and tubers can become shriveled in advanced cases. Early blight tuber lesions are not as prone to invasion by secondary organisms as compared to other tuber rots.

ii) Disease Cycle: *A. solani* can persist in crop debris, soil, infected tubers, or other solanaceous hosts. Spores are carried primarily by wind and infect potato leaves under ideal conditions of warm temperatures (68-86°F) and high moisture levels. Disease development is also favored by high relative humidity (about 95 percent). Spores are frequently produced after heavy dews and frequent rains and are blown in from mycelium growing on debris or on infected cultivated plants. The germinating spores penetrate susceptible tissue directly or through wounds and produce new conidia that are further spread by wind, splashing rain, etc. Periods of alternating wet and dry weather favor spore dispersal and disease spread. Early blight is often more prevalent on older, senescing tissues and particularly when plants have been predisposed by injury, poor nutrition, insect damage, or other type of stress. Because of the relative resistance of young to intermediate-aged potato plants, major outbreaks of disease do not occur until late in the season.

If sufficient moisture is present, spores can germinate and infect tubers. Infection of potato tubers occurs mostly through wounds inflicted during harvest. Immature tubers are particularly susceptible to infection by *A. solani.*

iii) Management : Use disease-free seed tubers for planting. Avoid cultivation of solanaceous crops nearby potato fields. Spray 1% Bordeaux mixture to control the disease. Use disease resistant cvs. like Kufri Sindhuri, Kufri Jyoti, Kufri Lalima for planting.

The disease is controlled primarily through the use of resistant cultivars, cultural practices, and foliar fungicides such as chlorothalonil, maneb, triphenyl tin hydroxide and mancozeb. Cultivars with levels of field resistance are available, however no immunity to disease exists in commercial potato cultivars or in their wild parents. Highly susceptible cultivars such as Red Norland, Norchip, and Superior should be avoided in locations where early blight is prevalent and disease pressure is high. Field resistance to foliage infection is associated with plant maturity. Late maturing cultivars are usually more resistant than early maturing cultivars; therefore, one should avoid planting early and late varieties in the same or adjacent fields.

Cultural practices, such as crop rotation, removal and burning of infected plant debris, and eradication of weed hosts helps reduce the inoculum level for subsequent plantings and susceptible crops. Because *A. solani* persists in plant debris in the field from one season to the next, rotation with non-host crops reduces the amount of initial inoculum. Other cultural control measures include avoiding irrigation in cool, cloudy periods, harvesting fully mature tubers and avoiding bruising or other mechanical damage and adjacent potato and tomato fields planted the previous season should be plowed immediately after harvest. Current recommendations include a combination of cultural and chemical measures to retard the development of early blight and reduce the impact of this disease on yield and tuber quality.

To minimize tuber infection after harvest, tubers should be stored under conditions that promote rapid suberization. Suberization of the wounded surface provides a significant degree of protection against the penetration of *A. solani*. The storage of tubers for three weeks at approximately 61°F with more than 90% relative humidity will facilitate rapid wound healing.

The most common and effective control method of early blight is the application of foliar fungicides. The protectant fungicides recommended for late blight control (i.e. maneb, mancozeb, chlorothalonil, and triphenyl tin hydroxide) are effective against early blight when applied at approxi-mately 7-10 days intervals. The application of foliar fungicides is not needed in plants at the vegetative stage, when they are relatively resistant. Accordingly, spraying should commence at the first sign of disease or immediately after bloom. Frequency of subsequent sprays should be determined according to the genotype and age-related resistance of the cultivar. Protectant fungicides should be applied initially at relatively long intervals and subsequently at shorter intervals as the crop ages.

Early season applications of fungicides before secondary inoculum is produced often have minimal or no effect on the spread of the disease. Early blight can be adequately controlled by relatively few fungicide applications if the initial application is properly timed. Regular inspection of fields after plants reach 30 cm in height is recommended in order to detect early infections.

Disease forecasting models can be used to predict the onset of early blight and specify when the initial fungicide application for a season should be applied. For example, the computerized forecasting system, modified FAST, was developed to identify periods when environmental conditions are favourable for the development of *A. solani* on potato and to provide an efficient fungicide application schedule. This forecasting model calculates the increased susceptibility of potato plants according to their physiological age. It determines the number of days (P-days) after potato emergence at which plants become susceptible to the disease. In general, early maturing cultivars should be sprayed at approximately 250 P-days, whereas fungicide applications for late maturing cultivars should commence at approximately 300 P-days.

3.1.2. Late Blight, *Phytophthora infestans*

The late blight of potato shot into prominence during Irish potato famine in 1844-45 which resulted in one million deaths due to starvation and another 1.5 million fled the country when entire potato crops rotted in the field or in storage because of infection by *Phytophthora infestans*. The losses are caused by the foliage infection which leads to premature death of the plant and consequently the reduction in tuber yield. The losses are also caused directly by tuber infection which leads to rottage of infected tubers in the field and stores (Bhattacharya *et al.*, 1990). The disease is capable of wiping out entire potato commercial fields within a week if the conditions are wet. Farmers who grow potatoes are at serious risk of losing their entire income for the season. Losses are more in hilly regions where crop is grown under rain-fed conditions as compared to plains. In India losses range between 21-74% in Shimla hills, 11-38% in Mukteshwar hills, 33-65% in Darjeeling hills and 32-39% in Ooty hills.

i) Symptoms: Late blight affects the foliage and tubers of potato (Fig. 3.2). The characteristic foliar symptoms on any susceptible plant are irregular to circular lesions. Initially, in wet weather, the lesions may appear water-soaked; in dry weather, the centres tend to dry out. The dark centre is often surrounded by yellow, chlorotic tissue or brown, collapsed tissue. These spots often occur at the tips or edges of leaves. If moist weather persists, the leaf spots enlarge rapidly to blight larger areas of the foliage (Fig. 3.2). Dark brown discolouration of leaf stalks and stems is also commonly encountered. The symptoms frequently develop first on lower leaves but can develop at any level in the canopy. The symptoms are usually most dramatic inside the canopy. During humid weather, a white fuzz or mildew-like growth appears on the leaves (upper or under sides) and on stems. This indicates that the fungus is producing spores.

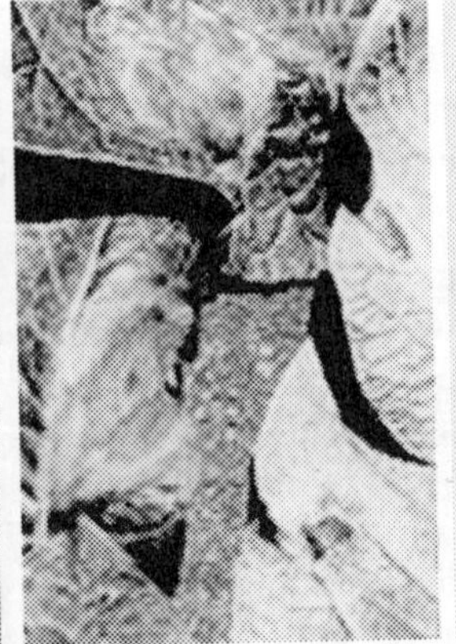
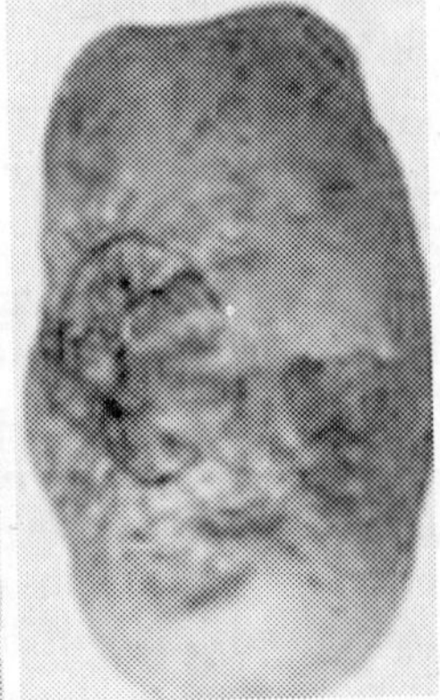

Fig. 3.2. Late blight on potato leaves and tuber.

Infected potato tubers are discolored by a reddish-brown, granular rot. The rot can be superficial or penetrate more deeply into the tubers (2.5 cm or more). Tubers must be cut open to observe this symptom. The rot continues to develop after the tubers are harvested. Further decay of the tubers is often hastened by secondary fungi and bacteria. The soft rot bacteria often turn tubers into a foul smelling liquid.

Tubers become infected when spores of the fungus wash down through the soil and come into contact with the fleshy organs.

ii) Epidemiology: The infected seed tubers and refuse piles in the temperate regions and cold stored infected tubers in the sub-tropical regions serve as the primary source of the inoculum.

The fungal growth and spores produced could only survive in living tissues. Infected potato tubers used for seed, infected tubers left in cull piles, and unharvested, infected tubers left in the ground were the sources of inoculum, or spores, for the disease from one growing season to another.

The development of the disease is favored by daytime temperatures of 21-27°C (70-80°F) and night time temperatures of 10-16°C (50-60°F). The presence of free moisture (such as dew, rain or fog) on the foliage also favors the disease. Free moisture is a prerequisite for spore germination and infection and thus is essential for disease development. As the foliage becomes further blighted, more spores are produced and spread of the disease to adjacent plants becomes more likely. As the season progresses, potato tubers can become infected as spores from the blighted foliage are washed down through the soil and contact developing tubers.

iii) Management: The effective management of late blight of potato can be achieved by reduction of both foliar and tuber infection by adopting three pronged attack i.e. by employing chemicals, cultural practices and host resistance in an integrated manner.

a) Cultural Methods: The most effective strategy for managing late blight is to avoid sources of inoculum (spores). Do not save potato seed from year to year. Use healthy certified potato seed. Many states that produce potato seed have seed certification programs to ensure that the seed meets certain standards for disease levels. Using certified seed also reduces the carry over of a number of potato viruses. Certified seed, however, is not a guarantee that late blight or other diseases will not be present. Examine your seed carefully before you plant, and plant only sound, blemish-free tubers. Examine tubers for late blight symptoms by cutting them open with a knife. Immediately sterilize or thoroughly wash the knife if late blight symptoms are found.

Use disease-free seed tubers for planting. Ridges should be made up high enough to cover all daughter tubers and reduce chances of their infection upon exposure.

Destroy any rejected seed tubers that you do not plant (bury them deeply in an area away from the field, burn them, or discard them in a plastic bag in the trash). Make sure any tubers put into compost piles are completely decomposed. Pull up and destroy potato plants that come up from tubers left in the field from the previous season.

If late blight becomes severe, remove diseased plants by digging them up. Destroy these plants immediately by one of the following: burying them deeply in an area away from the field, burning them, or by bagging them in a plastic bag and discarding the bag. These steps will help to avoid production of a larger number of spores. Harvest all potato tubers in the field. If late blight occurs when the tubers have already 'sized up', harvest the crop as soon as possible to avoid post-harvest tuber rot. Again, destroy diseased foliage and stems.

It is best to avoid wetting the leaves when you irrigate; soak the ground around the plants instead of spraying the plants with water. If this is not possible, water your garden mid-day so that the foliage dries off quickly. Eliminate weeds around the field so as to maximize air circulation around the plants. Additionally, earthing up the soil around the plants to provide a more substantial barrier against tuber infection from spores washing down through the soil.

The following cultural practices may be adopted to reduce the disease incidence:

- Use of disease-free seed.
- Removing/rogueing of ground keepers.
- Adjustment in planting time.
- High ridges and proper earthing up to avoid exposure of seed tubers.
- Water management – light or no irrigation when the blight is likely to appear.
- Killing of haulms at 75% crop damage.

b) Host Resistance: There are several potato varieties that show resistance to late blight. These varieties will slow down, but do not prevent, the development of the disease. Elba is currently the most resistant potato variety available. Potato varieties with moderate levels of resistance include: Kennebec, Sebago, Allegany, and Rosa.

In North Indian Plains, use disease resistant cvs. like Kufri Sindhuri, Kufri Jyoti, Kufri Badshah, Kufri Chipsona 1 and 2, Kufri Jawahar and Kufri Sutlej. In hills, use disease resistant cvs. like Kufri Jyoti, Kufri Giriraj in Himachal Pradesh hills, Kufri Kanchan and Kufri Jyoti in Darjeeling hills, Kufri Swarna and Kufri Thenamalai in Nilgiri hills.

c) Chemical Methods: Potatoes are susceptible to late blight at any time during the growing season. If the growing season is wet, late blight occurrence is very likely. Fungicides may be necessary to protect your plants from infection. Continuous fungicide coverage is necessary to protect plants from infection. Spray 1% Bordeaux mixture to control the disease.

When disease first appears, begin treatment with chlorothalonil, maneb or copper oxychloride. Additional pesticides that may be used later in rotation include Heritage or products containing the active ingredients *Bacillus subtilis* or basic cupric sulfate.

The following strategies can be adopted for chemical control:

- Application of prophylactic spray of contact fungicide and subsequently need based fungicide sprays.
- Application of 1 or 2 sprays of systemic fungicides mixtures (metalaxyl, ridomil, etc.) in areas where blight comes every year in epiphytotic form. First and last spray should be of contact fungicides (mancozeb, captafol, copper oxychloride, Bordeaux mixture) to avoid development of resistant strains of the fungus.

d) Biological Methods: Biocontrol agents like *Pencillium* sp. and *Trichoderma* sp. have been found effective.

3.1.3. Dry Rot or Black Scurf, *Rhizoctonia solani*

Rhizoctonia stem and stolen canker, commonly called black scurf because of its tuber phase, is an occasionally serious disease of potato found in all production areas of the world. This disease affects potato development from emergence to harvest. Under cold, moist soil conditions following planting, *Rhizoctonia* can affect emergence and the developing plant. The development of resting structures (sclerotia) causing the black scurf or "dirt that won't wash off" phase of this disease occurs after vine desiccation. Most reports of this disease claim it will limit total yield of a potato crop only under extreme conditions. *Rhizoctonia* can seriously limit marketable yield, however, by causing infected plants to produce an abnormally high percentage of tubers that are misshapen, knobby and cracked while also adversely affecting size distribution. *Rhizoctonia* will invariably decrease the amount of bruise-free potatoes during harvest due to the increased percentages of these misshapen and knobby tubers.

i) Symptoms: Perhaps the most readily apparent phase of *Rhizoctonia* disease is the black scurf or sclerotia present on tuber surfaces. These sclerotia can vary in size from very small, flat, superficial black specks to large, raised, irregularly shaped masses that can cover a major portion of the tuber. While black scurf can, under extreme conditions, affect the marketability of table stock/fresh pack potatoes, it is an extremely important seed-borne phase of the pathogen that deserves attention by all potato producers

Lesions characteristic of *Rhizoctonia* on stems and stolons are brown to black and sunken. These cankers can continue to expand and are capable of girdling stems and stolons of young developing plants (Fig. 3.3). *Rhizoctonia* infection of older plants very seldom lead to girdled stems that die. However, the health of these plants can be severely compromised and they can frequently become more susceptible to other diseases, particularly early blight.

Potato plants are most severely affected in the spring when underground sprouts can be killed prior to emergence. The secondary sprouts that develop are generally less vigorous and emerge much later causing irregular, uneven stands. Sprouts of severely affected plants that do not die are frequently stunted and remain so for the rest of the growing season. Infections of stolons that occur early in the growing season frequently result in pruning to tuber formation or abortion of tubers early in their development.

Fig. 3.3. Black scurf on collar region of potato.

Mid-season *Rhizoctonia* infections of potato plants will result in deep, sunken cankers on the main stem. The above ground portion of the plant will appear yellow with some purpling and upward curling of the foliage. These symptoms are frequently confused with leaf roll or purple top. A key in differentiating *Rhizoctonia* from these two diseases is the frequently formation of aerial tubers in the lower leaf axils and the brown, sunken lesions just below the soil surface.

ii) Epidemiology: *R. solani* can be either a soil-borne or seed-borne pathogen. The fungus survives in soil as mycelium in decomposing plant tissues. It also survives as sclerotia on tuber surfaces (seed-borne) or in the soil for extended periods. Populations of *R. solani* decline in the absence of a susceptible host although the rate of decline is affected by soil type, rotational crops and possibly the amount of organic matter present in the soil.

Disease development on emerging sprouts is favored by cold, wet soil conditions. Although these conditions may not increase disease incidence, disease severity is generally much greater. These conditions slow sprout development but favor germination of sclerotia and infection causing cankers to develop on young, under-developed tissues. Tuber-borne inoculum is very important in this phase of the disease while soil-borne inoculum is believed to generally more important in stem and stolon infections. However, tuber-borne inoculum is also important in the pruning of the first stolons developed by the plant. When stolons become pruned, the secondary stolons are generally weaker and much shorter in length. Tubers produced on these shorter stolons are generally the ones that become misshapen because they tend to grow around the plant stem. Cool, moist conditions, with moisture being the most critical factor, also favor disease development on stems and stolons. Disease development is optimum at around 65°F and decreases as soil temperatures increase.

iii) Management: Management of *Rhizoctonia* canker must be integrated since no single tactic is totally effective. First and foremost is crop rotation. In this regard it is important to consider both the number of years between potato crops and the rotational crops used in the system. Most research has shown that one potato crop in three or four is adequate to reduce soil inoculum to economic levels. However, soil type can potentially impact the rate of pathogen decline. Observations suggest that the rate of *Rhizoctonia* decline in sandy soils exceed that of loamy soils with more organic matter. Rotational crops such as sugar beet and dry beans, which are affected primarily by different AG's of *R. solani* tend to raise the general activity of *Rhizoctonia* in the soil thereby increasing the inoculum potential of soil-borne inoculum. Cereal crops such as oat and barley are generally regarded as beirg good rotations with potato for the reduction of soil-borne inoculum.

Since young developing potato crops are more susceptible to injury by *Rhizoctonia*, practices that encourage rapid emergence and plant development will reduce disease severity. Potatoes planted into soils with excessive amounts of residue from a previous crop, such as corn, tend to have serious *Rhizoctonia* problems. Soils with large amounts of residue remain cold for extended periods and more moist, thereby favoring disease development. Therefore, high amounts of crop residue should be dealt with prior to planting.

Cultural practices such as pre-cutting seed, shallow hilling or hill drag-off and delaying planting until soil temperatures are more than 60°F encourage emergence and will also reduce sprout pruning. It is also important to avoid high soil moisture during this period. Early irrigation, prior to emergence, should be kept to a minimum. Potato hills with steep sides and sharp peaks generally warm more rapidly than large, flat beds which can also reduce *Rhizoctonia* disease severity.

Seed selection is also important. Seed tubers free of obvious sclerotia is more desirable than seed with a significant amount of black scurf. Sclerotia coverage as little as 5% can significantly increase disease incidence and severity. In reality, however, much less black scurf sclerotia can significantly increase disease development. One black sclerotium per square inch, which is far less than 1% coverage, can significantly increase disease pressure under environmental conditions that favor disease development. For this reason, disease-free seed is the best option. Harvesting seed tubers within 2-3 weeks of vine desiccation will reduce the amount of sclerotial formation. Tubers left in the soil for more than 3 weeks after vine desiccation will generally have significantly higher levels of black scurf which can lead to deep discounts on the fresh market.

A number of fungicide seed piece treatments have been tested for their effectiveness in controlling the seed-borne *Rhizoctonia* inoculum. Of these, Maxim has consistently provided the best control of tuber borne *Rhizoctonia*.

Use disease-free seed tubers for planting. Seed treatment with acetic acid + Zn, boric acid or *T. viride* at 3.5 g/kg was found very effective. Combination of soil solarization of black scurf infested fields and seed treatment with *T. viride* further improved the disease control.

Soil application of mustard/groundnut cake at 2.5 MT/ha significantly reduced the number of black scurf diseased tubers and intensity of the disease.

Tuber damage and injury must be avoided during harvest and storage. Tubers should be washed to remove contaminated soil adhering to the tubers, followed by drying under shade reduces the infection substantially. In plains, tubers should be stored in cold stores. In country stores, tubers must be examined periodically and rotting tubers sorted out.

3.1.4. Wart, *Synchytrium endobioticum*

i) Symptoms: Tumors or warts appear on tubers (Fig. 3.4), stems and stolons. The wart pathogen is readily seed and soil-borne. Warted tubers used as seed are the chief means of disease spread. The wart pathogen may also spread through seed of wart immune varieties grown in infested land, contaminated soil carried on the feet of workers, animals or farm implements and manure containing diseased material.

Fig. 3.4. Wart on potato tubers.

ii) Management: Cultivate immune potato cvs. like Kufri Jyoti, Kufri Sherpa, Kufri Kanchan and Pimpernel. Practice long crop rotation (5 years or more). Rogue out infected plants along with tubers. Rogue out plants of susceptible varieties by burning and destroying warty tubers. Use disease-free seeds for planting.

Integrated Disease Management

- Deep summer ploughing to expose resting stages of pests to sunlight.
- Use disease-free seed material for planting.
- Grow resistant varieties.

- Adjustment of planting and harvesting time for avoidance of diseases. Optimum time for planting is 15th to 30th October in Western and Central Plains and 1st to 15th November in Eastern Plains.
- Covering the exposed tubers by high ridging.
- Rouging of diseased plants.

Table 3.6. Management practices for diseases of potato at each crop stage

Crop stage	Action	Reason
Before planting	Clean up crop area and remove weeds and self-sown potatoes.	Weeds and self-sowns harbour viruses and diseases which can be transferred to the new crop.
	Avoid continuous cropping with potatoes (minimum 2 years gap between crops).	Reduces the build-up of diseases in the soil.
Planting	Plant certified seed.	Certified seed is grown under strict hygiene regulations and will reduce the risk of introducing seed-borne diseases to the crop.
	Plant varieties which are less susceptible to powdery scab or common scab.	In paddocks where scabs have been present in the past, less susceptible varieties will reduce disease risk.
	Change planting depth according to relative risk of pests and diseases.	Shallow planting will help reduce *Rhizoctonia* infection.
	Plant in optimum conditions of soil temperature (16-20°C) and soil moisture.	High temperature and high moisture can lead to seed piece breakdown (especially in Coliban).Low temperature and high moisture favours infection by *Rhizoctonia*.
Hilling	Maintain and improve soil structure and create well-drained hills.	Poorly-drained soils which are easily saturated favour development of diseases like powdery scab, blackleg and phoma (gangrene).
	Maintain good soil cover throughout season.	Good soil cover acts as a barrier, preve-nting exposure of tubers to greening.
	Monitor diseases throughout season.	Control measures are only required when disease buid up reaches a critical level.
Irrigation / rain	Check soil coverage of tubers regularly throughout season.	Rolling and hilling prevents greening of tubers as well as protecting them from diseases attack.
	Manage irrigation to avoid diseases attack.	Paddocks with a history of powdery scab should not have saturated soil during tuber set.

	Avoid water stress.	Maintaining good crop health makes plants less susceptible to diseases attack (especially black dot).
Die-down / slashing	Maintain soil barrier using irrigation/rolling.	Prevents greening late blight.
Harvesting	Ensure tubers are mature before digging.	Mature tubers have a protective skin which reduces the risk of infection.
	Harvest as soon as possible after maturity.	Increased time in the ground increases risk of exposure to diseases (e.g. *Rhizoctonia*, silver scurf, and black dot).
	Handle tubers carefully during harvest.	Damage to tubers creates an entry point for disease (e.g. *Phoma*, dry rot, soft rot).
	Harvest under cool (12-18°C), moist conditions.	Hot and dry, or cold, conditions increase damage to tubers and the risk of rots.
Post harvest	If tubers were very wet at harvest, place them in a drying tent with circulating warm air until dry.	Free moisture on tubers increases the risk of rots and of seed piece breakdown.
	If tubers were cold at harvest or have just been removed from cool storage, warm them to at least 12°C before grading or cutting for seed.	Handling cold tubers is likely to damage them and increase the risk of disease.
	Handle tubers carefully during grading and inspection.	Damage to tubers increases risk of infection and can make them unsalable.
	If tubers are to be cool stored, cure them beforehand by holding them at 14-16°C and 85-95% relative humidity, with good airflow, for 10-14 days.	Curing heals skin wounds and reduces the risk of storage rots (e.g. Phoma, dry rot).
	Don't retain tubers for seed.	Non-certified seed increases chances of pests and diseases (especially viruses) developing in next season's crop.
	Store tubers at optimum temperature and humidity (3-4°C for seed; 4-10°C for ware; 10°C for processing tubers; 85-95% relative humidity for all).	Storing tubers under optimum conditions minimizes disease and prolongs storage life.

3.2. TOMATO

Tomato (*Lycopersicon esculentum*) is the most important and remunerative vegetable crop in India. It is being cultivated in 0.535 million hectares producing 9.362 million tonnes of tomatoes with an average yield of 17.5 tonnes per hectare (Bijay Kumar, 2007) (Tables 3.7 and 3.8). A rich source of minerals, vitamins and organic acids, tomato fruits provides 3-4% total solids, 15-30 mg/100 g ascorbic acid, 7.5-10.0 mg/100 ml titratable acidity and 20-50 mg/100 g fruit weight of lycopene. Tomato fruits contain 0.4% protein, 2.5% carbohydrates, 1.5 mg iron and 2 mg vitamin C in 100g of fresh weight. Fruits are good for people suffering from constipation, jaundice and indigestion.

Table 3.7. Major tomato producing countries of the world

Country	Area (in '000 ha)	Production (in '000 MT)	Productivity (in MT/ha)	% Share of world production
China	1305	31626	24.2	24.7
USA	167	11043	66.3	8.6
Turkey	260	10050	38.7	7.9
India	535	9362	17.5	7.3
Egypt	195	7600	39.0	5.9
Italy	139	7187	51.8	5.6
Spain	72	4810	66.5	3.8
Iran	139	4781	34.4	3.7
Brazil	61	3453	57.0	2.7
Mexico	119	2800	23.6	2.2
Others	1624	35281	21.7	27.6
Total	4615	127993		

Source: FAO (except Indian data) (Indian data source: Indian Horticulture Database 2006)

Table 3.8. Area, production and productivity of leading tomato growing states in India (2005-06)

State	Area (in'000 ha)	Production (in '000 MT)	Productivity (MT/ha)
Andhra Pradesh	76.5	1453.5	19.0
Orissa	100.4	1332.2	13.3
Karnataka	44.5	1188.1	26.7
Maharashtra	35.0	987.0	28.2
West Bengal	50.0	857.2	17.2
Bihar	46.0	727.2	15.8
Gujarat	29.3	650.0	22.2
Chhattisgarh	29.2	365.8	12.5
Madhya Pradesh	20.4	306.7	15.0
Tamil Nadu	22.0	277.7	12.6
Haryana	17.1	257.3	15.0
Others	64.1	959.1	15.0
Total	534.5	9361.8	17.5

(Source: Indian Horticulture Database 2006)

3.2.1. Damping-Off, *Pythium aphanidermatum*

i) Symptoms: In the pre-emergence phase of the disease, the young seedlings are killed even before they emerge out of the soil surface. However, this disease mostly occurs at post-emergence stage which is characterized by the toppling over of the infected seedlings anytime after they emerge from the soil until the stem has hardened sufficiently to resist invasion by the pathogen. Infestation usually occurs at ground level or through roots. The infected tissue appears soft and water-soaked. As the disease advances, the stem becomes constricted at the base of the plants that collapse later. Seedlings that are healthy looking one day may have collapsed the next morning.

ii) Management: Use disease-free seeds for raising seedlings. Treat the seeds with chlorothalanil or captan or metalaxyl at 2 g/kg. Pick rotten fruits and destroy. Spray 0.2% metalaxyl or mancozeb just before onset of monsoon and 1% Bordeaux mixture or 0.3% copper oxychloride at 8-10 days interval. Spraying of 0.2% cymoxanil, propineb mancozeb or chlorothalanil is also helpful. Application of neem cake along with *Trichoderma* spp. may reduce the soil inoculum.

Cultural practices like rabbing, thin sowing, use of light soil, light but frequent irrigation and use of well decomposed FYM should be adopted. *Streptomyces distaticus* was effective in suppressing damping-off of tomato. Success in suppression of damping-off in tomato was achieved through soil and seed application of *Trichoderma virens* and *T. longibrachiatum*.

Seed treatment with *Trichoderma viride* and *Pseudomonas fluorescens* and addition of neem cake to the nursery beds enhanced seed germination and seedling stand. Soil drenching with 1% Bordeaux mixture at 1 litre/m^2 helps in reducing the disease incidence.

Soil solarization using polythene sheet for 30 days during peak summer was most effective and resulted in least occurrence of damping-off (3.9%) when compared to control (30.64%). Seed treatment with *Trichoderma viride* resulted in lowest incidence of damping off (5.7%).

Delivery of *T. harzianum* through solid matrix priming (SMP) reduced the incidence of damping off in nursery. SMP is a process in which moistened seeds are mixed with an organic carrier and the moisture content is brought to a level just below that required for seed sprouting. SMP can further enhance the efficacy of the antagonist in treated seeds. *Trichoderma* spp. grow on seed surface during priming process and increase in numbers and other microbes may not easily dislodge it.

3.2.2. Early Blight, *Alternaria solani*

i) Symptoms: It not only affects leaves but also the fruits. Symptoms appear on all above ground parts of the plant causing pre and post emergence damping off in seed beds. Symptoms appear as small, isolated scattered pale-brown spots on leaves. They enlarge and develop into concentric rings in the necrotic tissue and induce yellowing leading to leaf blight (Fig. 3.5). Girdling of the stem at the base of the plant is also observed. The stem lesions are usually restricted to one side of a stem and become elongated and sunken. Similar symptoms also appear on the fruit causing fruit rot or stem end rot or blossom end rot. Under favourable conditions to the pathogen, losses may go up to 70%. It is more severe in June-July sown crop. In severe attacks, the entire plant may be defoliated, exposing the fruits to sun scald.

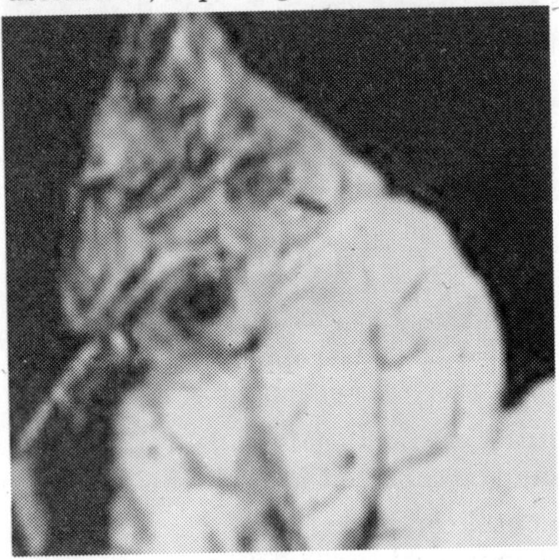

Fig 3.5. Early blight on tomato leaves.

ii) Management: The pathogen is soil as well as seed-borne. Use disease-free seed for raising seedlings. Healthy seeds should be collected only from disease-free fruits. Treat seeds with thiram 75 WP at 2.5 g/kg seed.

The crop debris and infected plant parts should be destroyed immediately after harvest by burning. Summer ploughing, avoiding susceptible crops adjacent to tomato, maintaining plant vigour, minimizing injury to plant are essential for checking the disease.

Give wider spacing of 90 x 60 cm in order to minimize the relative humidity in plant canopy. Remove the lower diseased leaves periodically and follow field sanitation.

Spray 0.2% chlorothalanil or 0.2% propineb or zineb or 0.3% copper oxychloride or 0.2% iprodion + carbendazim at 8-10 days interval starting at flowering stage.

3.2.3. Late Blight, *Phytophthora infestans*

The disease is more common in North-Western India but also appears in Uttar Pradesh and Bihar in the months of December-January.

i) Symptoms: Symptoms on leaves (Fig. 3.6), twigs and fruits are more common. Blight appears on foliage as light-pale green water-soaked area. The lesions are wet in the morning and dried, shriveled in daylight. It enlarges rapidly until the entire leaflets are killed. In moist and cloudy weather, a downy white fungus develops near the outer margin of the lesion on underside of leaves. Green and ripe fruits affected by the blight show olivacious, greasy appearing spots which may expand until the entire fruit is covered.

Fig. 3.6. Late blight on tomato.

ii) Management: Certified seeds should always be used. Infected crop debris and fruits should be collected from the field and burnt.

When environmental conditions are favourable for late blight incidence, preventive sprays of mancozeb at 0.25% at 5-7 days interval provide good control. Spraying of metalaxyl + mancozeb applied within 2 days of infection is very effective.

3.2.4. Buckeye Rot, *Phytophthora nicotianae* var. *parasitica*

i) Symptoms: This is a very serious disease and causes losses up to 92%. Spots with pale and dark brown concentric rings appear on fruits. These may be small or may cover a major portion of the fruit surface causing fruit rot (Fig. 3.7), thereby lowering the yield considerably. One infected fruit with latent infection rots the entire basket overnight.

Fig. 3.7. Buckeye rot on tomato fruits.

ii) Management: Ridge and furrow system of planting, staking plants at flowering stage and removal of basal leaves up to 15-20 cm height from ground at the start of monsoon reduce the disease incidence. The rotten fruits should be collected and destroyed. Spray 1% Bordeaux mixture or 0.2% propineb or dithionon or metalaxyl + mancozeb to reduce the disease incidence.

3.2.5. Septoria Leaf Blight, *Septoria lycopersici*

i) Symptoms: The plant may be attacked at any stage of its growth. Small, circular, water-soaked grey spots having dark border first appear on under surface of lower leaves. On enlargement, they develop dark brown margins and sunken white-grey centres (Fig. 3.8). Similar spots may develop on floral calyx and stem (rarely on fruits) causing severe defoliation resulting in yield losses and sun scalding of fruits.

ii) Epidemiology: The pathogen survives in the soil on infected plant debris and also in seeds. Low temperature, high moisture and humidity favour infection of the pathogen.

Fig. 3.8. *Septoria* leaf spot on tomato leaves.

iii) Management: Destroy infected plant material in the field. Treat the seeds with thiram 75 WP at 2 g/kg of seed. Spray chlorothalanil, propineb or mancozeb M at 0.2% or iprodion + carbendazim or mancozeb at 0.2% at 10-15 days intervals. Spraying of 0.4% copper oxychloride or 0.2% mancozeb or 0.1% carbendazim also gives effective control of the disease.

3.2.6. Wilt, *Fusarium oxysporum* f. sp. *lycopersici*

i) Symptoms: The wilt disease is becoming a major disease of concern in many parts of the country. The disease is characterized by yellowing and wilting of leaves and finally the entire plant wilts and dies prematurely (Fig. 3.9). Stem tissue often is discoloured throughout the plant. Vascular browning takes place in the root.

Fig. 3.9. Tomato plant infected by *Fusarium* wilt.

The fungus can persist in the soil for many years and is virulent at moderate temperature (26-28°C).

ii) Management: The infection could be minimized by crop rotation and planting resistant varieties. Incidence of wilt can be significantly reduced by the application of 0.1% carbendazim or 0.2% benomyl as soil drench (Sen and Kapoor, 1974).

3.2.7. Collar Rot, *Sclerotium rolfsii*

It has wide host range and causes severe problem on tomato and other vegetable crops.

i) Symptoms: The first symptom of the disease is observed as soft tissue necrosis of bark of the stem near soil line. White growth of cottony mycelium is clearly visible on the affected portion just below the soil surface. Later, dense silvery fungus growth along with white to light brown mustard-like sclerotia is observed on the same portion. Progressive drooping, yellowing or wilting of the entire plant is observed. Sometimes the plant is collapsed soon after infection. The disease is soil-borne where the pathogen survives in the form of sclerotia.

ii) Management: Long rotations with cereals viz., corn, sorghum and millets will reduce the disease incidence. All the weeds from the field should be removed. Use of infested FYM, soil, compost should be avoided in the field. Summer ploughing for 3 times and irrigating the field destroys the germinating sclerotia. Avoid dense planting and maintain proper aeration near collar region.

Soil drenching in the evening with 0.3% copper oxychloride or 0.25% captafol/captan near the collar region is effective to check the disease.

The disease was effectively managed under greenhouse and field conditions by soil + root-dip treatment and seed + soil treatment with *T. harzianum;* and seed + soil treatment with *B. subtilis.*

3.2.8. Powdery Mildew, *Leveillula taurica*

i) Symptoms: The leaves show white powdery patches, on both sides of leaves (Fig. 3.10). At later stages of infection, leaves turn yellowish brown and wither. If the infection is on lower side of the leaves, yellow patches are visible on the upper side.

Fig. 3.10. Powdery mildew on tomato leaves.

ii) Management

- At the onset of the disease, application of 0.075% hexconazole or 0.1% dinocap or 0.3% wettable sulphur or 0.1% carbendazim reduces the disease.
- The ergosterol biosynthesis inhibitor fungicides like tridimefon/propiconazole/triforine are more effective in reducing the disease intensity.

Integrated Disease Management

- Crop rotation with French beans reduces bacterial wilt.
- Soil solarization of nursery beds for 2-3 weeks using transparent polythene sheets (100-160 gauge) kills soil borne pathogens.
- Raising nursery in raised seed beds (10-15 cm) provides good drainage and thereby prevent damping off.
- Use of bacterial wilt resistant varieties (Arka Abha, Arka Alok, Arka Shreshta, Arka Abhijit, Megha, Shakthi, Sonali).
- Seed treatment with *Trichoderma viride*/*T. harzianum*/ *Pseudomonas fluorescens* at 20 g/kg of seeds to prevent damping off.
- Deep summer ploughing helps in exposing pathogens and their resting stages to sun's heat.
- Collection and destruction of diseased plants/plant parts from the field from time to time.
- Soil application of FYM enriched with *Trichoderma viride* at 1 tonne/ha.

- Deep off-season/summer ploughing helps in exposing resting stages of the diseases.
- Green manuring with sunnhemp in July – August.
- Soil solarization using transparent polythene sheets on nursery beds for 2-3 weeks which helps in killing soil-borne pathogens.
- Seed treatment with *T. viride* at 4 g/kg of seed to prevent soil-borne infection of fungi and nematodes.
- Soil application of *T. harzianum* at 50 g/m^2.
- Giving wider spacing of 90 x 60 cm to reduce the incidence of early blight and powdery mildew.
- Spray chlorothalonil at 2 g/litre of water to control early and late blight diseases.

3.3. BRINJAL

Brinjal (*Solanum melongena*) is a widely grown vegetable crop in Asian countries. It is being cultivated in 0.553 million hectares producing 9.136 million tonnes of brinjal with an average yield of 16.5 tonnes per hectare (Bijay Kumar, 2007) (Tables 3.9 and 3.10). In India, it is adapted to a wide range of climatic conditions from north to south and east to west. In hilly regions, it is grown only in summer. Brinjal is used in a variety of culinary preparations. Pickles and industrially processed foods are also produced. It is rich in Vitamins A and B and good for diabetic patients.

Table 3.9. Major brinjal producing countries of the world

Country	Area (in '000 ha)	Production (in '000 MT)	Productivity (in MT/ha)	% Share of world production
China	952	17028	17.9	53.5
India	553	9136	16.5	28.7
Egypt	43	1000	23.3	3.1
Turkey	35	930	26.6	2.9
Iraq	23	439	19.1	1.4
Japan	11	396	34.7	1.2
Italy	12	339	27.8	1.1
Sudan	15	272	18.7	0.9
Indonesia	43	252	5.9	0.8
Philippines	21	188	8.8	0.6
Others	120	1856	15.4	5.8
Total	1829	31836		

Source: FAO (except Indian data)

(Indian data source: Indian Horticulture Database 2006)

Table 3.10. Area, production and productivity of leading brinjal growing states in India (2005-06)

State	Area (in'000 ha)	Production (in '000 MT)	Productivity (MT/ha)
West Bengal	152.9	2757.4	18.0
Orissa	127.9	1853.9	14.5
Bihar	53.7	1031.0	19.2
Gujarat	62.7	982.6	15.7
Andhra Pradesh	26.2	549.4	21.0
Maharashtra	32.0	521.6	16.3
Karnataka	14.7	326.0	22.1
Chhattisgarh	19.4	266.4	13.7
Madhya Pradesh	14.8	221.9	15.0
Assam	13.0	171.5	13.2
Haryana	11.5	148.7	12.9
Others	24.5	305.8	12.5
Total	553.3	9136.4	16.5

(Source: Indian Horticulture Database 2006)

3.3.1. Damping-Off, *Pythium aphanidermatum*

i) Symptoms: The pathogen causes death of plants in the nursery beds.

ii) Management: Soil solarization using polythene sheet for 30 days during peak summer was most effective and resulted in least occurrence of damping-off (1.5%) when compared to control (30.64%). Seed treatment with *Trichoderma viride* resulted in lowest incidence of damping off (4.7%).

3.3.2. Root Rot, *Rhizoctonia solani*

i) Symptoms: Generally early transplanted brinjal crop is much affected during the months of August-September. Lesions start on stem near collar region at or below the soil level and expand downward into the roots. Initially the bark become wet, soft with macerated tissue. Later on, drooping and wilting of the plant is observed. The disease is more common in poorly drained soil and the fields having prolonged excessive moisture.

ii) Management: Field sanitation and crop rotation are most important for this soil-borne disease. All infected plants should be uprooted and burnt. Summer ploughing and fallowing should be carried out to reduce the soil inoculum. Long crop rotation, particularly with sugarcane should be followed. Proper drainage should be maintained in the field.

Seed treatment with 0.25% captan greatly reduced the inoculum from the contaminated seeds. When initial symptoms are noticed in the field, drenching the collar region with 0.1% carbendazim is recommended to reduce the disease incidence.

3.3.3. Collar Rot, *Sclerotinia sclorotiorum*

i) Symptoms: Infection takes place on leaves, twigs, flowers and fruits. Water-soaked lesions develop and the infected tissue is macerated by the pathogen. White mass of mycelium grow on the surface and later embedded sclerotia are formed. These sclerotia become black after drying. The pathogen is soil-borne.

ii) Management: Deep summer ploughing and crop rotation with flooded paddy is beneficial for reducing the soil inoculum. Crop debris and infected plant parts should be collected from the field and burnt without shattering and shedding of sclerotia in the field. Dense planting should be avoided to reduce the shade and humidity in plant canopy.

Spraying of 0.1% carbendazim should be initiated just prior to flowering stage. Alternate spray of 0.25% mancozeb can also bè given.

B. subtilis, T. viride and *T. virens* were effective.

3.3.4. Phomopsis Blight and Fruit Rot, *Phomopsis vexans*

i) Symptoms: In the seed beds, the disease appears as damping-off. After transplanting, the leaves show clearly defined, circular, grey to brown spots with light coloured centre. The affected leaves turn yellow and die. The lesions on stem are dark brown becoming grey in the centre. Mostly the stem base is attacked and is characterized by constriction of the base or a grey dry rot. Pale, sunken spots develop on the fruit and may progress to cover the entire fruit surface. The internal portion of the fruit rots. If the fungus enters through the calyx, the whole fruit is mummified due to dry rot.

ii) Epidemiology: The fungus is seed-borne and also survives between crop seasons on infected plant debris in the field. The fungus requires wet weather and high temperatures for rapid development.

iii) Management

The following practices may be adopted to manage the disease:

- Use disease-free seeds for planting.
- Collect seeds from only healthy plants and blemish-free fruits.
- Deep ploughing in summer should be practiced.
- Treat seeds in hot water at 50°C for 30 min.
- Spray foltaf or copper oxychloride both at 0.3% concentration.
- Since the pathogen attacks only brinjal, 3 years crop rotation shall be effective for the management of the disease.
- The infected crop debris should be collected and burnt in the field.

3.3.5. Leaf Spot, *Cercospora melonginae*

i) Symptoms: Leaf spots are circular to irregular in shape, usually large and brown to grayish brown in colour (Fig. 3.11). Infected leaves are shed.

ii) Epidemiology: The pathogen survives in infected plant debris. Moderate temperature, cool and humid weather conditions favour the disease development.

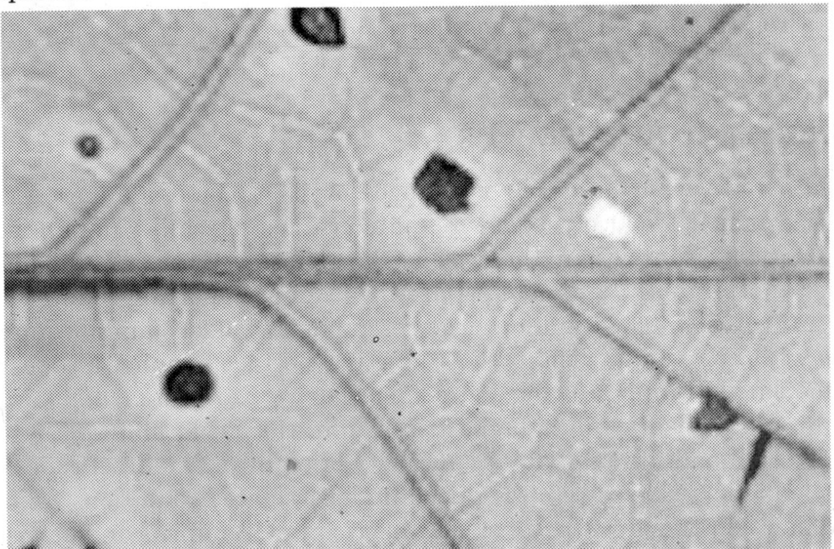

Fig. 3.11. *Cercospora* leaf spot on brinjal leaf.

iii) Management: Spraying 0.1% carbendazim is effective for the management of the disease.

Integrated Disease Management

Nursery

- Deep off-season/summer ploughing to expose resting stages of diseases to sunlight.
- Soil solarization for 2-3 weeks during summer using 400 gauge thick polythene sheet.
- Soil application of *Trichoderma harzianum* mixed with FYM.
- Seed treatment with 0.25% captan or carbendazim.
- Seed treatment with *Trichoderma viride* at 4 g/kg of seeds to prevent seed and soil borne infection of fungi and nematodes.
- Raising nursery on raised beds (10 cm above ground level).
- Use of bacterial wilt resistant varieties (Arka Nidhi, Arka Keshav, Arka Neelkant, Swarna Shree, Swarna Mani, Swarna Shyamali, Surya, Pant Rituraj, Pusa Purple Cluster).

Main field

- Avoid transplanting before last week of August for prevention of *Phomopsis* stem blight disease.

- Spraying 0.1% carbendazim after flowering.
- Destruction of root/collar rot, little leaf and bacterial wilt affected plants from time to time.
- Crop rotation with non-host crops like sorghum, wheat, marigold or onion to reduce nematode population in soil.

3.4. CHILLI AND BELL PEPPER

Chilli (*Capsicum annuum*) is valued for its diverse commercial uses. India is a major producer, exporter and consumer of chilli. Indian chillies reach over 90 countries in the world. Bangladesh, Bahrain, Canada, Sri Lanka, Saudi Arabia, USA and UAE are the leading importing countries. It is being cultivated in 0.834 million hectares producing 0.847 million tonnes of chillies with an average yield of 1.0 tonne per hectare. Andhra Pradesh has been the largest chilli-growing state followed by Karnataka and Maharashtra. Productivity of chilli is highest in Andhra Pradesh, followed by Arunachal Pradesh and Punjab. The productivity of chilli in Andhra Pradesh and Punjab is higher since the crop is raised under irrigated conditions compared with Maharashtra and Karnataka. Chilli is rich in vitamins A and C.

Bell pepper (*Capsicum annuum*), also known as capsicum or sweet pepper, is a popular vegetable in India. It is relatively a new entrant into our country. It is mainly cultivated in Himachal Pradesh, Uttar Pradesh, parts of Gujarat, Maharashtra, Karnataka, Ranchi region of Bihar and hilly regions of Tamil Nadu. It grows well in summer season in hills and cooler season in the plains.

3.4.1. Damping-Off, *Pythium aphanidermatum*

i) Symptoms: The pathogen is responsible for death of seedlings in nursery beds.

ii) Management: Soil solarization using polythene sheet for 30 days during peak summer was most effective and resulted in least occurrence of damping off (1.28%) when compared to control (30.64%). Seed treatment with *Trichoderma viride* resulted in lowest incidence of damping off (4.7%).

Delivery of *T. harzianum* through solid matrix priming (SMP) reduced the incidence of damping off in nursery. SMP is a process in which moistened seeds are mixed with an organic carrier and the moisture content is brought to a level just below that required for seed sprouting. SMP can further enhance the efficacy of the antagonist in treated seeds. *Trichoderma* spp. grow on seed surface during priming process and increase in numbers and other microbes may not easily dislodge it.

Integration of soil solarization, application of neem cake and *T. viride* gives effective control of the disease.

3.4.2. Phomopsis Blight and Soft Rot, *Phomopsis vexans, Diaporthe vexans*

i) Symptoms: Infected seeds may cause damping-off in the nursery. Leaves of seedlings show clearly defined circular, light brown spots mostly on lower leaves. Centre of the spots is slightly lighter in colour. Old spots are having numerous black dot-like pycnidia. These spots are papery, often crack and sometimes with shot holes. Almost similar symptoms are observed on the leaves of transplanted crop in the main field. Sometimes irregular big lesions also are observed on leaves, particularly in cool and humid weather. Stem lesions start from basal part of nodal portion as grey dry rot with constriction. This leads to drying of few twigs or partial wilting of the portion of the plant. Pale to light brown sunken spots develop on the old fruits. Individual spots expand and coalesce to cover the entire fruit which leads to fruit rot.

ii) Management: Deep summer ploughing, seed treatment in hot water at 50°C for 30 minutes, crop rotation with non-solanaceous crops and collecting and destruction of crop debris shall keep the disease under check. Seeds should be collected from healthy fruits and disease-free field. Seed treatment with 0.25% carbendazim is essential to eliminate any seed inoculum. At least 3 year crop rotation is required with non-host crops to manage the disease. Protective sprays of 0.1% carbendazim at 10 days interval check the disease initiation in the field. Use tolerant varieties like Ramnagar Giant, Pusa Hybrid-5, KS-24 and Green Long.

3.4.3. Wilt, *Fusarium oxysporum*

i) Symptoms: Symptoms include leaf chlorosis, vascular discoloration, and wilting of chilli pepper plants. High temperature and high moisture are conducive to symptom development.

ii) Management: Reducing saturated conditions in soil is an important step in managing the disease. If economically justified, soil fumigants and solarization may be used to reduce pathogen populations in soil.

Seedling root-dip treatment and soil application of non-pathogenic *Fusarium* (Fo 52) significantly suppressed the wilt incidence. The combination of Fo 52 with *P. fluorescens* gave maximum inhibition of wilt.

3.4.4. *Phytophthora* Root Rot, Fruit Rot and Leaf Blight, *Phytophthora capsici*

i) Symptoms: This is a disease of rainy season, characterized by small, water-soaked spots appearing on fruits leading to complete rot of fruits. On leaves, water-soaked bleached spots appear, resulting in blighting. It causes damage to plants and affect fruit yield. Infection of the root and lower portion of the stem leads to plant wilting (Fig. 3.12), which is the most conspicuous symptom of the disease. In furrow-irrigated fields, there is a row-delimited pattern of wilted plants (Fig. 3.12). Over time, wilting is displayed over the entire field. *Phytophthora capsici* produces specialized swimming cells, known as zoospores, which enable spread of the pathogen from plant to plant.

Fig. 3.12. Wilting of chilli plant and a field severely affected by *Phytophthora* root rot (Courtesy: D. Lindsey).

ii) Management: Management of *P. capsici* consists of reducing soil water saturation; reducing populations of the pathogen through crop rotation, soil solarization, and application of organic amendments; and restricting pathogen dispersal through mulching and judicious use of fungicides such as metalaxyl and mefenoxam (Ristaino and Johnston, 1999).

Use of disease-free seeds for raising seedlings, collection and destruction of rotten fruits and soil application of neem cake + *Trichoderma viride* reduce the disease incidence. Minimum plant mortality (16.7%) was recorded when the plants were subjected to *T. viride* (PDBC TV 31) and *T. harzianum* (PDBC TH 10) followed by *T. harzianum* (PDBC TH 18) and *T. viride* (PDBC TV 23) where the mortality was 30%. In fungicidal treatment, 33.3% plant mortality was recorded while in pathogen check, it was highest (46.7%).

Treat seeds with chlorothalanil or captan or metalaxyl at 2 g/kg. Spray metalaxyl or mancozeb at 0.2% just before onset of monsoon and 1% Bordeaux mixture or copper oxychloride at 0.3% at 8-10 days interval. Spray of cymoxanil, propineb mancozeb or chlorothalanil at 0.2% is also helpful to fruits.

3.4.5. Anthracnose or Dieback, *Colletotrichum dematium*

i) Symptoms: This is characterized by small grey spots on leaves. It causes necrosis of twigs, lower branches or the entire top of the plant may wither away. Black dots (acervuli) are formed all over the necrotic surface of the affected twig. Later dark brown to black sunken spots having pink spore masses appear on fruits (Fig. 3.13). It may cause losses up to 25% under suitable weather conditions. Fruit rot extends to the seed cavity making it internally seed-borne.

Fig. 3.13. Anthracnose on chilli fruit.

ii) Management: Disease-free seeds should be collected from healthy fruits. Avoid apical damage during transplanting and also in the main field particularly at flowering stage. Well drained and porus soil reduces the disease incidence. Pick rotten fruits and destroy.

Treat the seed with 0.25% carbendazim or thiram 75 WP at 2 g/kg seed. Seedlings should be sprayed with 0.1% carbendazim before transplanting. Spray propineb or mancozeb at 0.25% or carbendazim or thiophanate methyl at 0.1% or 0.3% copper oxychloride at fortnightly intervals in the main field. Collect diseased plants and burn. To control post-harvest infection, spraying with benomyl/thiophanate methyl at 1 g/litre helps in reducing the disease incidence.

If it is a seed crop, spraying with metalaxyl mancozeb or cymoxanil mancozeb or Al-Fosetyl at 0.2% at fortnightly intervals gives very good control.

Cv. Arka Lohit is tolerant to anthracnose disease.

3.4.6. Leaf Blight, *Alternaria alternata, Cercospora capsici*

i) Symptoms: Symptoms of *Cercospora* leaf blight are observed as circular spots with light grey to white grey centres with dark brown margins (Fig. 3.14). In severe cases, small spots coalesce and lead to defoliation. *Alternaria* leaf spots are covered with a dark grayish brown to black spores. In moist condition, sporulation of both the pathogens was observed on the spots. Disease leads to severe defoliation and affects flowering and fruiting in infected plants.

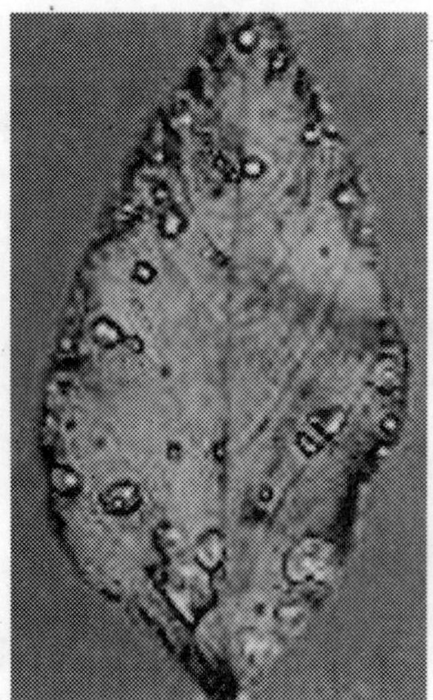

Fig. 3.14. Cercospora leaf spot on chilli leaf.

ii) Management: Selection of disease-free seed is important to check the primary infection. Plant vigour should be maintained by spraying 0.2% tricel after fourth picking of green chillies. Foliar spray of 0.2% chlorothalanil is a must to protect the crop from the disease.

3.4.7. Powdery Mildew, *Leveillula taurica*

i) Symptoms: The symptoms appear as whitish powdery patches on the leaves (Fig. 3.15), stem and petiole and rarely on fruit. Infected leaves at later stages show yellowish patches. The affected plants yield poor quality fruits. The disease occurs at all stages of crop growth under protected cultivation. The delay in initiating plant protection measures increases severity of the disease.

Fig. 3.15. Powdery mildew on chilli leaf.

ii) Management: Since the spread of the disease is fast within the polyhouse, care should be taken to destroy the initial inoculum by application of fungicides (0.1% dinocap or 0.075% hexaconazole or 0.3% wettable sulphur) to prevent further spread of the inoculum.

Baycor at 0.2%, Devisulphur at 0.3% and Bayleton at 0.05% have been reported to be very effective for the management of this disease (Keshwal and Choubay, 1983).

3.4.8. *Verticillium* Wilt, *Verticillium dahliae, V. albo-atrum*

i) Symptoms: Verticillium wilt of chilli is caused by *Verticillium dahliae* and *V. albo-atrum.* Visual symptoms of the disease include stunting, defoliation, and wilting, with discoloration of the vascular system. In the field, the disease first appears in a single plant or a small group of wilting plants (foci) scattered throughout the field (Fig. 3.16), depending on the distribution of the pathogen in soil. The disease gradually spreads from these foci, and this expansion may cover the entire area of plantings. Infected plants may exhibit physiological changes such as reduction in photosynthesis, increased plant transpiration, and respiration.

Fig. 3.16. Early stage of *Verticillium* wilt, and scattered plants of chilli affected by *Verticillium* wilt (Courtesy: D. Lindsey).

ii) Management: Application of chloropicrin and metham-sodium fumigants, and use of organic amendments (Lazarovits *et al.*, 2000), may decrease *Verticillium* wilt through reduction of pathogen populations in soil.

3.4.9. *Rhizoctonia* Root Rot, *Rhizoctonia solani*

i) Symptoms: *Rhizoctonia* root rot caused by *Rhizoctonia solani* generally affects seedlings. But, *R. solani* can also infect mature plants and induce root rot (Fig. 3.17), which leads to wilting and death of chilli plants. *Rhizoctonia solani* persists in soils and organic debris.

ii) Management: Management of this fungal pathogen relies on using high-quality seed treated with fungicides such as thiram or captan, and avoiding saturated soil conditions. In fields with high pathogen populations, control measures may include the use of soil fumigants and soil solarisation.

Fig. 3.17. Chilli plant infected by *Rhizoctonia solani*, with a reddish-brown discoloration at the soil line level of the stem and root (Courtesy: P. Bosland).

IDM

- Crop rotation with French beans reduces bacterial wilt.
- Crop rotation with cereals, sesame and mustard, so also intercropping with marigold, onion, garlic reduce nematode infestation.
- Soil solarization of nursery beds for 2-3 weeks using transparent polythene sheets (100-160 gauge) kills soil borne pathogens and nematodes.
- Raising nursery in raised seed beds (10-15 cm) provides good drainage and thereby avoid damping-off.
- Seed treatment with *Trichoderma viride/T. harzianum/ Pseudomonas fluorescens* at 20 g/kg of seeds to prevent damping-off and nematodes.
- Deep summer ploughing helps in exposing pathogens and their resting stages to sun's heat.
- Collection and destruction of diseased plants/plant parts from the field from time to time.
- Soil application of neem cake at 250 kg/ha as basal dose to control nematodes.
- Soil application of FYM enriched with *Trichoderma viride* at 1 MT/ha.

Table 3.11. Disease management in chilli and bell pepper

Chemical	Time of application (days after planting)
Wettable sulphur 3 g + Teepol 0.5 ml/litre of water	15
Wettable sulphur 3 g + Teepol 0.5 ml/litre of water	25
Wettable sulphur 3 g + Teepol 0.5 ml/litre of water	40

Carbendazim 1 g + Teepol 0.5 ml/litre of water	70
Metalaxyl and Mancozeb (ridomil) 2 g + Teepol 0.5 ml/litre of water	80
Metalaxyl and Mancozeb (ridomil) 2 g + Teepol 0.5 ml/litre of water	90
Metalaxyl and Mancozeb (ridomil) 2 g + Teepol 0.5 ml/litre of water	110

Repeat sprays at 130 days after planting, if necessary.

3.5. ONION AND GARLIC

Onion (*Allium cepa*) is one of the most important commercial vegetables. The bulbs are consumed both in mature and tender stages. Even the leaves and stalks of inflorescence are used as vegetables. It is being cultivated in 0.695 million hectares producing 9.248 million tonnes of onions with an average yield of 13.3 tonnes per hectare (Bijay Kumar, 2007) (Tables 3.12 and 3.13).

Table 3.12. Major onion producing countries of the world

Country	Area (in '000 ha)	Production (in '000 MT)	Productivity (in MT/ha)	% Share of world production
China	901	19040	21.1	30.2
India	695	9248	13.3	14.7
USA	65	3346	51.2	5.3
Turkey	77	2070	26.8	3.3
Pakistan	128	1765	13.8	2.8
Russia	123	1759	14.3	2.8
Iran	50	1685	33.7	2.7
Egypt	43	1302	30.6	2.1
Brazil	58	1138	19.5	1.8
Mexico	40	1115	28.1	1.8
Others	1179	20500	17.4	32.6
Total	3359	62968		

Source: FAO (except Indian data) (Indian data source: Indian Horticulture Database 2006)

Table 3.13. Area, production and productivity of leading onion growing states in India (2005-06)

State	Area (in'000 ha)	Production (in '000 MT)	Productivity (MT/ha)
Maharashtra	154.0	1878.8	12.2
Gujarat	84.9	1876.6	22.1

Bihar	49.8	1011.7	20.3
Karnataka	139.0	1008.5	7.3
Andhra Pradesh	39.4	670.0	17.0
Madhya Pradesh	40.0	639.8	16.0
Rajasthan	44.5	427.2	9.6
Haryana	15.5	317.7	20.5
Uttar Pradesh	21.2	283.1	13.3
Orissa	26.8	243.9	9.1
Tamil Nadu	29.2	233.8	8.0
Others	50.8	657.3	12.9
Total	695.1	9248.4	13.3

(Source: Indian Horticulture Database, 2006)

Garlic (*Allium sativum*) is mainly used for flavouring vegetables and meat dishes. It is carminative, stimulant and cures digestive disorders. It has been found to be very effective in curing diabetes and high blood pressure. It is being cultivated in 0.107 million hectares producing 0.436 million tonnes of garlic with an average yield of 4.1 tonnes per hectare. Madhya Pradesh, Gujarat, Orissa, Maharashtra, Uttar Pradesh and Rajasthan are major garlic -growing states. More than 50% production comes from Madhya Pradesh and Gujarat only.

3.5.1. Damping-Off, *Pythium* sp., *Rhizoctonia solani, Fusarium* sp.

i) Symptoms: Cold, wet soils often encourage the development of damping-off symptoms very early in the seedling's growth. Seedlings may fall over and die as a result of breakdown of plant tissues at the soil line. Sometimes damping-off occurs before the seedling even emerges. The disease is usually caused by *Pythium, Rhizoctonia* or *Fusarium* fungi, either alone or in combination. Damping-off can occur in the field, or in the nursery if conditions are too wet.

ii) Management : Use disease-free seeds for sowing. Seed treatment with *T. harzianum* at 5 g/kg and sowing in solarized soil helps in reducing the disease incidence. Overcrowding of seedlings and water stagnation in the nursery beds should be avoided. Drenching the soil with 1% Bordeaux mixture or 0.2% copper oxychloride reduces the disease incidence.

3.5.2. Basal Rot, *Fusarium oxysporum* f.sp. *cepae*

Basal rot is caused by the soil-borne fungus *Fusarium oxysporum* f. sp. *cepae* and generally occurs when soil temperatures are very warm (optimum 29°C).

i) Symptoms: The early symptoms in the field are yellowing of leaves and tip dieback (Fig. 3.18). As the disease progresses, the whole plant may collapse and, if the plant is pulled, it often comes out without any roots attached since they have decayed. The basal plate of the onion becomes pinkish-brown and secondary bacterial rots may develop in the affected area (Fig. 3.18). If infection occurs late in the season, the symptoms may not show up until the onions are in storage.

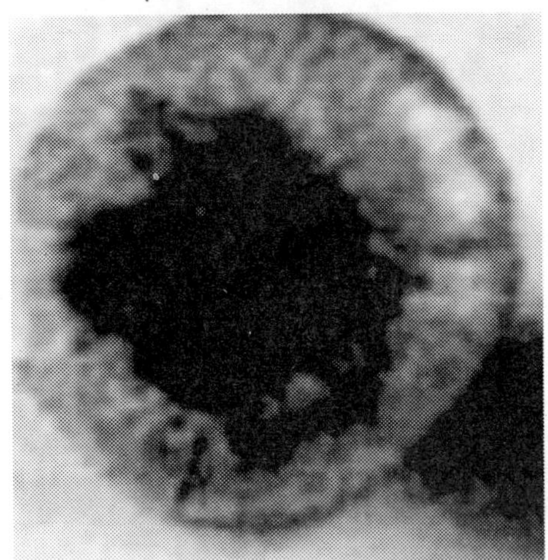
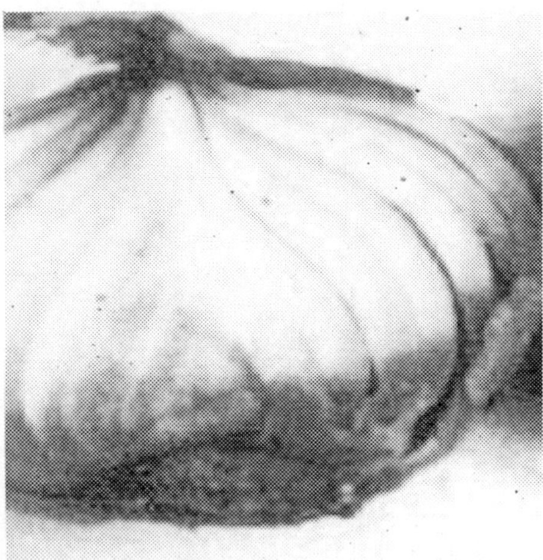

Fig. 3.18. Basal rot on onion bulbs.

ii) Management: Mixed cropping and crop rotation reduces the incidence of the disease. Losses can be reduced with a 4-year rotation out of onions. Seedling dip in antagonists like *Pseudomonas cepacis* and *T. viride* gives protection to seedlings. Soil solarization using polythene sheet of 750 gauge in summer season for 20-30 days reduces the infectious propagules, which in turn reduces the disease incidence. *T. viride* at 1250 g + 50 kg FYM if applied in soil before planting gives good control of basal rot in seed crop. Onion cvs. Arka Lalima, Arka Niketan, Arka Pragati are reported to be resistant/tolerant to basal rot.

3.5.3. Purple Blotch, *Alternaria porri*

i) Symptoms: The disease is severe in Maharashtra, Gujarat, Tamil Nadu, Rajasthan, Uttar Pradesh and Punjab causing damage up to 50% (Thind and Jhooty, 1984). The disease is more damaging to the seed crop. The disease appears on leaves as purplish spots with dark or red purplish margin and a yellow halo that extends upwards and downwards to some extent (Fig. 3.19). The seeds do not develop or remain shriveled on the infected seed stalks and heavily infected seed stalks are prone to breaking. The spots on leaves rapidly enlarge and the leaves are gradually shed.

ii) Epidemiology: The temperature range of 21-30°C and high humidity favoured disease development. The fungus survives from one season to another as mycelium and spores in leaves and stalks left as plant debris, as chlamydospores and is also seed borne in nature.

iii) Management: Heat treatment of onion bulbs at 35°C for 8 hr before planting gives effective control of the disease. Summer sown crop escapes the disease infection. Spraying of 0.2% mancozeb gives effective control. Onion cvs. Arka Kalyan, Arka Kirtiman, Arka Lalima and Agrofound Dark Red are reported to be tolerant to purple blotch. Crop rotation with coarse cereals also reduces blotch incidence.

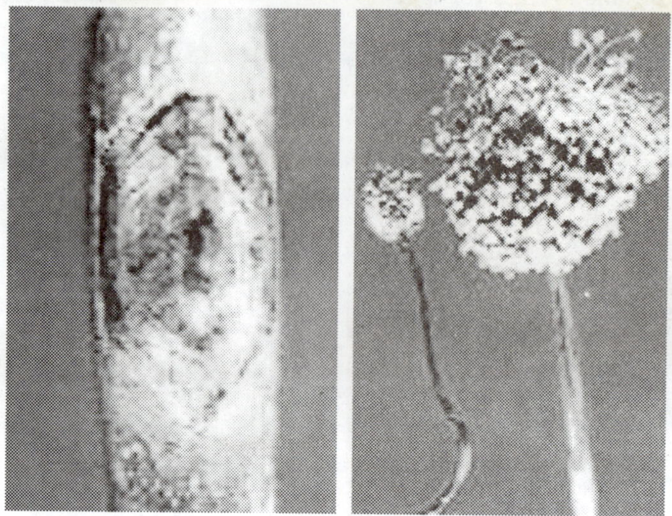

Fig. 3.19. Purple blotch on onion leaf and inflorescence.

3.5.4. Blight, *Stemphyllium vesicarium*

i) Symptoms: The disease causes blight in northern parts of India and mainly damages radical leaves of the seedling at 3-4 leaf stage. But it has also been recorded on the inflorescence. The disease is serious in other parts of the world. The perfect stage of the pathogen is *Pleospora allii*.

Lesions are initially light yellow to brown and water-soaked. They elongate, often reaching the leaf tips, and become dark brown to black. The disease can become serious following periods of more than 24 hours of rainy weather.

ii) Management: Summer ploughing of infested field reduces inoculum potential of blight. Field sanitation, collection and burning of crop residues minimizes infection rate.

The disease can be controlled by 4 sprays on bulb crop and 5 sprays in seed crop with 0.25% mancozeb along with sticker Triton-X.

3.5.5. Downy Mildew, *Peronospora destructor*

The pathogen always attacks the oldest leaves. This disease can rapidly destroy the foliage and thus prevents the formation, ripening and storage of onion bulbs. The disease is very damaging to onion seed crop when the flower stem is destroyed before the seeds ripen.

i) Symptoms: The first sign of downy mildew is a purple-grey, velvety growth on otherwise green leaves (most easily seen in the early morning) (Fig. 3.20). The disease often starts in patches and is favoured by cool (less than 22°C) and humid weather. Diseased leaves quickly turn pale-green, then yellow, and collapse and die. The pale-green and yellow phase is characterized by oval-shaped lesions that often become infected with other diseases such as purple blotch or bacterial infections. Several cycles of sporulation and infection can occur.

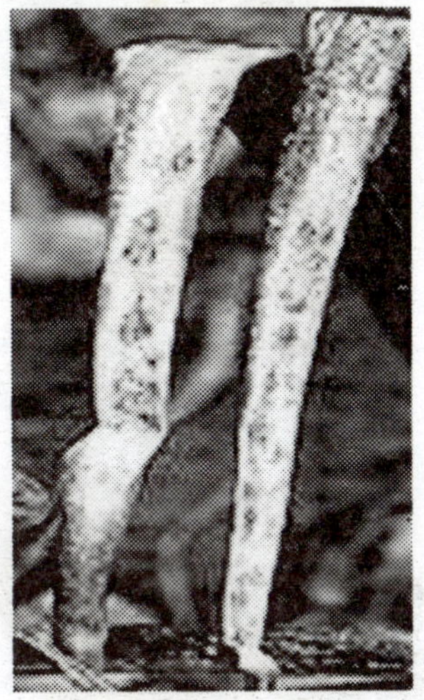

Fig. 3.20. Downy mildew infection on onion leaves.

Under favourable conditions, the diseased areas become covered with white to purplish fruiting bodies. Often, the leaves fold over at the affected areas and the leaf tip withers. Bulbs of such plants do not attain full growth and are often soft and immature. The fungus over winters in infected crop residues and also persists from one year to another as mycelium in infected bulbs in storage.

ii) Management: Regular fungicidal sprays with difolatan or cumin or ridomil all at 0.1% are recommended for the control of downy mildew disease (Mirakhur *et al.*, 1977; Mir, 1987).

3.5.6. Smut, *Uromyces cepulae*

i) Symptoms: The fungus causes considerable losses at the seedling stage. The first symptom of the disease appears on the cotyledons and young leaves in the form of longitudinal blisters which are black and contain the fruiting bodies of the fungus. The seedlings often die before emergence. If the plant survives, the disease becomes systemic and they remain at a vegetative stage for the whole growing season. The bulbs also become covered with blackish lesions (Fig. 3.21).

The pathogen is soil-borne and the disease is more prevalent in temperate region. The onion is susceptible to this disease for a very short time i.e., between germination and seedling emergence. A cool, wet spring increases the incidence of smut infection because the onion seedlings grow slowly and the flag leaf is in the soil for a longer period. Similarly, planting onion seeds too deep will also make them more likely to be infected. Smut spores survive in the soil for many years, and even long crop rotations may not reduce disease incidence. The disease is spread when contaminated soil or set onions are transferred to smut-free areas.

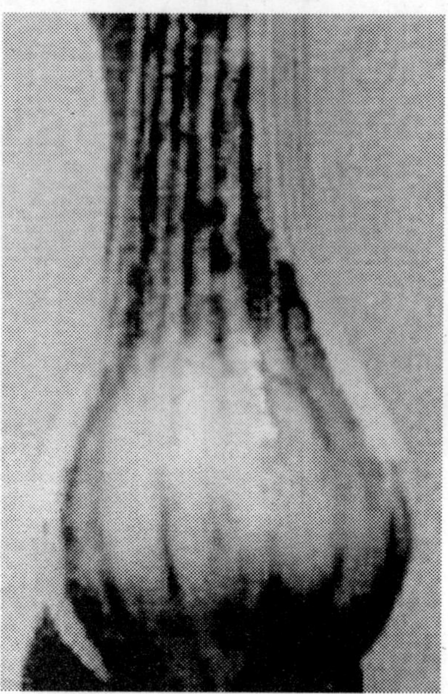

Fig. 3.21. Onion smut.

ii) Management: Seed treatment with thiram or captan at 3 g/kg of seed checks soil-borne infection. Thiram and captan both at 0.2% may be applied to the soil before sowing the seeds. The soil may be treated with formalin to eliminate the pathogen in the soil. Growing onions from transplants avoids the disease.

3.5.7. White Rot, *Sclerotium cepivorum*

White rot, caused by the soil-borne fungus *Sclerotium cepivorum*, is a very destructive disease that begins in the field and can carry over into storage.

i) Symptoms: The first above-ground symptoms are yellowing and dieback of the leaf tips, followed by a collapse of the affected leaves; however, these symptoms alone could easily be confused with other types of damage (i.e., onion maggot). When the bulbs and roots are examined, a white, fluffy mold and soft rot will be observed. Masses of tiny black sclerotia can also be seen within this mold (Fig. 3.22). These sclerotia remain in the soil for many years. Infected bulbs can rot in storage boxes and stain other bulbs. White rot typically develops in patches in the field and is less of a problem when soils are warm (higher than 24°C) and dry.

ii) Management : Removal and destruction of infected plants reduces the inoculum potential in the soil. Early planting of onion and garlic is an important management strategy to avoid white rot in areas with infected soil. Soil solarization for 8-11 weeks during summer using transparent sheet resulted in reduction of *S. cepivorum* to undetectable levels in upper 20 cm of soil. Application of *T. harzianum, T. viride, T. virens* and *Bacillus subtilis* to the solarized soil effectively controlled the disease.

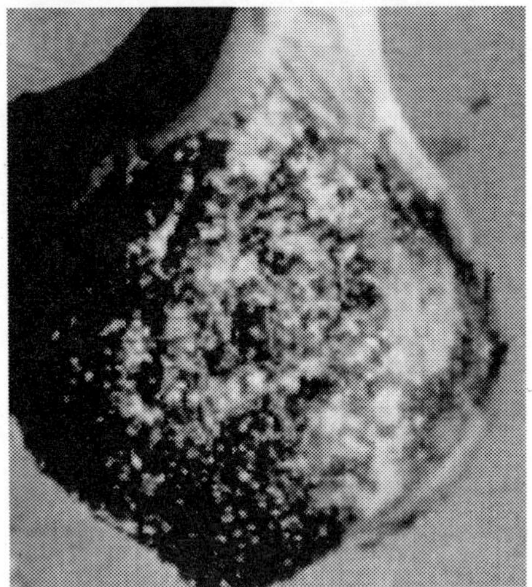

Fig. 3.22. White rot of onion.

3.6. OKRA

Okra (*Abelmoschus esculentus*) is an annual vegetable crop grown in tropical and sub-tropical regions. It is being cultivated in 0.376 million hectares producing 3.684 million tonnes of okra with an average yield of 9.8 tonnes per hectare (Bijay Kumar, 2007) (Table 3.14). Tender green fruits are cooked in curry and soup. The root and stem are used for clearing cane juice in preparation of 'gur'. High iodine content of fruits helps to control goitre while leaves are used in inflammation and dysentery. It is a good source of Vitamins A, B and C. The fruits are also used in cases of renal colic, leucorrhoea and general weakness. The dry seed contains 13-22% good edible oil and 20-24% protein. The oil is used in soap, cosmetic industry and as vanaspati while protein is used for fortified preparations. The crushed seed is fed to cattle for more milk production and the fibre is utilized in jute, textile and paper industry.

Table 3.14. Area, production and productivity of leading okra growing states in India (2005-06)

State	Area (in'000 ha)	Production (in '000 MT)	Productivity (MT/ha)
West Bengal	65.8	718.9	10.9
Bihar	56.2	714.1	12.7
Orissa	71.4	618.8	8.7
Gujarat	41.0	350.3	8.5
Andhra Pradesh	25.6	332.7	13.0

Chattishgarh	18.1	184.7	10.2
Maharashtra	29.0	182.7	6.3
Assam	9.6	119.7	12.5
Uttar Pradesh	8.8	110.2	12.5
Haryana	14.0	93.7	6.7
Karnataka	7.7	63.2	8.2
Others	28.9	194.9	6.7
Total	376.1	3684.0	9.8

(Source: Indian Horticulture Database 2006)

3.6.1. Leaf Spot, *Cercospora abelmoschi*

i) Symptoms: The fungus causes no definite leaf spot but grow as sooty to dark olivaceous ectoparasite covering the entire leaf lamina. The disease is initiated as small, isolated fungal growth on lower surface of leaves, but very soon it spreads to both sides without any definite area (Fig. 3.23). The pathogen severely reduces the photosynthetic area. Infected leaves fall down very quickly soon after drooping and rolling. The disease causes severe loss in seed crop as well as late sown crop grown for fruits. Sometimes light brown lesions are also observed after washing out of spores from leaves due to rainfall.

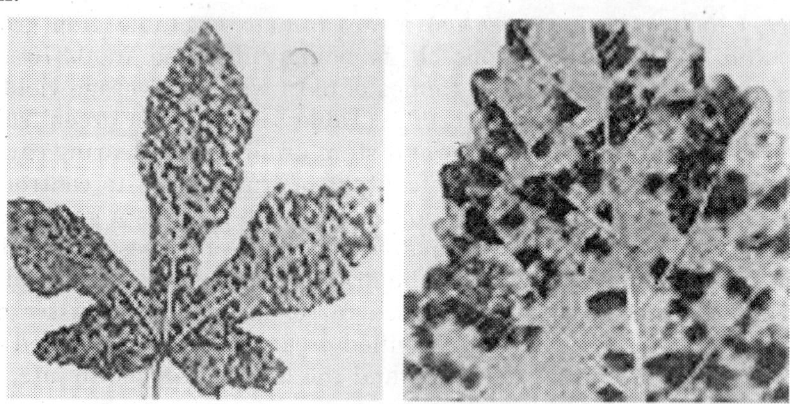

Fig. 3.23. *Cercospora* leaf spot on okra.

ii) Management: Sowing of okra should be completed during first week of July. The defoliated leaves should be collected and burnt in the field to reduce the inoculum at source.

Three sprays of 0.05% bitertanol or 0.025% tridimefon or 0.8% Bordeaux mixture at 500 liters/ha starting from the appearance of symptoms at weekly intervals is effective for the management of leaf spot disease.

3.6.2. Powdery Mildew

i) Symptoms: The powdery mildew disease usually appears at the end of last harvesting of the fruits. The disease is characterized by the appearance of tiny, white to dirty grey spots. They become powdery as they enlarge. The superficial powdery mass may ultimately cover the entire green surface of the host (Fig. 3.24). The effect of severe infection is premature defoliation and death of the plant parts.

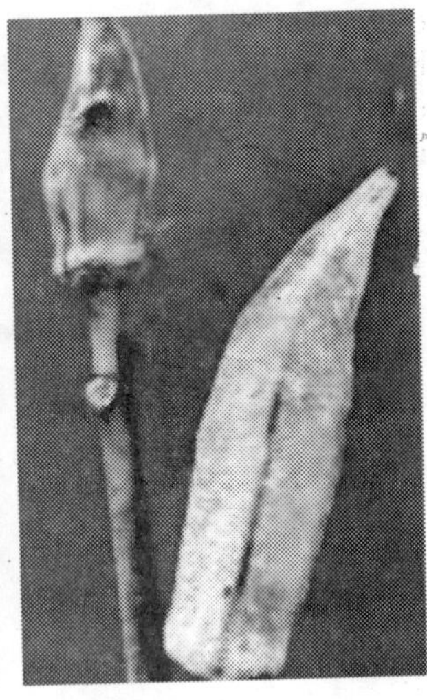

Fig. 3.24. Powdery mildew on okra leaves and fruits.

ii) Management: Spraying of 4% NSKE and 1% neem oil effectively controlled the disease.

Integrated Disease Management

- Deep off-season/summer ploughing is useful to expose nematodes and resting stages of pathogens.
- Using YVMV resistant varieties (Arka Abhay, Arka Anamika, Parbhani Kranti, Varsha Uphar) for planting.
- Seed treatment with *Trichoderma viride* at 4 g/kg of seeds.
- Intercropping with onion or cowpea is useful.
- Crop rotation with non-host crops such as sesame, marigold, onion or mustard reduces nematode infestation.
- Destruction of YVMV affected plants from time to time.

Table 3.15. Disease management in okra

Days after sowing	Management	Pests and Diseases controlled
25	Mancozeb 2 g/l of water*	Leaf spot diseases
40	Wettable sulphur 3 g/l of water	Powdery mildew
60	Wettable sulphur 3 g/l of water	Powdery mildew

3.7. CABBAGE AND CAULIFLOWER

Cabbage (*Brassica oleracea* var. *capitata*) is an important winter vegetable of cole group. It is a rich source of Vitamins A, B and C and also contains minerals. It is being cultivated in 3.121 million hectares producing 67.315 million tonnes of cabbages with an average yield of 21.568 tonnes per hectare (Bijay Kumar, 2007) (Table 5.1). It covers about 4% of total area under vegetables. India comes next to China in cabbage production. In India, it is being cultivated in 0.267 million hectares producing 5.922 million tonnes of cabbage with an average yield of 22.1 tonnes per hectare (Bijay Kumar, 2007) (Tables 3.16 and 3.17).

Table 3.16. Major cabbage producing countries of the world

Country	Area (in '000 ha)	Production (in '000 MT)	Productivity (in MT/ha)	% share of world production
China	1719	33817	19.5	50.2
India	267	5922	22.1	8.8
Russia	166	4051	24.5	6.0
Korea	42	2603	61.8	3.9
Japan	53	2287	42.9	3.4
Ukraine	71	1475	20.9	2.2
Poland	39	1382	35.6	2.1
Indonesia	64	1293	20.1	1.9
USA	30	1100	36.9	1.6
Romania	55	1012	18.5	1.5
Others	615	12375	20.1	18.4
Total	3121	67315	21.5	

Source: FAO (except Indian data) (Indian data source: Indian Horticulture Database 2006)

Table 3.17. Area, production and productivity of leading cabbage growing states in India (2005-06)

State	Area (in'000 ha)	Production (in '000 MT)	Productivity (MT/ha)
West Bengal	74.7	1982.7	26.5
Bihar	33.7	931.6	27.6
Orissa	36.6	578.3	15.8
Assam	24.3	480.3	19.8
Maharashtra	15.0	381.0	25.4
Gujarat	19.0	329.2	17.3
Uttar Pradesh	7.0	209.9	30.0
Haryana	12.3	172.2	14.0

Karnataka	6.4	134.2	20.8
Chattishgarh	6.8	114.0	16.8
Madhya Pradesh	3.8	75.0	20.0
Others	27.7	533.1	19.2
Total	267.4	5921.6	22.1

(Source: Indian Horticulture Database 2006)

Cauliflower (*Brassica oleracea* var. *botrytis*) is the most popular winter vegetable among cole crops. It is rich in Vitamin C. It is being cultivated in 0.292 million hectares producing 5.260 million tonnes of cauliflowers with an average yield of 18.0 tonnes per hectare (Bijay Kumar, 2007) (Tables 3.18 and 3.19). With the development of new varieties, it is now being grown in non-traditional areas like Andhra Pradesh, Tamil Nadu and Kerala.

Table 3.18. Major cauliflower producing countries of the world

Country	Area (in '000 ha)	Production (in '000 MT)	Productivity (in MT/ha)	% share of world production
China	363	7369	20.3	40.5
India	292	5260	18.0	28.9
USA	69	1288	18.6	7.1
Spain	25	442	17.4	2.4
Italy	18	431	23.7	2.4
France	27	375	13.8	2.1
Mexico	25	359	14.2	2.0
Poland	14	239	17.0	1.3
UK	19	219	11.7	1.2
Pakistan	11	206	18.2	1.1
Others	130	1992	15.3	11.0
Total	995	18180		

Source: FAO (except Indian data) (Indian data source: Indian Horticulture Database 2006)

Table 3.19. Area, production and productivity of leading cauliflower growing in India (2005-06)

State	Area (in'000 ha)	Production (in '000 MT)	Productivity (MT/ha)
West Bengal	65.6	1666.2	25.4
Bihar	59.8	938.5	15.7
Orissa	45.1	638.9	14.2
Maharashtra	13.0	328.9	25.3
Haryana	21.0	302.8	14.4
Gujarat	16.1	277.8	17.2
Assam	16.7	247.8	14.8

Madhya Pradesh	8.6	137.4	16.0
Uttar Pradesh	6.7	134.4	20.0
Punjab	5.6	131.4	23.5
Chattishgarh	9.6	123.5	12.9
Others	24.0	332.7	13.9
Total	291.8	5260.1	18.0

(Source: Indian Horticulture Database 2006)

3.7.1. Damping-Off, *Pythium aphanidermatum, Rhizoctonia solani*

i) Symptoms: The fungus attacks in the seedling stage causing damping-off as well as wire stem. The pathogen is responsible for bottom rot, head rot and foot rot in older plants.

Rhizoctonia solani attacks the seedlings at the bottom of the young stem (hypocotyl) or at the soil line. The tissue becomes water-soaked and rapidly collapse causing the seedlings to topple over and die. At later stages, plants are brownish to black just above and below soil line. The stems are usually somewhat smaller than normal but tough and woody.

ii) Epidemiology: The fungus survives in soil. Secondary spread takes place by irrigation water and during cultural operations.

iii) Management: Well drained nursery beds, use of well decomposed organic manure, wide spacing between the rows, lesser density of seedlings and light but frequent irrigations helps in reduction of damping-off. Good control of damping-off was obtained through soil incorporation of wheat bran preparation of *T. harzianum* in commercial nursery. Seed treatment with 0.2% carbendazim or benomyl or *Trichoderma* spp. was found to be as effective as soil treatment for managing the disease. Soil application of 0.1% carbendazim or 0.2% rovrol is effective.

Soil drenching with 0.6% Bordeaux mixture commencing 15 DAS and subsequent 2 more drenchings at 10 to 15 days interval gives effective control of the disease.

3.7.2. Downy Mildew, *Peronospora parasitica*

The disease occurs generally in nursery and to a lesser extent in the main field.

i) Symptoms: The pathogen infects stem, leaves and flowers. The disease is characterized by the appearance of purplish brown spots on the under surface of the leaves. On the upper surface of the leaves, the lesions are tan or yellow. Thin grayish-white downy fungal growth occurs on the under surface of leaves (Fig. 3.25). Infected leaves dry up, shrivel and fall off. Flower parts are deformed. The stem show dark brown and sunken lesions or streaks which later develop the downy growth of the fungus. Heads formed are very small in size.

Fig. 3.25. Downy mildew on cabbage leaves.

ii) Epidemiology: The pathogen survives on diseased plant debris. Low temperature, high humidity and high moisture favour the disease development.

iii) Management

- Removal of lower leaves in morning and burning them.
- Seeds should be cleaned properly so that fragments or parts of the pod which may contain oospores are not present.
- Crop rotation with non-cruciferous crops is effective in reducing the disease incidence.
- Spraying with ridomil MZ or aliette or mancozeb or metalaxyl + mancozeb each at 0.2% or copper oxychloride at 0.3% gives effective control of the disease.
- Cauliflower cvs. Kuwari-17, Kuwari-8 and First Early Laxmi are reported to be moderately resistant to downy mildew.

3.7.3. White Rot or Stalk Rot, *Sclerotinia sclerotiorum*

i) Symptoms: Leaves when infected turn light green and later yellow and are shed prematurely. Black soft rot often appears on petioles and midrib touching the soil and are covered with fluffy growth of the fungus. Stem at the ground level shows girdling. Fluffy mycelium of the fungus grows on these infected plant parts with numerous sclerotia embedded. Cool (20-25°C) and humid (95-100% RH) conditions are ideal for disease initiation and spread (Sharma and Sharma, 1985).

ii) Management: Avoiding excess moisture with less frequent irrigation and reducing the plant populations by following wider spacing will create microclimate less congenial for disease development. Application of FYM or saw dust reduces the disease incidence. Soil amendment with sunflower inflorescence, rape seeds and pine needles reduce the disease incidence (Gupta *et al.*, 1986). Follow cabbage/cauliflower-paddy rotation. Remove diseased leaves and destroy them.

Application of benomyl and MBC both at 0.1% was found effective in reducing the disease incidence (Gangopadhyay, 1984). Soil application of *T. harzianum* (TH 4) and *Aspergillus niger* (Sanjeevini and Kalisena) were highly effective.

3.7.4. *Alternaria* Black Spot, *Alternaria brassicae*

i) Symptoms: On leaves several small, dark brown zonate spots are produced expanding rapidly to form circular lesions up to 1 cm in diameter. The enlargement of lesions may lead to formation of concentric circles which coalesce in the centre (Fig. 3.26). In humid weather, the fungus may cause bluish growth in the centre of the spots which hampers the photosynthetic activity of the plant thereby hampering the overall productive potential of the plants. Sometimes cauliflower heads are also infected by the pathogen. The heads show browning, starting at the margin of the individual flowers or flower clusters.

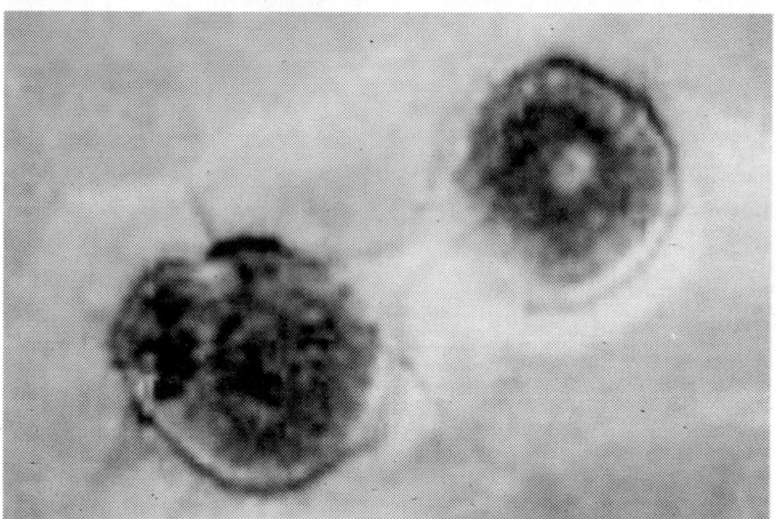

Fig. 3.26. *Alternaria* leaf spot on cabbage.

ii) Epidemiology: The fungus is seed-borne and survives in diseased crop debris. The disease is favoured by hot moist weather.

iii) Management: Seed treatment with hot water maintained at 50°C for 30 min is effective. Giving wider spacing of 60 x 50 cm and removal of basal leaves from time to time help to reduce the disease incidence. Seed treatment with *T. viride* and *P. fluorescens* and soil treatment of nursery beds with neem cake enhanced seed germination and seedling stand. Spraying with captan or copper oxychloride both at 0.3% is effective in controlling the disease.

3.7.5. Club Root, *Plasmodiophora brassicae*

i) Symptoms: Roots enlarge to form spindle-shaped growth (Fig. 3.27). The pathogen is responsible for temporary flagging of leaves on bright days leading to their wilting. Secondary invasion of soft rot bacteria follows. This helps in accumulation of toxic materials in plants.

Fig. 3.27. Club root affected cabbage seedlings and roots.

ii) Management: Trap cropping with rayosak among susceptible brassicas in the first year and tori sarson and radish among resistant brassicas in the second year with 30 days optimum trapping period were found most effective in reducing the disease intensity and increasing the yield.

Integrated Disease Management

Nursery

- Deep off-season ploughing.
- Soil solarization for 2-3 weeks during summer using 400 gauge thick transparent polythene sheet.
- Green manuring with sunnhemp in July – August.
- Hot water treatment of seeds at 50°C for 30 min or seed soaking in 100 ppm streptocycline solution for 30 min to control black rot disease.
- Raising nursery on raised beds (10 cm above ground level) for good drainage, to avoid damping-off.
- Seed treatment with *T. harzianum/T. viride* at 4 g/kg to prevent infection of soil/ seed borne fungal diseases.
- Apply 1 kg FYM enriched with 50 g of *T. harzianum/T. viride*/m^2 nursery bed.

Main Field

- Giving wider spacing of 60 x 50 cm to reduce the incidence of *Alternaria* black spot, black rot and downy mildew.
- Removal of basal leaves to reduce the incidence of *Alternaria* black spot, black rot and downy mildew.
- Spray 0.25% mancozeb or 0.2% chlorothalanil to control foliar diseases.

3.8. PEA

Pea (*Pisum sativum*) is an important vegetable crop widely grown throughout the world. As a cool season crop, it is extensively grown in temperate zone; but restricted to cooler altitudes in the tropics and winter season in the sub-tropics. A rich source of proteins (25%), amino acids and sugars (12%), green peas are an all time favourite vegetables. It is rich in Vitamins A,B,C and minerals. It is being cultivated in 0.314 million hectares producing 2.299 million tonnes of peas with an average yield of 7.3 tonnes per hectare (Bijay Kumar, 2007) (Tables 3.20 and 3.21).

Table 3.20. Major Peas Producing Countries of the World

Country	Area (in '000 ha)	Production (in '000 MT)	Productivity (in MT/ha)	% Share of world production
India	314	2299	7.3	29.5
China	241	2210	9.2	28.3
USA	86	859	10.0	11.0
France	31	360	11.7	4.6
Egypt	28	290	10.4	3.7
UK	32	133	4.2	1.7
Turkey	17	122	7.0	1.6
Algeria	24	110	4.5	1.4
Hungary	14	85	6.2	1.1
Netherlands	5	84	16.5	1.1
Others	303	1248	4.1	16.0
Total	1094	7799		

Source: FAO (except Indian data) (Indian data source: Indian Horticulture Database 2006)

Table 3.21. Area, production and productivity of leading peas growing states in India (2005-06)

State	Area (in'000 ha)	Production (in '000 MT)	Productivity (MT/ha)
Uttar Pradesh	139.0	1146.8	8.3
Madhya Pradesh	19.4	212.9	11.0
Himachal Pradesh	16.3	177.0	10.9
West Bengal	21.7	129.4	6.0
Jammu & Kashmir	7.3	112.2	15.4
Punjab	17.2	103.3	6.0
Haryana	11.1	84.9	7.6
Uttarakhand	9.6	60.9	6.3
Chattishgarh	7.2	50.8	7.1

Bihar	8.1	50.5	6.2
Orissa	4.8	42.0	8.7
Others	52.4	128.0	2.4
Total	314.0	2298.7	7.3

(Source: Indian Horticulture Database 2006)

3.8.1. White Rot, *Sclerotium rolfsii*

i) Symptoms: It is a ubiquitous fungus with a broad host range. Initially wet rotting of bark is observed. Entire bark of the plant near collar region rots. The characteristic symptoms include white, cottony fungus growth is observed on affected portion as well as on parts in contact with soil. Gradually this hyphal mat is converted into small, mustard-like sclerotia that survive in the soil.

ii) Management: Summer ploughing followed by irrigation in summer helps in reducing the initial inoculum of the fungus. Field sanitation by burning all plant debris after harvest is effective. Crop rotation with low land paddy, wheat/barley is effective in reducing the initial inoculum. Follow proper spacing between rows and plants. Avoid excessive irrigation from pre-flowering stage onwards. Stake medium to tall cultivars. Avoid injury to stem during picking. The late sown crop in November escapes the disease to a great extent.

Soil drenching with 0.3% copper oxychloride followed by 0.1% carbendazim near the collar region in evening is recommended. Soil application of *Trichoderma* spp. supplemented with organic matter and green manuring are beneficial.

3.8.2. White Mold, *Sclerotinia sclerotiorum*

i) Symptoms: The disease is commonly found on pea which is responsible for complete killing of plants. Symptoms observed include wet and soft rotting of the tissues. White fungus growth is observed on the rotted portion. Embedded sclerotia covered with white mycelium are formed on the infected portion as well as on inner portion of the pith and fruits.

ii) Management: All infected parts must be carefully collected and burnt before drying of the plants. Deep ploughing and crop rotation with low land paddy with continuous stagnation of water reduces the sclerotial population. Close planting and dense canopy of plants should be avoided. Alternate sprays of 0.1% carbendazim and 0.25% mancozeb is essential at 7 days interval to check the disease. Reduction in white mold of pea was reported under field conditions through the application of *Trichoderma harzianum*.

3.8.3. Powdery Mildew, *Erysiphe polygonum*

i) Symptoms: Mealy white patches appear on both sides of leaves, stems, branches, pods and tendrils (Fig. 3.28). These patches originate as minute discoloured specks from which a powdery mass radiates on all sides and covers a large area of aerial parts. It causes considerable damage and may result in 20-30% losses in pod number and quality. If it attacks early, plants fail to bear fruit or the pods get chaffy.

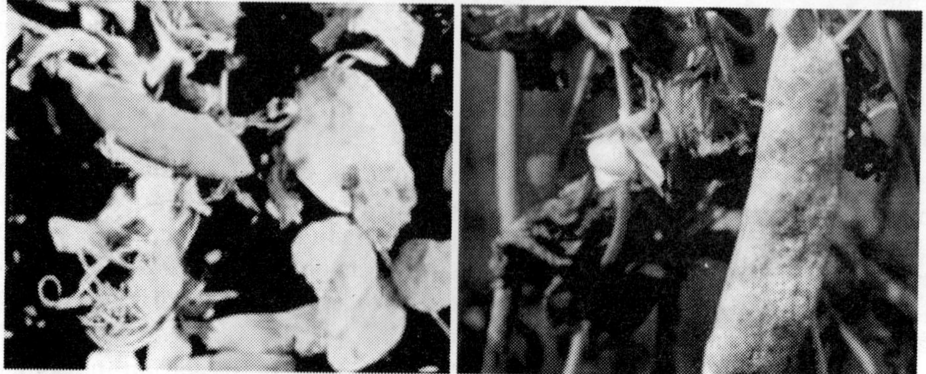

Fig. 3.28. Powdery mildew on pea leaves and pod.

ii) Management: Avoid delay in sowing which should be completed by first fortnight of October. Infected plant debris must be burnt after harvesting of pods. Spray 0.01% penconazole or dinocap or karathane or carbendazim or baycor and repeat the spray after 10-15 days, if necessary. Use tolerant cvs. like Ageta, Organ, Sugar and NDVP-4 which are also vegetable types (Pandey *et al.*, 1999).

In organically grown pea, biopriming of pea seeds with *Trichoderma harzianum* + *Pseudomonas fluorescens* was most effective in improving the seedling stand. Foliar application of bioagents significantly reduced the incidence of powdery mildew disease. Application of *P. fluorescens* + *T. harzianum* through seed and foliar spray was most effective in increasing the yield of organically grown pea (Table 3.22).

Table 3.22. Effect of different biocontrol agents applied through seed, FYM and/or foliar application on powdery mildew disease and yield of pea

Treatment	Seedling stand/m^2	Powdery mildew incidence (%)	Yield (kg/ha)
Seed biopriming with *Trichoderma harzianum* + sprays with *Pseudomonas fluorescens*	56	13	355
FYM colonized with *Trichoderma harzianum* + sprays with *Pseudomonas fluorescens*	63	16	352
Seed biopriming with *Trichoderma harzianum* + FYM colonized with *T. harzianum* + sprays with *T. harzianum*	69	12	388
Control	52	36	301
CD at 5%	6	10	53

3.8.4. Wilt, *Fusarium oxysporum* f. sp. *pisi*

i) Symptoms: Infection leads to drooping and wilting of plants at early stage of plant growth (Fig. 3.29). Vascular discolouration of stem and reddish appearance in the pith extends towards roots. The roots turn black and rot. Plant growth is checked, foliage turns yellow and downward curling of stipules and leaflets takes place. The entire plant wilts and the stem shrivels. Epinasty of affected plants is a characteristic symptom.

Fig. 3.29. Fusarium wilt of pea.

ii) Management: Summer ploughing should be carried out to reduce the soil inoculum. Field sanitation and crop rotation are most important for controlling this soil-borne pathogen. All infected plants should be uprooted and burnt. Proper drainage should be maintained in the field.

Treat seeds with thiram at 0.3% or carbendazim at 0.2% before sowing and avoid early sowing. In badly infected areas, treat the soil with 0.1% carbendazim. Soil amendment with green manure and application of *Trichoderma harzianum* is most beneficial.

3.8.5. Rust, *Uromyces pisi, U. fabae*

i) Symptoms: It occurs in severe form in cooler regions of India. The disease is characterized by yellow spots bearing asci in clusters with curling symptoms. Pustules present a light appearance on both surfaces of leaves and other parts. Teleuto pustules are dark brown or almost black in colour. In cool and wet climate of Northern India, the entire plant may be killed.

ii) Management: Burning of infected crop debris will reduce the inoculum. Early bloom stage of crop 4 weeks before harvest is most critical period for rust and must be protected from the infection. Fungicide spray should be immediately started when average of 2 spots per leaf are observed in the field. The best fungicide is 0.025% flusilazole sprayed at weekly interval. Spraying of hexaconazole or bitertanol or tridimefon or tebuconazole or karathane or carbendazim or baycor all at 0.05% at 10-15 days interval is also effective. The disease can also be controlled by sprays of calixin, bayleton and mancozeb. Pea cvs. NDVP-3 (vegetable type), Batri Brown and Pea-9 are reported to be resistant to rust disease (Katiyar and Ram, 1987).

3.9. FRENCH BEAN

French bean (*Phaseolus vulgaris*), also known as common, haricot, kidney, string, salad, runner or snap bean, is an important leguminous vegetable. It is consumed as tender pods, shelled green beans and dry beans. It is a nutritious vegetable which contains proteins (1.7g), calcium (50 mg), phosphorus (28 mg), iron (1.7 mg), carotene (132 mg), Thiamine (0.08 mg), Riboflavin (0.06 mg) and Vitamin C (24 mg/100 g of edible pods). It is largely grown in hilly areas of Himachal Pradesh, Jammu and Kashmir and north-eastern states during summer, winter and autumn crop in parts of Uttar Pradesh, Maharashtra, Karnataka and Andhra Pradesh.

3.9.1. Anthracnose, *Colletotrichum lindemuthianum*

i) Symptoms: Most characteristic symptoms appear on immature pods, leaves (Fig. 3.30) and cotyledons. Stems are also infected. Black, sunken cankers with light coloured or grey central area appear on pods. The central portion of the spot shows pinkish masses of the fungus in wet weather. It causes more damage in cool and temperate regions.

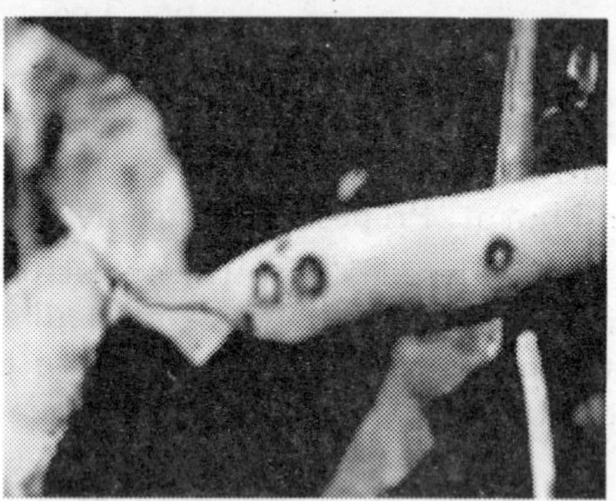

Fig. 3.30. Anthracnose on French bean leaf and pod.

ii) Management: Use disease-free seed and treat seed with carbendazim at 2 g/kg seed before sowing. In disease prone areas, spray right from the beginning 0.2% chlorothalanil or zineb or 0.1% mancozeb.

3.9.2. Collar Rot or Web Blight, *Rhizoctonia solani*

i) Symptoms: At emergence, the infection on hypocotyl and stem results in damping-off symptoms, but afterwards, elongated, sunken, reddish brown lesions are produced on the stem at ground level. On the foliage, circular to irregular brown spots having distinct borders appear. Leaf scald symptoms are common. In rainy season, the disease assumes serious proportions. Under continuous wet weather, along with high temperature, extensive damage results.

ii) Management: Treat seed with 0.3% thiram 75 WP or 0.2% carbendazim. On disease appearance, spray 0.1% carbendazim. Soil application of *Trichoderma* spp. along with FYM before sowing the seed helps to control soil-borne problems.

Integration of *Glomus mosseae* with *T. viride* gave total protection.

3.9.3. Rust, *Uromyces phaseoli*

i) Symptoms: Rust disease occurs throughout India where beans are grown. Disease symptoms are mostly confined to leaves but young stem and branches are also attacked. The reddish brown rust pustules appear abundantly on the lower surface of leaves (Fig. 3.31). When the leaves become thoroughly infected, these shrivel and fall off from the plant. In severe cases, most of the vines are defoliated. Uredospores of the rust pathogen are the main source of infection.

ii) Management: Cultural operations like good field sanitation, destruction of infected plant material and crop rotation are important in minimizing the disease incidence. Fungicidal sprays with 0.2% mancozeb and 0.05% bayleton are effective for disease control.

Fig. 3.31. Rust on leaf of French bean.

T. viride and *T. koningii* as 2% foliar sprays were effective both in French bean and cowpea.

3.9.4. Angular Leaf Spot, *Isariopsis grisicola*

i) Symptoms: Reddish-brown lesions appear mostly on lower surface of leaves. Later, symptoms also appear on pods which results in defoliation.

ii) Management: Follow at least 2-year crop rotation with non-host crops in infected areas. On appearance of disease, spray chlorothalanil/dithionon/propineb/quantaf/carbendazim at 0.1%.

3.10. COWPEA

Cowpea (*Vigna unguiculata*) also known as southern pea and black-eye pea, is one of the most important vegetables. It is cultivated for its long, green or purplish pods to be cooked as vegetable or for dry seeds used as pulse. Its foliage is used as fodder or green manure. In India, cowpea is grown almost throughout the country.

3.10.1. Leaf Blight, *Pseudocercospora cruenta, Ccercospora cruenta*

i) Symptoms: *Pseudocercospora cruenta* is more destructive pathogen than *C. cruenta*. Symptoms appear as black, sooty and mouldy growth of fungus observed on all foliar parts. Initially the spore masses are olivaceous dark brown and spread very fast in the field. Quick defoliation is characteristic feature of the disease which leads to naked stem and twigs on the plant. However, *C. cruenta* is less destructive and form small circular brown spots with reddish margin.

ii) Management: Management practices include crop rotation with non-host crops and field sanitation by burning the infected crop debris. Avoid delay in planting particularly during kharif season. Hexaconazole or biter-tanol or triadimefon or tebuconazole all at 0.05% are highly effective (Pandey and Pandey, 2002).

3.11. PIGEONPEA

3.11.1. Wilt, *Fusarium udum*

i) Symptoms: The disease is characterized by slow wilting of the plant. The symptoms can be observed after a month of sowing or at the flowering or pod formation stage. The affected plants become yellow in colour followed by drooping and finally the whole plant dries up. The symptoms resemble as if the plant is suffering from the drought. The disease can be diagnosed whenever the affected stem is cut and opened where the browning of the xylem vessels could be clearly seen. It has been recorded that wilting can be partial or total. However, it has been observed that partial wilting is mainly associated with lateral root infection, whereas, tap root infection may cause complete wilting.

ii) Management: *Bacillus subtilis* brought about a marked reduction in the incidence of pigeon pea wilt. Amendment of soil with roots of certain leguminous crops (sweet clover), molasses and oil cakes (groundnut cake) markedly increased the antibiotic (bulbiformin) production by *B. subtilis* and a reduction of 88% in the incidence of wilt. Seedlings gained resistance to *Fusarium* infection when the seeds were bacterized with *B. subtilis* before sowing. It was suggested that bulbiformin became systemic in the plant and provided a protective zone around the roots of pigeon pea seedlings.

Trichoderma harzianum effectively suppressed *Fusarial* wilt of pigeon pea.

3.12. CLUSTER BEAN

Cluster bean (*Cyamopsis tetragonoloba*) is also popularly called as 'guar' in India. Its tender pods being edible and are used in preparation of curry. The green pods are rich in Vitamins A and C. Since it is draught tolerant, it can be grown in summer. It is also grown as a forage and green manure crop. Its seeds are reported to be good cattle feed.

3.12.1. Root Rot, *Sclerotium rolfsii*

i) Symptoms: It is a ubiquitous fungus with a broad host range. Initially wet rotting of bark is observed. Entire bark of the plant near collar region rots. The characteristic symptoms include white, cottony fungus growth is observed on affected portion as well as on parts in contact with soil. Gradually this hyphal mat is converted into small, mustard-like sclerotia that survive in the soil.

ii) Management: Control of root rot of cluster bean was achieved with *Streptomyces nigrifaciens.*

3.13. AMARANTHUS

3.13.1. Damping-off, *Pythium aphanidermatum*, *Rhizoctonia solani*, *Aphanomyces* sp.

i) Symptoms: Damping-off is a serious disease in seedlings. The disease is caused by *Pythium aphanidermatum, Rhizoctonia solani* and *Aphanomyces* sp. Seeds may rot in the soil before emergence (pre-emergence damping-off) or seedlings may exhibit stem canker above the soil line and/or root necrosis. Affected seedlings eventually wilt (post-emergence damping-off) (Fig. 3.32). The disease is favoured by high soil water content and low soil temperatures. Also dense planting without sufficient aeration enhances disease development.

Fig. 3.32. Damping-off of amaranthus seedlings.

ii) Management

- Use disease-free seeds.
- Avoid over watering.
- Avoid dense sowing and planting.
- Provide good drainage.
- Spray dithiocarbamates.

3.13.2. Stem Rot/Blight, *Choanephora cucurbitarium*

i) Symptoms: Stem rot is caused by the fungus *Choanephora cucurbitarium*. It causes wet rot of stems and leaves. Affected plant parts have hairy appearance (silk-like threads) consisting of fungal spores. Infection is predisposed by injuries. During rainy season it can cause heavy defoliation. The disease is spread by air currents and infected seeds. Warm, and moist conditions favour disease development.

ii) Management

- Plant certified disease-free seeds.
- Avoid dense planting to allow sufficient aeration.
- Practice good field sanitation.
- Wider spacing reduces the rate of infection.

3.13.3. White Rust, *Albugo bliti*

i) Symptoms: White rust pathogen forms white blisters on the leaves (Fig. 3.33).

Fig. 3.33. White rust on amaranthus leaf.

3.13.4. Downy mildew, *Peronospora* spp.

i) Symptoms: Downy mildew causes discolouration and drying of the leaves (Fig. 3.34).

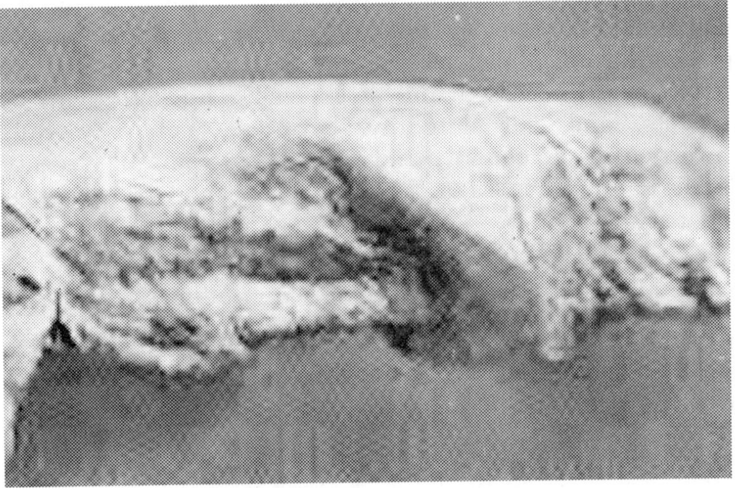

Fig. 3.34. Downy mildew on amaranthus leaf.

3.14. CURRY LEAF

3.14.1. Leaf Spot, *Cercospora* sp.

i) Symptoms: The pathogen causes leaf spots on curry leaves (Fig. 3.35).

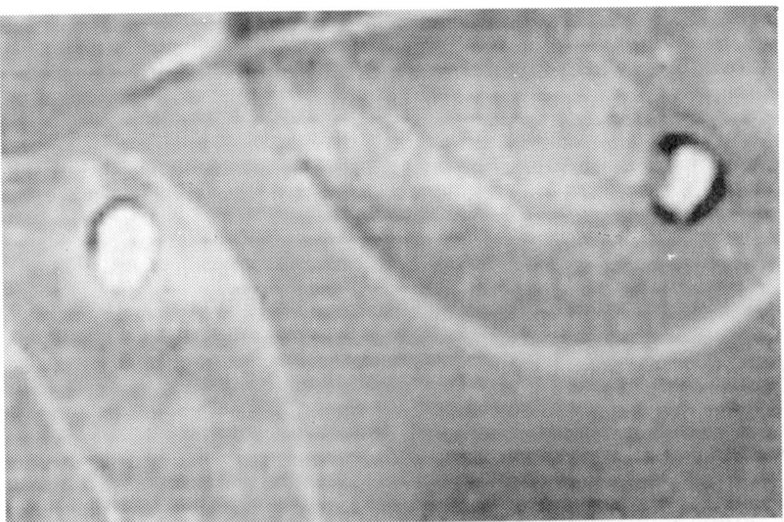

Fig. 3.35. *Cercospora* leaf spots on curry leaf.

ii) Management: Spraying carbendazim at 0.25 ml/litre of water gives effective management of leaf spot disease.

3.15 PUMPKIN

Pumpkin (*Cucurbita moschata*) occupies a prominent place among vegetables owing to its high productivity, nutritive value, good storability, long period of availability, better transport qualities and extensive cultivation in sub-tropical and tropical parts of the world. Fruits are rich in Vitamin A. It is also consumed as processed and stock feed. The fruits can be used in preparing sweets, candy, fermented into beverages. Yellow or orange-fleshed pumpkins are rich in carotene. In India, it is grown mainly in Assam, West Bengal, Tamil Nadu, Karnataka, Madhya Pradesh, Uttar Pradesh, Orissa and Bihar.

3.15.1. Downy Mildew, *Pseudoperonospora cubensis*

i) Symptoms: This is the most important disease in successful cultivation of cucurbitaceous crops. Characteristic symptoms are angular yellow spots often restricted by veins appearing on upper surface, giving purplish growth on the lower surface. In moist weather, such leaves die quickly. Young fruits and tips of vines are affected and fail to grow.

ii) Management: Removal of badly infected leaves, followed by spraying fungicides prevents spread of the disease. Spray 0.3% copper oxychloride or 0.2% mancozeb or propineb or chlorothalanil or dithionon.

If it is a seed crop, the spray of 0.2% metalaxyl mancozeb or cymoxanil mancozeb or AL-Fosetyl at fortnightly interval gives very good control.

3.15.2. Anthracnose, *Colletotrichum lagenarium*

i) Symptoms: This disease is serious on fruits, young stems and vines. Small, yellowish or water-soaked areas that enlarge rapidly and turn brownish appear on the foliage of most cucurbits. Growing leaves may be destroyed by the disease and coalescing spots may cause death of the entire leaf. On fruits, the spots are roughly circular, sunken, and water-soaked with dark borders. In the centre of these spots, point-wise blank stromata bearing pink spore masses are produced in humid weather.

ii) Management: Spray 0.25% mancozeb or chlorothalanil or dethionon or 0.1% carbendazim at 15 days interval, especially during rainy season.

3.15.3. Powdery Mildew, *Erysiphe cichoracearum, Sphaerotheca fuligenea*

i) Symptoms: This disease is widely occurring on musk melon, bottle gourd, cucumber, pointed gourd, water melon and pumpkin. The disease becomes rampant during dry and humid weather. The disease appears as mealy-white to dirty grey spots on the foliage (Fig. 3.36) and green stems. Later, these become powdery and ultimately cover the entire green surface of the plant, causing premature withering, drying and defoliation. Under severe infection, the fruits may also be covered with white powdery mass. Young fruits turn pale brown and drop off. The quality of fruit is adversely affected.

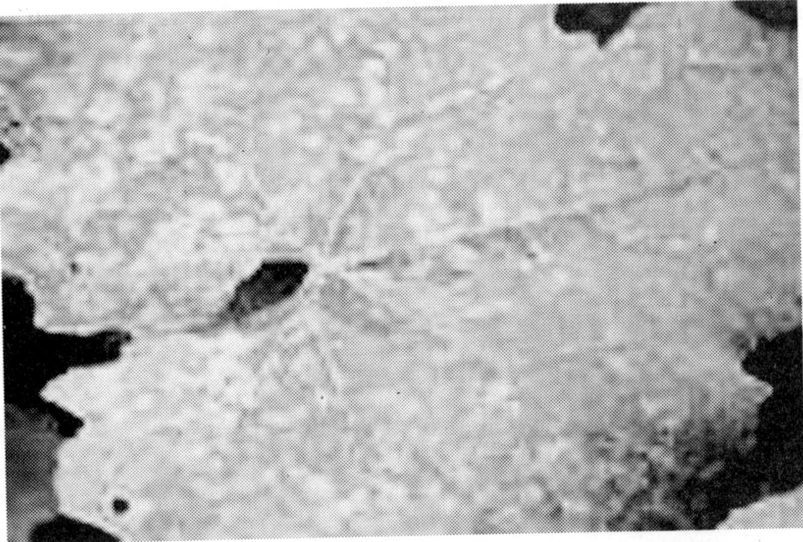

Fig. 3.36. Powdery mildew infection on pumpkin leaf.

ii) Management: Destruction of infected plant material and weeds harbouring the pathogen reduces the disease incidence. For effective control of the disease, 0.025% penconazole, 0.1% tridemorph, 0.15% benomyl, 0.05% karathane, 0.5% elosal, 0.2% wettable sulphur, 0.1% carbendazim have been recommended.

Muskmelon cvs. Arka Rajhans, Florigold, Gulfstream, Jacumba, Edisto and Cantaloupe No. 45 are reported to be resistant to powdery mildew disease.

Integrated Disease Management

- Ploughing after green manuring in the month of June and July.
- Soil application of FYM enriched with *Trichoderma harzianum* at 3-5 MT/ha within a week of ploughing of sunnhemp.
- Always prefer to grow all the cucurbits on bower system particularly in kharif/rainy season.
- Seed treatment with 0.25% carbendazim or *T. harzianum/T. viride* at 4g/kg of seed.
- Restrict the use of exotic hybrids in view of their susceptibility to gummy stem blight.
- Soil application of neem cake enriched with *T. harzianum* to minimize gummosis, root rot and collar rot.
- Give 2 foliar sprays of 0.25% mancozeb at initial infection of downy mildew.
- Spray 0.1% carbendazim or 0.25% chlorothalanil for controlling foliar diseases such as anthracnose and *Cercospora* leaf spot.
- Always use summer crop for disease-free seed production.

- Give enough spacing to prevent overlapping of the crop.

3.16. CUCUMBER

India is considered to be the home of cucumber (*Cucumis sativus*). Cucumber is cultivated for fresh consumption or as pickling cucumber for preservation, marinated with vinegar, salt, dill or other spices. It is an important salad crop cultivated both in north and south and lower as well as higher hills in India.

3.16.1. Powdery Mildew, *Sphaerotheca fuliginea, Erysiphe cichoracearum*

i) Symptoms: This is a fungal disease that is spread by spores carried by the wind. White to gray fungal deposits are observed on the leaves and stems. As the mildew spreads, the leaves may begin to drop off. White powdery coating is produced on the leaves. Affected plants turn yellow and wither.

ii) Management: Make sure to keep the field free of debris and destroy any infected plant parts. Crop rotation with paddy in low lying areas is very effective against powdery mildew. Application of *Ampelomyces quisqualis* to the foliage, starting shortly after colonies of *S. fuliginea* began to appear, efficiently suppressed powdery mildew and increased yields of marketable cucumbers. Water spray of *A. quisqualis* gave acceptable control of powdery mildew on cucumber in the field if applied with 2% paraffin oil.

Both *A. quisqualis* and *Sporothrix flocculosa* are effective in controlling powdery mildew on cucumber under high humidity conditions, especially in greenhouses. The pycnidial hyperparasite, *A. quisqualis* has sticky conidia that are splash dispersed, as are the spores of yeast like antagonist, *S. flocculosa*. Biocontrol of powdery mildew therefore, depends on maintaining high humidity and occasional water sprays to ensure splash dispersal of antagonist spores. *Tilletiopsis washingtonensis* and *T. pallescens* when applied at 1 x 10^8 spores/ml, could significantly reduce powdery mildew on cucumber in greenhouse.

3.16.2. Downy Mildew, *Pseudoperonospora cubensis*

i) Symptoms: This devastating disease is caused by the fungus *Pseudoperonospora cubensis*. The disease is most important in the Northeast on muskmelon, pumpkin, and winter squash, though all cucurbits are susceptible. Because the fungus does not over winter in the North, the disease is reintroduced each season by airborne conidia blown into the area from southern states. If appropriate conditions are present, plantings can be infected so rapidly that they appear to have been frosted. Foliar symptoms vary depending on how quickly infection occurs. Upper surfaces of leaves may show angular pale-green areas bounded by leaf veins that give the impression of mosaic. These areas change to yellow angular spots and become necrotic as the disease progresses. When leaves are infected by a heavy spore shower, small individual necrotic and chlorotic flecks appear on the upper surface. In a moist environment, sporulation occurs on the lower leaf surface. The chlorotic flecks appear as light brown areas when held up to the light. Areas with active sporulation give the undersurface a purplish or dark brown sooty appearance. The fungal growth consists of large lemon-shaped spores called sporangia that are borne on branched structures called sparangiophorus. The sporangia are wind-borne and can survive over long distances when the air is moist. They are also readily spread to adjoining plants by splashing rain.

Although most downy mildew fungi are considered cool, wet weather diseases, cucurbit mildew can infect plants over a wide temperature range (50-80° F) with the optimum at 61-72°F. Periods of heavy dew lasting into midmorning are sufficient for infection to occur. Mid-August is a critical time to scout for mildew because appropriate environmental conditions usually exist then in the Hudson Valley and on Long Island. Practice cleanliness in the garden by removing garden debris. Try and keep the foliage as dry and possible.

ii) Management : Because disease development is dependent upon outside inoculum, rotation cannot prevent the disease. Select resistant varieties if they are available. Downy mildew can be controlled with the application of protectant and systèmic fungicides.

Integrated Disease Management

- Seed treatment with *P. fluorescens* at 5 g/kg of seeds.
- Soil application of enriched FYM with *T. harzianum* at 1 MT/ha.
- Soil application of fresh neem cake at 250 kg/ha immediately after germination and repeated at the time of flowering.
- Spraying of *P. fluorescens* 2 times (20 and 45 DAP) to induce disease resistance.

3.17. WATERMELON

Watermelon (*Citrullus lunatus*) is an important cucurbitaceous summer vegetable grown for ripe fruits. Though it can be grown in garden land, it is a major river-bed crop of Uttar Pradesh, Rajasthan, Gujarat, Maharashtra and Andhra Pradesh. An excellent desert fruit, it is relished by rich as well as poor. The fruit juice makes an excellent refreshing and cooling beverage after adding a pinch of salt and black pepper. The fruits contain 92% water, 0.2% protein, 0.3% minerals and 7.0% carbohydrates in a 100g edible flesh.

3.17.1. Anthracnose, *Colletotrichum* sp.

i) Symptoms: Sunken spots with raised margins are produced on leaves and fruits (Fig. 3.37).

ii) Management : Spraying carbendazim at 2 g/litre of water gives effective management of the disease.

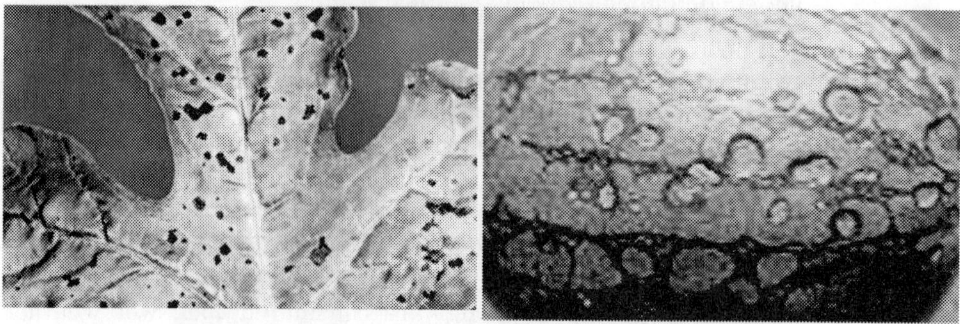

Fig. 3.37. Anthracnose on water melon leaf and fruit.

3.17.2. Powdery Mildew, *Erysiphe cichoracearum*

i) Symptoms: White powdery coating is produced on the leaves (Fig. 3.38). Affected plants turn yellow and wither.

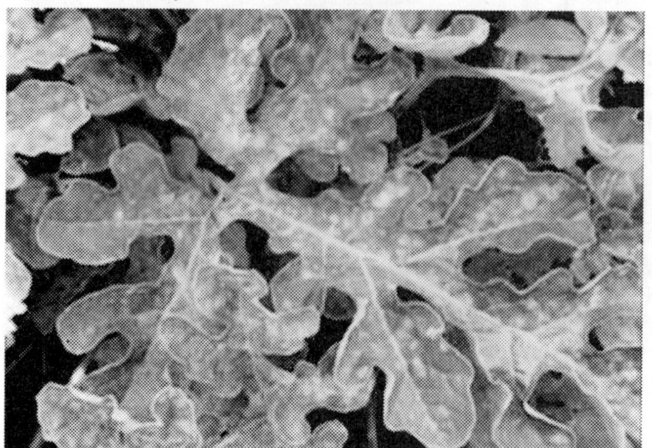

Fig. 3.38. Powdery mildew on water melon leaf.

ii) Management : Spraying carbendazim at 2 g/litre of water is effective for the control of the disease.

3.18. MUSK MELON

3.18.1. Alternaria Leaf Blight, *Alternaria cucumerina*

i) Symptoms: Alternaria leaf blight caused by the fungus *Alternaria cucumerina* is a common disease of muskmelon (cantaloupe) but can also infect watermelon, squash, and cucumber. Leaf blight usually occurs from midseason or when the canopy closes, can reduce late season fruit production, and can result in poor quality if the disease is severe. Symptoms first appear on the upper surface of crown leaves as small spots 1 to 2 mm in diameter although other lesions can be more than 10 mm in diameter (Fig. 3.39). These larger spots can show a target-like pattern of rings which is typical of most *Alternaria* spp.

Fig. 3.39. Altérnaria leaf blight on musk melon.

Leaf blight pathogen over winters as mycelium or chlamydospores in diseased plant debris and can probably survive for more than a year. Conidia are produced in the spring and act as the primary inoculum. Conidia produced on infected plants provide inoculum for repeating secondary cycles during the season.

ii) Management : A 2-year rotation out of cucurbits is the easiest way to break this cycle. There are few Alternaria-resistant varieties. Protectant fungicides should be applied in mid-July when vines run and fruit have been set. Infection and disease development are favoured by lengthy periods of high relative humidity (18 hours) over a broad range of temperatures (68 to 90° F).

3.18.2. Wilt, *Fusarium solani*

i) Management: *T. harzianum* and *T. viride* when applied as seed treatment (5 g/kg) followed by soil application along with FYM/neem cake at 1 kg/basin were effective. Soil application of neem cake/mustard cake at 2% reduced the wilt by 80% and 65%, respectively.

3.19. RADISH

Radish (*Raphanus sativus*) is a root vegetable which can be grown as an intercrop with other crops. Its leaves are rich in Vitamins A and C. It can be eaten raw and also cooked. It has some medicinal value. Radish is cultivated throughout India. West Bengal, Bihar, Uttar Pradesh, Punjab, Assam, Haryana, Gujarat and Himachal Pradesh are major radish-growing states.

3.19.1. White Rust, *Albugo candida*

i) Symptoms: Prominent white pustules appear on the surface of leaves, stem and inflorescence. Marked swelling and distortion of the affected plant parts are common. Due to hypertrophy and hyperplasia of the tissues, the floral parts show swellings and distortions. Infected plants are stunted in growth.

ii) Epidemiology: The pathogen survives in infected plant debris and on other hosts. Low temperature, high humidity and high moisture favour the disease development.

iii) Management: Spraying with mancozeb or Phytoalexin/Macimin all at 0.3% is effective.

3.20. BEET ROOT

3.20.1. Seedling Blight, *Rhizoctonia solani*

i) Management: Spraying 0.2% mancozeb gives effective control of seedling blight disease.

3.20.2. Leaf Spot, *Cercospora* sp.

i) Symptoms: The pathogen causes small circular spots in large number on older leaves. The centre of the spot turn brown which later turn grey with red margin.

ii) Management: Spraying of 1% Bordeaux mixture as soon as the disease incidence is noticed gives effective control of the disease.

3.20.3. Storage Rot, *Sclerotium rolfsii*

i) Management: *T. harzianum* was highly effective in suppressing *S. rolfsii* in beet root under field conditions. The conidial mixture of 2 bioagents *T. pseudokoningii* (effective in reducing the mycelial growth) and *T. virens* (good colonizer of sclerotia) was effective in reducing the incidence of storage rot.

3.21. CARROT

Carrot (*Daucus carota*) is another root vegetable rich in Vitamin A and some minerals. It is grown all over India. Main carrot growing states are Uttar Pradesh, Assam, Karnataka, Andhra Pradesh, Punjab and Haryana. It is taken raw as well as in cooked form. It is made into pickles and sweet meat. Carrot juice is a rich source of carotene (a precursor of Vitamin A) and contains appreciable quantities of thiamine and riboflavin. It is sometimes used for colouring butter and other foods. Green carrot leaves are highly nutritive, rich in protein, minerals and vitamins and used as fodder and also for preparation of poultry feed.

3.21.1. *Cercospora* Leaf Blight, *Cercospora carotae*

i) Symptoms: Lesions are formed on leaves, petioles, stems and floral parts. On leaves, brown to black spots appear and later may shrivel and break (Fig. 3.40). Floral parts when attacked also shrivel. The pathogen survives on and in the seed and also in diseased plant refuse in the soil. The fungus is active under moderately warm conditions.

Fig. 3.40. Cercospora leaf blight on carrot leaf.

ii) Management: Seed treatment with hot water at 50°C for 15 min is effective to manage the disease. Crop rotation and destruction of infected plant material in the field will minimize infection. Spraying of 0.2% foltaf or 0.3% copper oxychloride is effective to control leaf blight.

3.21.2. *Alternaria* Leaf Blight, *Alternaria dauci*

i) Symptoms: Dark brown to black spots appear mostly along the leaf margin and petioles (Fig. 3.41). In severe cases, leaflet is shriveled and killed. The pathogen survives in and on the seed and also in diseased material in the soil. The disease is severe under high moisture conditions.

Fig. 3.41. *Alternaria* leaf blight on carrot leaf.

ii) Management: Seed treatment with hot water at 50°C for 15 min is effective to manage the disease. Crop rotation and destruction of infected plant material in the field will minimize infection. Spraying of 0.2% foltaf or 0.3% copper oxychloride is effective to control leaf blight.

3.21.3. Powdery Mildew, *Erysiphe cichoracearum*

i) Symptoms: White powdery growth appears on leaves and petioles causing the leaves to turn brown and wilt (Fig. 3.42).

Fig. 3.42. Powdery mildew on carrot leaves.

ii) Management: Spray 0.2% wettable sulphur fungicides 4 weeks after sowing and repeat the spray at 7 and 10 weeks after sowing.

3.22. MUSHROOMS

Mushrooms have achieved significant importance due to their nutritive and medicinal values. Mushrooms, the edible fleshy fungi, represent one of the few valuable sources of nutritive and palatable food. They are rich in proteins (35-48%), vitamins and minerals. They are weight reducers and useful for diabetic patients. Though, these fungi have been naturally occurring and being utilized in India since time immemorial, the commercial cultivation is only of recent times. It was only in 1964 that the first successful crop of mushroom was grown at Solan in Himachal Pradesh. Since then, research on various aspects of mushroom cultivation has been strengthened and today it is one of the important cottage industries, not only in the state of Himachal Pradesh but its commercial cultivation has also been initiated in most of the other states and union territories of India.

At present, it is estimated that 40,000 tonnes of mushrooms are produced annually. The major share (85%) is contributed by button mushroom (*Agaricus bisporus* and *A. bitorquis*). Oyster mushroom (*Pleurotus* spp.) is popular in southern states and Maharashtra and milky mushroom (*Calocybe indica*) in Tamil Nadu, Andhra Pradesh and Karnataka (Table 3.23).

Table 3.23. State-Wise Production of Cultivated Mushrooms

State	Production in MT	Type of mushroom
Tamil Nadu	10000	*Agaricus / Volvariella*
Maharashtra	7000	*Agaricus / Pleurotus*
Haryana	4000	*Agaricus*
Punjab	4000	*Agaricus*
Utter Pradesh	4000	*Agaricus / Pleurotus*
Andhra Pradesh	2500	*Agaricus / Pleurotus*
Madhya Pradesh	2000	*Agaricus / Pleurotus*
Goa	1000	*Agaricus*
North Eastern States	1000	*Agaricus / Lentinus /*
Orissa	700	*Pleurotus / Volvariella*
Karnataka	600	*Agaricus / Volvariella*
Himachal Pradesh	500	*Agaricus*
Jammu & Kashmir	400	*Agaricus*
West Bengal	250	*Agaricus / Pleurotus*
Kerala	100	*Pleurotus / Volvariella*
Rajasthan	50	*Pleurotus*
Others	1900	*Agaricus / Pleurotus*
Total	40000	

Source : NCMR&T, Solan (1996-97)

3.22.1. Wet Bubble, *Mycogone perniciosa*

It is one of the serious diseases causing severe crop losses. It is the commonest and geographically most widespread disease of mushrooms. This disease is occurring in almost all the countries wherever mushrooms are grown.

i) Symptoms: The most characteristic symptom is the development of distorted masses of mushroom false tissue, which are initially white and fluffy but turn brown with age and decay. These distorted masses are known as 'Sclerodermoid masses' and they may be as big as 10 cm in diameter. Small amber to dark brown drops of liquid are seen on the surface of the undifferentiated tissue under high moisture/RH conditions. The name of the disease is derived from this typical condition. In addition to distortion symptom, small fluffy white patches of mycelium may occur on the surface of the casing, following the infection of a developing mushroom below the casing surface.

The base of the stalk shows brown discolouration when mature fruit bodies are attacked, and a white fluffy mycelial overgrowth appears on the affected portions of the mushroom. The stipe sometimes develops brown streaks, and the cap develops symptoms which appear as a sector of affected and stunted gills, covered by the contamination of the casing medium.

ii) Epidemiology: The pathogen produces two types of spores: single celled, thin walled and relatively short lived conidia; and two celled aleurospore (terminal chlamydospore) with very thick-walled terminal cell for longer survival under adverse conditions. The conidia are light and air borne. Both aleurospores and conidia are spread by water splash.

This fungus is commonly found in soils, and the disease outbreak often follows unrestricted soil movement on the farm. The primary source of the pathogen on most farms is through contaminated casing. Under temperate climate, spores may survive in crop debris and hence can contaminate subsequent crops. Once the pathogen is established in the crop beds, the main means of secondary spread is through water splash and by excess water running off the beds. Pickers and workers may also carry and spread the pathogen on their hands, tools, boxes, clothing, etc.

iii) Management

The most important and viable means of control of wet bubble disease is the elimination of the primary source of the pathogen by paying strict attention to hygiene at the farm. The casing medium should be freed from spores of *Mycogone perniciosa* which can be achieved by either steam pasteurization or formalin treatment of the casing medium and its careful storage in a clean room free from dust and debris from cropping rooms.

The following practices may be adopted to manage the disease:

- Disinfect tools/clothes of workers regularly to prevent the spread of inoculum.
- Avoid over-watering of cropping beds.
- Carefully remove all affected mushroom under a powerful light source.
- Pickers should use gloves and disinfectants at regular intervals.
- All diseased mushrooms should be removed before the crop is picked.
- Cover the affected mushroom with common salt or pots (push the pots well into the casing) to prevent the spread of the disease.
- Watering should be done only after all the diseased mushrooms have been removed.

Benzimidazole fungicides (benomyl, carbendazim and thiobendazole) are quite effective in controlling the disease. Prochloraz manganese also gives good control of wet bubble disease when mixed in casing soil. Dithiocarbamates (zineb and maneb) also control the disease when inoculum levels are low.

3.22.2. Dry Bubble, *Verticillium fungicola* var. *fungicola*

This is the most serious fungal disease of mushrooms causing serious economic losses. This fungus is a mycoparasite with necrotrophic behaviour and kills the host fungus on colonization. It is geographically most widespread disease of white button mushroom in the world. This disease is reported from U.K., U.S.A., Ireland, Germany, The Netherlands, Italy, India, Canada, Australia, etc. If this disease is left uncontrolled, the pathogen can totally destroy the mushroom crop within 2 to 3 weeks.

i) Symptoms: When the mushrooms are infected at an early stage of their development (primordial/pinhead stage), the typical symptoms are small undifferentiated mass of tissue (up to 2.5 cm in diameter). When mushrooms are affected at later stage of development, they are often imperfectly formed with partially differentiated caps, or with distorted shaggy stipes and with tilted caps. Such affected mushrooms are covered with a fine grey white mycelial growth, discoloured, dry and do not rot as in wet bubble. When fully grown mushrooms are affected, these show small pimple-like outgrowths on top of pileus, or blue grey spots (1-2 cm in diameter) on the cap surface. Such spots have yellow or bluish-grey halo around them.

The growth of the pathogen on mushrooms is in the form of fine, grayish brown felt. The infected mushrooms are dry at first, but may undergo bacterial rot if not removed soon from beds. The diseased sporophores when cut longitudinally, often showed a grey-brown zone of variable thickness immediately below the surface, others were completely discoloured internally. The cavities found in sporophores were lined with spore-bearing conidiophores.

ii) Epidemiology: Contaminated casing and not the compost is the most common initial source of inoculum on a mushroom farm. Air-borne spores can also serve as primary source of inoculum, or spores carried by flies, mites and pickers. The conidia of *Verticillium*, produced in clusters surrounded by sticky mucilage, help in effective dispersal of spores to new sites from infected beds. The mucilage enables the spores to get attached to dust particles, flies, mites, debris and pickers. Watering is another important means of dispersal, as excess water running off the infected beds carry spores to the lower beds and floor.

The optimum temperature for disease development is about 20°C. The time from infection to symptom appearance is 10 days for distortion symptoms and 3-4 days for cap spotting at 20°C. The pathogen grows best at 24°C.

iii) Management: The most effective way to control this disease is the prevention of entry of this pathogen into crop beds. Strict attention to hygiene will go a long way in keeping this disease at bay in mushroom houses. Ensure that casing soil is free from infection of *V. fungicola*. Pasteurize or treat the casing medium with steam/formalin and store in a clean area, free from dust and debris. The affected mushroom should be removed as quickly as possible from the beds to avoid the danger of dispersal of inoculum through insects/pests, tools, pickers, etc. The affected mushrooms can also be covered with common salt or pots. Fly and mite infestation should be controlled to prevent dissemination of the inoculum. The diseased beds should be effectively cooked out with steam or chemically.

Prochloraz manganese gives good control of dry bubble disease.

3.22.3. *Trichoderma* Disease, *Trichoderma harzianum*

Trichoderma has been reported as one of the serious fungal pathogen on mushrooms. As high as 30% crop losses have been reported by the attack of this fungus on mushrooms. It is most commonly associated with mushroom farms and thrives best in high humidity and on carbohydrate materials. Both these parameters are excellent for the growth of *Trichoderma.*

i) Symptoms: Green patches of sporulation, the main symptom of attack by *Trichoderma,* are clearly visible in bags, trays or base of shelves. When mature, these spores turn dark green and become less visible. These are then washed off into the casing below and the mushrooms emerging through this casing are badly spotted, often distorted and with a very short shelf life (Seaby, 1989).

There is drastic fall in yield with the appearance of blank patches on the bed, and casing becomes densely covered with red pepper mites or contaminated by nematodes.

ii) Epidemiology: The spores of *Trichoderma* are transmitted to freshly peak heated compost by wind, attached to spent compost dust particles, by mites who have fed on *Trichoderma* mycelium in infected compost, on the hands and clothing of workers and by machinery. The infection persists in wooden shelf by colonization of wood which is most of the times wet. When new compost is filled, it gets contaminated from the fungus from the wood, unless effective cook out has been carried out at termination of cropping.

iii) Management: The most effective and economic control measures are by giving maximum attention to farm hygiene. Benomyl spray is also reported to be effective but spot treatment of affected areas with common salt is often used in the mushroom farms. Cook out of the compost beds at termination of cropping with steam or formalin treatment reduces the infection. Clean all tools and equipment with formalin solution periodically. Use foot dips effectively while entering the mushroom house. Control mites effectively to prevent dissemination of infection.

Table 3.24. Fungal disease Management in Vegetable Crops

Crop	Pest	Symptoms	Pesticide Dose/litre	Remarks
Tomato	Buck-eye rot	Light brown zonate spots on fruits	Staking + Copper oxychloride-3 g	Rainy season when plants are green
	Early blight	Dark brown spots on leaves elyptical on stem	Bordeaux mixture-1% Copper oxychloride-3 g Zineb-2 g	September-December
	Late blight	Water soaked brown spots on leaves and stem	Bordeaux mixture-1% Copper oxychloride-3 g Mancozeb-2 g	November-January
	Blight	Small irregular spots on leaves	Copper oxychloride-3 g Carbendazim/ Benomyl-1 g	Rainy season
	Powdery mildew	White mealy growth on leaves & Calyx	Wettable sulphur-2 g Dinocap-1 g	September-February

Brinjal	Powdery mildew	White mealy growth on leaves & Calyx	Wettable sulphur-2 g Dinocap/Tridemorph/ Myclobutanil-1 ml	September-February
	Fruit rot	Light brown zonate spots	Copper oxychloride-3 g	Rainy season
	Leaf spot	Reddish brown spots on leaves	Carbendazim-1 g Thiophanate Methyl/ Bitertanol-1 g	Rainy season
Chili/ Capsicum	Powdery mildew	White mealy matter on leaves & Calyx	Wettable sulphur-2 g Carbendazim-1 g	During cloudy days
	Anthracnose	Dark brown with grayish centre	Carbendazim-1 g Thiophanate Methyl-1 g	Rainy season
	Leaf spot	Light to dark brown zonate spots	Copper oxychloride-3 g Zineb-2 g Chlorothalonil/Bitertanol/Thiophanate Methyl-1 g	Rainy season
	Leaf spot	As in tomato	Zineb-2 g Ziram-2 g Copper oxychloride-3 g	Rainy season, November-December
French Bean	Anthracnose	As in grapes	Copper oxychloride-3 g Carbendazim-1 g Benomyl-1 g Thiophanate Methyl-1 g Mancozeb-2 g	Rainy season
	Angular leaf spot	Light brown angular spots on leaves	Mancozeb-2 g Carbendazim-1 g Thiophanate Methyl-1 g	September-October
	Powdery mildew	As in tomato	Wettable sulphur-2 g	Cloudy weather
	Rust	Dark brown pustules on the lower surface of leaves leading to defoliation	Mancozeb-2 g Triadimefon/Bitertanol-1 g	Rainy season, December-January
Pea	Powdery mildew	White mealy growth on leaves, petiole and pods	Wettable sulphur-2 g Tridemorph-0.5 ml Dinocap-1 ml Myclobutanil-1 g	During cloudy weather
	Rust	Light yellow deposit on leaves and pods	Mancozeb-2 g Tridemorph/Triadimefon/ Bitertanol-1 ml	October-January
Cluster Bean	Powdery mildew	As in peas	Wettable sulphur-2 g Dinocap-1 ml	September-December
	Blight	Dark brown zonate spots on leaves	Mancozeb/Iprodione-3 g	Rainy season, November-December

Cowpea	Leaf spot	Brick red spots on leaves surrounded by yellow halo	Carbendazim-1 g Thiophanate Methyl-2 g Mancozeb-2 g	Rainy season, November-December
	Powdery mildew	As in peas	Wettable sulphur-2 g Dinocap/Myclobutanil-1 ml	Cloudy weather
	Anthracnose	Dark brown sunken spots	Copper oxychloride-3 g	Rainy season
	Rust	Dark brown rusty pustules on lower surface of leaves	Mancozeb-2 g Triadimefon/Bitertanol-0.5 g	October-January, Rainy season
Cucurbits	Anthracnose	Dark brown sunken spots. Infects leaves and stem parts	Carbendazim-1 g Thiophanate Methyl-1 g	September-February
	Powdery mildew	As in other crops	Wettable sulphur-2 g Dinocap/Myclobutanil-1 ml	Cloudy weather
	Downy mildew	Spots on upper surface of leaves with light spots on lower surface	Metalaxyl /Mancozeb-2 g AL-Phosetyl-2 g	Rainy season & November/ December months
Cabbage/ Cauliflower	Downy mildew	As in cucurbits	Metalaxyl /Mancozeb-2 g Al-Phosetyl-2 g Copper oxychloride-3 g	Rainy season & November/ December months
	Wire stem		Carbendazim-1 g	Soil application
	Leaf spot	Dark brown zonate spots on leaves	Captan/ Chlorothalonil/ Iprodione-2 g Copper oxychloride-3 g	Cooler months
	Black rot	Blackening of stem leading to rotting and foul smell	Streptocycline (seed soaking for 25 min) -0.1 g	Seed treatment
Radish	White rust	Milky white pustules on lower surface of leaves corresponding to the blisters on the upper surface	Mancozeb-2 g Potassium salt of phosphoric acid-3 ml	As and when required
	Downy mildew	As in other crops	Mancozeb/Metalaxyl / AL-Phosetyl-2 g	As and when required
	Blight	As in other crops	Mancozeb/ Chlorothalonil / Iprodione-2 g	As and when required
Onion	Purple blotch	Light brown to purple blotches on leaves	Mancozeb-2 g Chlorothalonil/Iprodione-3 g	During rabi crop

	Stemphylium blight	Purple brown spindle shaped patches on leaves and stem.	Mancozeb-2 g Chlorothalonil/Iprodione-3 g	During rabi crop
	Downy mildew	Pale olive patches on the leaves, later become curled and dry off.	Metalaxyl /Mancozeb/ AL-Phosetyl-2 g	During rabi crop
Carrot	Leaf blight	Light brown spots on the leaves and petiole	Copper oxychloride-3 g	November-January
	Powdery mildew	As in other crops	Wettable sulphur-2 g Dinocap /Myclobutanil-1 ml	October-February
Okra	Powdery mildew	As in other crops	Wettable sulphur-2 g Dinocap/Myclobutanil-1 ml	During cloudy weather
	Leaf spot	Light brown to dark spots with grey centre on the leaves	Carbendazim-1 g Chlorothalonil-2 g Mancozeb-2 g	Rainy season
	Leaf blight	dark brown to black zonate spots on leaves.	Mancozeb/ Chlorothalonil-2 g	Rainy season
Seedling diseases of Vegetables		Black girdling of stem near the base (Ground level)	Carbendazim--1 g Thiophanate Methyl(Soil application)/Iprodione-2 g	Seed treatment before sowing
	Damping off	Damping off of plants	Copper oxychloride (Soil application)-3 g Metalaxyl/Mancozeb-2 g	Rainy season
	Sclerotium	White thread like mycelium with mushroom coloured bodies	Mancozeb/ Iprodione/ (Soil application)-2 g	February to June
	Macrophomina	Charcoal coloured lesions turning black later on and girdling of stem	Only soil applications Carbendazim-1 g Iprodione/Chlorothalonil-2 g	February to June

Table 3.25. *Trichoderma* spp. and other biocontrol agents for the management of diseases of vegetable crops

Crop	Disease / Pathogen/s	Potential biocontrol agent	Mode of application
French bean	Seedling rot, *Pythium* sp., *Sclerotinia sclerotiorum, Botrytis cineria, Rhizoctonia solani*	*T. koningii*	Seed treatment
	Root rot, *R. solani*	*T. viride, T. harzianum, T. virens, T. hamatum*	Seed treatment-4 g/kg, Furrow appln.-130-160 kg/ha
	Dry root rot, *Macrophomina phaseolina*	*T. harzianum*	Furrow appln.-130-160 kg/ha
	Wilt, *Fusarium oxysporum*	*T. harzianum*	Soil appln.- 160 kg/ha

	Wilt, *Sclerotium rolfsii*	*T. viride*	Seed treatment-4 g/kg, Forrow appln.- 130-160 kg/ha
Pea	Seed & Collar rot, *Pythium* sp., *R. solani*	*T. harzianum, T. hamatum*	Seed treatment
	White rot, *Sclerotinia sclerotiorum*	*T. harzianum*	Soil treatment
	Wilt, *Fusarium oxysporum* f. sp. *pisi*	*T. viride, T. harzianum*	Soil treatment
Cabbage	Blight, *Alternaria brassicola*	*T. longibrachiatum, T. virens*	Seed treatment
	Damping-off, *R. solani*	*T. viride, T. harzianum, T. koningii*	Seed treatment
Cauliflower	Damping-off, *R. solani, Pythium aphanidermatum*	*T. harzianum, Aspergillus niger*	Seed & soil treatment
Okra	Wilt, *Pythium* sp.	*Aspergillus niger*	Soil treatment
	Stalk rot, *Sclerotinia sclerotiorum*	*Aspergillus niger*	Soil treatment
Tomato	Damping-off, *Pythium aphanidermatum*	*T. viride, T. harzianum, T. koningii, T. pseudokoningii*	Seed & soil treatment
	Pythium spp.	*Trichoderma viride / T. harzianum* at 4 g/l.	Soil treatment
	Wilt, *Fusarium oxysporum* f. sp. *lycopersici*	*T. viride, T. harzianum, Aspergillus niger*	Seed treatment Seed & soil treatment
		P. fluorescens	Seed treatment Seed & soil treatment
	Wilt, *Sclerotium rolfsii*	*T. viride*	Seed treatment-4 g/kg, Forrow appln.- 130-160 kg/ha
	White rot, *S. sclerotiorum*	*T. viride, T. harzianum*	Soil treatment
		Pseudomonas fluorescens	Soil treatment
	Root rot, *R. solani*	*T. viride, T. harzianum*	Soil treatment
	Grey mold, *Botrytis cineria*	*T. harzianum*	Foliar spray
Potato	Black scurf, *R. solani*	*T. viride, T. virens, T. harzianum, Epicoccum purpurescens*	Tuber treatment
	Charcoal rot, *Macrophomina phaseolina*	*Aspergillus niger*	Soil treatment
Bell pepper	*Phytophthora capsici*	*T. viride, T. harzianum*	Fruit treatment
Chilli	Damping-off, *Pythium aphanidermatum*	*T. harzianum*	Seed & soil treatment

	Wilt, *Sclerotium rolfsii*	*T. harzianum*	Soil treatment
	Fruit rot & Dieback, *Colletotrichum capsici*	*T. viride, T. harzianum, T. pileatus, T. koningii, T. pseudokoningii, T. hamatum, T. longibrachiatum*	Seed & seedling treatment
		P. fluorescens	Seed & seedling treatment Foliar spray
	Wilt, *F. oxysporum*	*E. purpurescens, T. viride, T. harzianum,* Non-pathogenic *F. oxysporum*	Soil treatment
		P. fluorescens	Soil treatment
	Stem rot, *S. sclerotiorum*	*T. viride, T. harzianum*	Soil treatment
	Blight, *Phomopsis vexans*	*T. viride, T. harzianum*	Seed treatment
	Macrophomina phaseolina	*P. fluorescens*	Soil treatment
	Powdery mildew, *Oidium taurica*	*P. fluorescens*	Foliar spray
Brinjal	Damping-off, Wilt, *Phytophthora* sp., *Pythium aphanidermatum, F. solani*	*T. viride, T. harzianum, T. koningii*	Seed & soil treatment
	Collar rot, *Sclerotinia sclerotiorum*	*T. viride, T. virens*	Soil treatment
	Collar rot, *Sclerotinia sclerotiorum*	*Bacillus subtilis*	Soil treatment
		T. viride, T. harzianum	Soil treatment
	Root rot, *R. solani*	*P. fluorescens*	Soil treatment
	Blight, *Pythium vexans*	*P. fluorescens*	Seed treatment
Radish	Seedling rot, Damping-off, *Pythium* sp., *R. solani*	*T. harzianum, T. hamatum*	Seed treatment
Beet root	Wilt, *Sclerotium rolfsii*	*T. viride*	Seed treatment-4 g/kg, Forrow appln.-130-160 kg/ha
	Seed rot, *R. solani*	*T. harzianum*	Seed treatment
Cucurbits	Anthracnose, *Colletotrichum capsici*	*E. purpurescens*	Foliar spray
	Powdery mildew, *Sphaerotheca fuligena*	*Ampelomyces quisqualis*	Foliar spray
	Charcoal rot, *Macrophomina phaseolina*	*E. purpurescens*	Soil treatment
Cucumber	Wilt, *Fusarium oxysporum* f. sp. *cucumerinum, R. solani*	*Aspergillus niger*	Soil treatment

	Root rot, *R. solani*	*T. viride*	Seed treatment-4 g/kg, Forrow appln.- 130-160 kg/ha
	Powdery mildew	*Ampelomyces quisqualis*	Foliar spray
Bottle gourd	Wilt, root rot, *Fusarium oxysporum, F. solani*	*Aspergillus niger*	Soil treatment
Water melon	Wilt, *Fusarium oxysporum* f. sp. *solani, F. o.* f. sp. *niveum*	*T. viride, Aspergillus niger*	Seed & soil treatment
Musk melon	Wilt, *F. oxysporum, F. solani, R. solani*	*T. harzianum, Aspergillus niger*	Seed treatment
Onion	Leaf blight, *Botrytis cineria*	*Gliocladium roseum*	Foliar application
	Tip blight, *Alternaria* sp.	*P. fluorescens*	Foliar application

Table 3.26. Field efficacy of antagonists on diseases of vegetable crops.

Antagonist	Crop	Disease	Pathogen	% Control
T. harzianum	Cauliflower	Sclerotinia rot	*Sclerotinia sclerotiorum*	72
		Damping off	*Pythium aphanidermatum*	83
	Cabbage	Sclerotinia rot	*Sclerotinia sclerotiorum*	70
		Damping off	*Pythium aphanidermatum*	80
	Tomato	Sclerotinia rot	*Sclerotinia sclerotiorum*	73
		Damping off	*Pythium aphanidermatum*	85
	Chilli	Dieback	*Colletotrichum capsici*	61
		Fruit rot	*Colletotrichum capsici*	64
T. viride	Cauliflower	Sclerotinia rot	*Sclerotinia sclerotiorum*	70
		Damping off	*Pythium aphanidermatum*	80
	Cabbage	Sclerotinia rot	*Sclerotinia sclerotiorum*	68
		Damping off	*Pythium aphanidermatum*	78
	Tomato	Sclerotinia rot	*Sclerotinia sclerotiorum*	71
		Damping off	*Pythium aphanidermatum*	83
	Chilli	Dieback	*Colletotrichum capsici*	60
		Fruit rot	*Colletotrichum capsici*	62

Table 3.27. Vegetable crop varieties resistant to fungal diseases.

Vegetable crop	Disease/s	Resistant varieties
Potato	Late blight	Kufri Sutlej, Kufri Badshah, Kufri Jawahar (in plains), Kufri Jyothi, Kufri Giriraj, Kufri Kanchan, Kufri Meghad (in hills)
Tomato	Fusarium wilt	HS-110, Sel-28.
	Powdery mildew	Arka Asish.
Brinjal	*Phomopsis* blight	Pusa Bhairav.

Chilli	Powdery mildew	Arka Suphala (T). Arka Harita, Arka Meghana.
French bean	Rust	Arka Bold, Swarna Priya, Swarna Latha, Arka Anoop.
	Rust, *Alternaria* leaf spot	Arka Bold.
Cowpea	Bacterial blight	Pusa Komal.
Pea	Powdery mildew	Pusa Pragati, JP-4, JP-93, FC-1, PRS-4, NDVP-4.
	PM, Rust	Arka Ajit, Arka Karthik, Arka Sampoorna.
	Fusarium Wilt	JP Batri Brown 3, JP Batri Brown 4.
Cluster bean	Powdery mildew, *Alternaria* leaf spot	Gomah Manjari
Pigeonpea	*Fusarium* wilt	Maruti
Muskmelon	Powdery mildew	Arka Rajhans, Pusa Madhuras (MR), RM-43.
	PM, Downey mildew	Punjab Rasila.
	Fusarium wilt	Pusa Madhuras, Durgapura Madhu, Arka Jeet, Punjab Sunehari (MR), Harela.
Watermelon	PM, DM, Anthracnose	Arka Manik
Ridge gourd	PM, DM	Swarna Uphaar.
Bottle gourd	Blossom end rot	Arka Bahar (T).
Cucumber	PM	Swarna Poorna.
	PM, DM	Poinsette
Cabbage	Black leg	Pusa Drum Head.
Cauliflower	Curd blight	Pusa Shubhra, Pusa Synthetic.
	DM	Pusa Hybrid-2.
Carrot	PM	Arka Suraj.
Onion	Purple blotch, Basal rot	Arka Pitamber, Arka Kirtiman, Arka Lalima.
	Purple blotch, *Alternaria pori*	Arka Kalyan.
Garlic	Purple blotch, *Stemphylium* disease	Agri-found White (G-41), G-1, G-50, G-51
Amaranth	White rust	Arka Arunima, Arka Suguna (MR).
Palak	*Cercospora* leaf spot	Arka Anupama.

MR – Moderately Resistant T - Tolerant

Chapter 4

ORNAMENTAL CROPS

India is known for growing of traditional flowers such as jasmine, marigold, chrysanthemum, tuberose, crossandra and aster. Commercial cultivation of cut flowers- rose, orchids, gladiolus, carnation, gerbera, anthurium and lilies-has also become popular.

With the liberalization of Indian economy, the domestic floricultural trade, which hitherto remained an unorganized sector, became an organized sector with the active participation of corporate giants. The establishment of state-of-art hi-tech floricultural units in early 90's for intensive cultivation of cut flowers contributes about Rs. 300 crores of valuable foreign exchange today. Various government agencies are actively involved in the promotion of this sector and are highly committed to resolve the teething problems. The world demand for floricultural products is approximately US $ 40 billions and cut flowers contribute nearly 60% (US$ 25 billions) to the global trade. India's share in the world flower market is about 0.7% which is contributed mainly from open cultivation.

Floriculture is estimated to cover an area of 126,000 ha with production of 694,000 MT of loose flowers during 2006. The area under cut flowers has increased significantly in recent years and the production has reached 2762 million numbers (Table 4.1). Another 500 ha of climatic control greenhouse are available for growing quality flowers for export. Many corporate houses (more than 100) have set up 100% export oriented units with an investment of more than Rs. 1000 crores. India is bestowed with natural advantages like favourable weather conditions, soil, cheap labour which are likely to make a real mark in international trade. With this, India is emerging as one of the powerful forces in world flower trade.

Table 4.1. State-wise area and production of flowers (1999-2000)

State / UT's	Area (in ha)	Production (in MT)	
		Loose	Cut
Andhra Pradesh	18087	126609	---
Delhi	3450	21007	---
Haryana	2250	40500	760
Himachal Pradesh	133.3	976	181
Jammu & Kashmir	42	7	17.4
Karnataka	20801	124440	---
Madhya Pradesh	1387	832	---
Maharashtra	6600	28336	---
Orissa	175	868.5	9.3
Tamil Nadu	18120	142250	---
U.P. Plains	160	---	160
West Bengal	13227	14644	5678
Others	4174.7	8723.5	1.0
Total	**88607**	**509193**	**6806**

Floriculture is an intensive type of agriculture wherein the income per unit area is much higher than any other agricultural product if it is done in a scientific way. The area used for production of ornamental plants is low, but the cash returns are high because of the intensive cropping. The flowers grown in large areas include jasmine, rose, marigold, chrysanthemum, crossandra, champak, tuberose, gladiolus and aster. New crops like lilies, tulips, anthurium, carnation, gerbera, etc., are also finding potential place in the foreign market. With the rapid development in urban planning and hotel industry, there has been an increasing demand for cut flowers and ornamentals. The cut flowers followed by live plants, cut foliage, seeds and bulbs have got an immense potential for export to Western countries where the cost of production is comparatively high. India exports roses, gladioli, carnations, chrysanthemums, lilies and jasmines to Gulf, Germany, France, England, Netherlands and Singapore. Our share in the export market has registered a sharp increase from Rs.14.45 crores in 1991-92 to Rs.299.41 crores in 2005-06.

4.1. ROSE

Rose (*Rosa* spp.) is the most ancient and popular flower grown the world over. In India, it is commercially grown for cut flowers and rose oil. Rose flowers without stems and loose flower petals are used in traditional markets for making garlands, for offering in temples, while the florist shops sell cut roses with stems mainly for bouquets and floral arrangements. In recent times, about 60 units have been established under joint ventures around Bangalore, Pune, Nasik, Mumbai, Hyderabad, Gurgaon (Haryana), Chandigarh and Saharanpur (Uttar Pradesh) for growing roses in greenhouses for export of cut flowers to Japan, Major protected rose growing areas of include Ambala (Punjab), Bangalore (Karnataka), Hyderabad (Andhra Pradesh), Kodaikanal (Tamil Nadu), Nasik (Maharashtra), New Delhi and Pune (Maharashtra). The Netherlands, Germany and other European countries. Besides, the Damask rose (*Rosa damascena*) and Edward rose (*R. bourboniana*) are cultivated for rose *attar* and other products – *gulkand, gulabjal,* and *pankkurj*. It is cultivated in about 6,000 ha area in the states of West Bengal, Uttar Pradesh, Gujarat, Haryana, Jammu and Kashmir, Tamil Nadu, Karnataka, Rajasthan, Madhya Pradesh, Andhra Pradesh and Punjab.

As is the case with other plants, maintaining the health of roses makes them less susceptible to diseases and better able to withstand disease attack. Cultivars that are more resistant to certain diseases should be used. Humidity encourages fungal disease, as does poor air circulation. Avoid planting roses too closely; avoid overhead watering; and prune to maintain a fairly open structure.

4.1.1. Grey Mold or Botrytis Blight, *Botrytis cinerea*

i) Symptoms: *Botrytis* causes water-soaked lesions on leaves (Fig. 4.1), which spreads very fast. On flowers, the infection occurs as small water-soaked spots resulting in immature fading and drying of petals, which drop ultimately. The infection is generally seen on inner whorls of petals. During the favourable conditions, the infection spreads very fast resulting in complete blighting of the flowers. In roses, grey mold coincides with the progress of senescence. The fungus affects all growth stages of the crop starting from planting to post-harvest. Humid and wet conditions of the greenhouses favour the growth and multiplication of the fungus. A temperature range of 10-25°C and a RH of above 85% are highly favourable for disease development. During cold storage and transshipment, the condensation on the flowers caused due to temperature fluctuations leads to flower blight. High RH besides promoting prolific sporulation of the fungus is also known to predispose the rose plants for grey mold infection by impeding the uptake of calcium by the plants. Application of calcium to rose buds reduces the severity of post-harvest *Botrytis* blight. The onset of post-harvest senescence of the flowers has a direct correlation with the severity of the disease in storage and transit.

ii) Management: Many fungicides belonging to benzimidazoles (carbendazim, benomyl at 0.1%), dicarboximides (iprodion) and sterol biosynthesis inhibitors are commonly used for controlling the disease. *Botrytis* has developed resistance to all the important groups of fungicides that were effectively used earlier. Chemicals with anti-senescent activity render the plants more tolerant.

Application of growth regulators like gibberellic acid, paclobutrazol and methyl jasmonate are reported to suppress grey mold in roses. Inhibitors of ethylene biosynthesis such as amino-oxyacetic acid and carbon dioxide also reduced the disease incidence.

Since, *Botrytis* sporulates readily on wounded and senescent tissues; these should be thoroughly removed from the greenhouses and destroyed. Roses under photo-selective plastic have shown a lower disease incidence compared to commercially used plastic. Occurrence of free moisture should be reduced by proper ventilation. Polyfilms that block UV rays and allow blue light are known to reduce the sporulation of the grey mold fungus to a large extent.

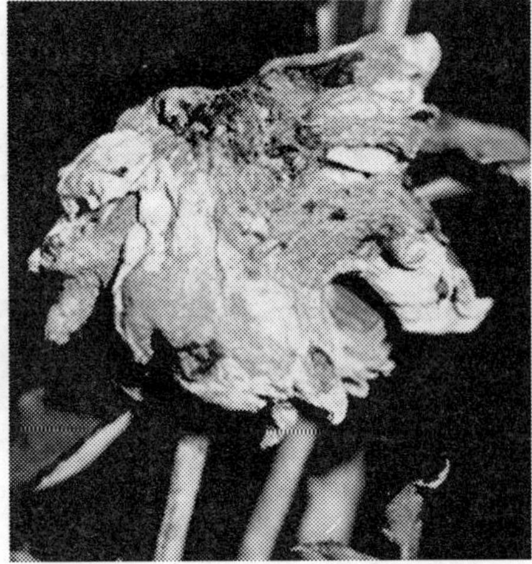

Fig 4.1. Grey mold on leaves and flowers of rose.

Biological agents like *Gliocladium roseum, Myrothecium verrucaria, Trichoderma harzianum* and *Pencillium* spp. are known to be effective in controlling grey mold.

4.1.2. Powdery Mildew, *Podosphaera pannosa*

i) Symptoms: It is a debilitating disease. Though it does not kill the plant, it renders the foliage and flowers unsightly. This is the most important disease on roses grown in both polyhouses and in the open. Grayish white powdery patches are seen on the tender leaves (Fig. 4.2), pedicels and flower buds. Younger leaves of 7-10 days old are more susceptible than older leaves. Infected leaves get malformed, flower buds fail to open and plants exhibit wilting symptoms. The infection occurs on the abaxial side of leaves. The luxuriant growth of the fungal mat on leaves results in greater water loss and reduced photosynthetic efficiency.

The fungus has a very short life cycle and spreads at an alarming rate. Unlike in garden roses where the infection occurs mostly in winter months, the disease occurs throughout the year on greenhouse roses. Favoured by warm and dry weather and widely varying day and night temperatures, the fungus grows well at 18-25°C and optimum germination of conidia occurs at 21°C. A night temperature of 15.5°C and RH of 90-99% promote optimum production of conidia and their germination. The fungal spores are spread by wind.

The fungus survives in the form of mycelial mat called "Pannose mycelium" on buds, stems and thorns. Water stress predisposes the plants to infection by the fungus.

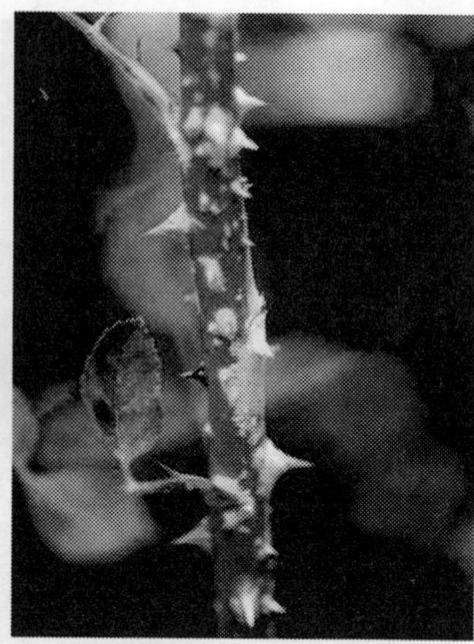

Fig. 4.2. Powdery mildew on leaves and stem of rose.

ii) Management: Since the disease spreads very fast, the spray schedule should begin before the onset of favourable conditions for fungal growth. As the fungus grows on external surfaces of plants even contact fungicides such as sulphur are quite effective in controlling the disease. However, when the days are hot, application of wettable sulphur should be avoided because of phytotoxicity. Early detection followed by sulphur fumigation is most effective and economical under greenhouse conditions. The most commonly used contact and systemic fungicides are 0.1% carbendazim (Bavistin), 0.1% tridemorph (Calixin), 0.1% dinocap (Karathane), 0.1% triadimefon (Bayleton), 0.1% bitertanol (Baycor), 0.05% hexaconazole (Contaf) and 0.05% myclobutanil (Systhane). Among the more recent compounds, strobilurins such as kresoxym methyl and azoxystrobin which are surface systemic protectants are effective in controlling powdery mildew disease.

Foliar sprays of some inorganic salts namely, sodium bicarbonate, dipotassium hydrogen orthophosphate and potassium dihydrogen orthophosphate at 4.5 g/litre gave better control of powdery mildew compared to fungicides. Other desirable aspects of these inorganics are a) their rapid absorption and extreme mobility in the plant tissue, b) nutrient value, c) low mammalian toxicity and environmental safety, and d) low cost.

Some antitranspirants such as polyterpenes and hydrocarbon wax emulsions are also known to be effective in controlling the disease.

The cultural practices that keep the disease under check include minimal use of nitrogenous fertilizers and irrigation through sprinklers or mistifiers.

Among the bioagents, two antagonistic fungi namely, *Ampelomyces quisqualis* and *Pseudozyma flocculosa* can control powdery mildew when their spore suspension are sprayed onto the foliage.

4.1.3. Black Spot, *Marssonina rosae* (syn. *Diplocarpon roseae*)

i) Symptoms: This disease is more of a problem in open-field cultivation. The characteristic symptom of the disease is the appearance of leaf spots which are coal-black, circular or irregular with black margins and yellow halos (Fig. 4.3). The spots develop on either side of the leaf. Symptoms first appear on older leaves on the lower portion of plant causing severe defoliation, reducing yield and the size of flowers. On close examination with a hand lens, the spots show small, black blister-like fruiting bodies of the fungus. Infected leaves soon drop off from the plant. Continual defoliation will cause weakness, dieback or death of the plant. Some very susceptible species may have stems affected with a considerable reduction in plant vigour.

The pathogen can survive in the canes. Spores are spread by wind and water-splash.

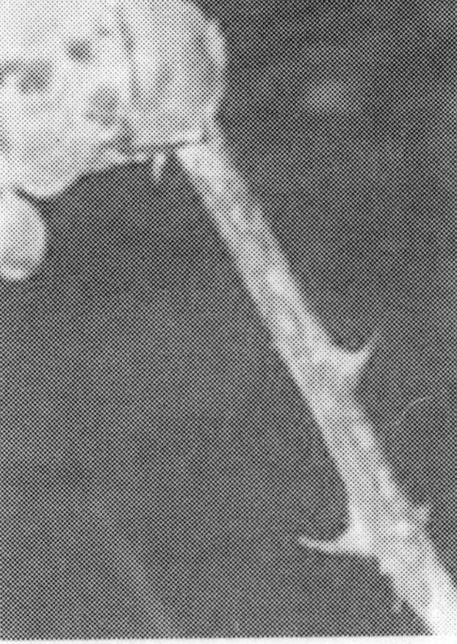

Fig. 4.3. Black spots on rose leaves and stem.

ii) Management: Cultural practices like wider spacing, pruning the dense central portion of the canopy to allow good air circulation contribute to disease control. The debris from infected plants should be removed and burnt to minimize the inoculum potential.

It is desirable to give a dormant spray of lime-sulphur when the leaves begin to emerge. Spraying with 0.2% captan/chlorothalanil/ dithionon/ propineb at 15 days interval can control the disease. Foliar application of chlorothalanil/benomyl/carbendazim/propiconazole/triforine is also effective. Lower surfaces of the leaves should be particularly covered with the fungicides.

4.1.4. Dieback, *Diplodia rosarum, Colletotrichum* sp.

i) Symptoms: Die-back generally occurs after pruning. The infection occurs at the pruned ends of stems or through wounds caused due to harvesing of flowers or other operations. Damage caused by digger wasps also facilitates the infection. Old plants are more prone to die-back. The infected stems dry from tip downwards, often extending to the main stem and thus killing the entire plant. Browning and rotting of infected stem extends inside the stem even beyond the infected lesions on the stem surface. The fruiting bodies of the fungus are seen as small eruptions around the bases of thorns on the infected stems (Fig. 4.4). Plants defoliated due to black spot and downy mildew are more susceptible because of depletion of carbohydrates in the stems. The fungus *Alternaria alternata* is also reported to cause die-back in *Rosa damascene*.

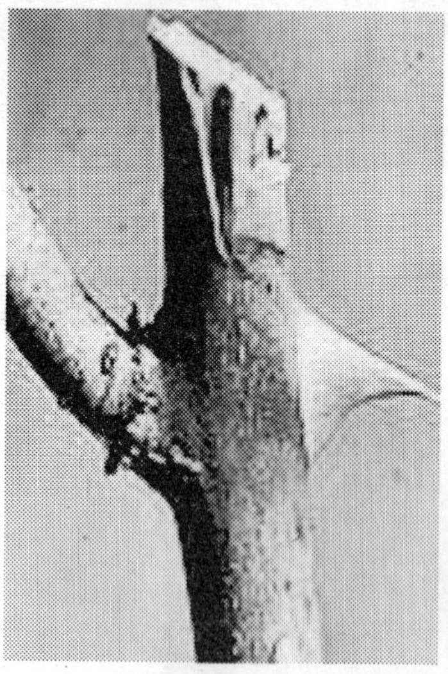

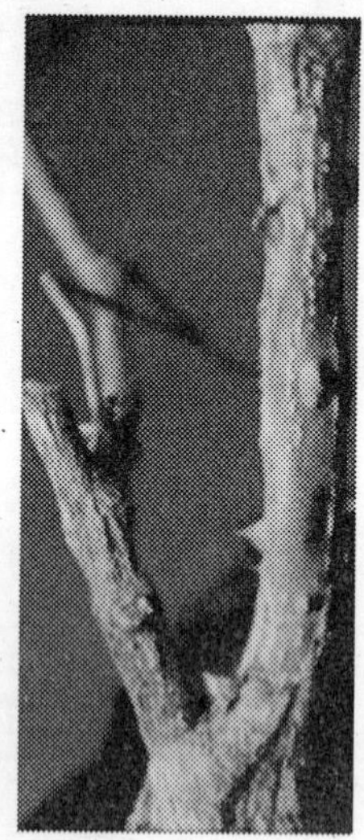

Fig. 4.4. Die-back on stem of rose.

ii) Management: Pruning the infected stems 5-10 cm below the margin of infection and pasting the pruned ends with Bordeaux paste or rose paint [copper carbonate, red lead (Pb_3O_4) and linseed oil at 4: 4: 5 ratio] or 0.1% carbendazim or 0.2% copper oxychloride gives effective control of the disease. To avoid digger wasp damage, any of the common insecticides may be mixed with these fungicides. While pruning, a slant cut is advised in order to avoid formation of water droplets on the cut surface since the fungus requires water film for spore germination and infection. Use of sprinklers and mistifiers enhances the spread of the disease.

Cvs. White Sign, Red Gold, Bhim, Summer Queen, Quebec, Resolute, Samba and Blue Moon are reported to be resistant.

4.1.5. Downy Mildew, *Peronospora sparsa*

i) Symptoms: Initially the symptoms appear as pinkish brown irregular spots on tender leaves. The fungus causes chlorotic blotches on the upper surface of the matured leaves, which ultimately become necrotic (Fig. 4.5). The corresponding areas on the lower surface of the leaves show sparse growth of the fungal mycelium and sporangia that are not easily noticed. The infection on flower buds occurs very early and leads to total drying of flowers. Tender canes on infection show die-back type of symptoms. The infection results in total defoliation and severe flower blight. The infected flowers get malformed and fail to open.

Fig. 4.5. Downy mildew on leaves of rose.

Downy mildew is very common on greenhouse roses because of the prevalence of high humidity and low temperatures. The disease spreads very fast resulting in total defoliation of plants. The fungus survives through special perennating structures called oospores. Humid, wet and cool conditions favour the spread of the disease. RH of about 85% and temperature around 18°C are optimal for multiplication of this fungus. Under these conditions the fungus is capable of explosive spread through the production of innumerable zoospores which are the dispersal units of the fungus. Since *Peronospora* is an obligate pathogen, movement of the planting material from infected nurseries and polyhouses is the main source of spread of the fungus to new areas.

ii) Management: The disease can be effectively brought under control by spraying 0.3% fosetyl-AL (Aliette), 0.2% metalaxyl-MZ (Ridomil-MZ) or 0.2% cymoxanil (Curzate-M). Strict phytosanitation is a must in order to keep the disease under check as the fungus can survive in the infected plant parts. All the infected flowers, stems and fallen leaves should be collected and destroyed. Anti-resistant strategies like mixing or alternating fungicides with different modes of action should be followed in order to overcome the fungus developing resistance to fungicides. Reducing the humidity in polyhouses effectively checks the spread of the disease. In nurseries, overcrowding of seedlings should be avoided by providing enough space between them. Irrigation should be so timed so as to facilitate quick drying of the foliage.

4.1.6. Rust, *Phragmidium mucronatum*

i) Symptoms: Most rusts are host-specific. The fungus *Phragmidium mucronatum* that causes rust on roses does not infect any other plants. In spring and early summer, orange powdery pustules develop on the undersides of leaves (Fig. 4.6), while upper surfaces become speckled yellow. These leaves drop prematurely. In late summer, black or brown spore masses appear with the orange pustules; these are the teliospores that can over winter. The fungus spreads rapidly and can cause severe defoliation. The spores are spread by wind.

Fig. 4.6. Rust pustules on rose leaves.

ii) Management: To control rose rust, remove and destroy diseased material. When new growth occurs, spray bushes with bitertanol (such as Baycor), mancozeb, lime sulphur, triforine, or zineb.

4.1.7. Anthracnose, *Sphaceloma rosarum*

i) Symptoms: The fungus *Sphaceloma rosarum* causes anthracnose, which is sometimes confused with black spot on rose leaves. The differences are that anthracnose spots are smaller and initially have sharply defined margins (Fig. 4.7). As the spots enlarge, the centres become ash grey and drop away, leaving the black margins intact. The leaves become red or yellow, and may drop prematurely, but not to the same degree as with black spot. As the spots enlarge, the centre becomes gray and may fall out resulting in a shot-hole appearance.

ii) Management : To control anthracnose, remove and destroy diseased material. Spray bushes with chlorothalonil, mancozeb, triforine, or zineb.

Fig. 4.7. Anthracnose leaf spots on Rose.

4.1.8. Cane Canker, *Coniothyrium* spp., *Botryosphaeria* spp.

i) Symptoms: Apart from *Botrytis* (see above), two of the commonest fungi to cause cane canker are *Coniothyrium* and *Botryosphaeria* species. These diseases are more commonly seen on plants stressed through poor management practices such as inadequate nutrition or irrigation. Water-borne spores enter mainly through pruning wounds or other injuries. The disease can affect any part of the stem, initially producing yellow or red spots on the bark. These enlarge into brown cankers with dark margins, and the bark cracks and becomes sunken. Tiny black fruiting structures are sometimes seen. Canker can encircle the stem or travel down it, killing the plant.

ii) Management : To treat canker, prune stems back to healthy buds and destroy diseased material. After pruning, apply mancozeb to protect wounds from infection.

4.1.9. Root Rots, *Phytophthora cinnamomi, Pythium* spp.

i) Symptoms: Roses affected by root rots wilt and may eventually die. The root ball may be water soaked and brown and none or few new white roots may be seen. *Phytophthora* root rot may be caused by more than one species of the pathogen. Common sources of *Phytophthora cinnamomi* (jarrah dieback) are infected soil or irrigation water that has been in contact with the soil, for example, dam water. Other species of *Phytophthora* may be imported in potting mixes.

Pythium affects plants in a similar manner but is more often a sign that the plants are being stressed in some other way, for example, by high salinity, low pH, or water logging.

These root diseases will often show up first after a few warm days when plants wilt because their water conducting mechanisms are disrupted. Sometimes the disease affects only one side of the plant.

ii) Management : The *Phytophthora* disease may be suppressed with the use of phosphorous acid. The best treatment for *Pythium* is to eliminate the causes of stress.

4.2. CARNATION

Carnation (*Dianthus caryophyllus*) flower is valued for its excellent keeping quality, wide array of colours and forms, and ability to rehydrate after continuous transportation. Moderate climatic control measures that are economical can deliver quality carnations at the internationally competitive prices year round. In India, Carnation and gerbera are commercially grown mostly under cover around big cites like Solan, Shimla, Mandi, Kullu, Chandigarh, Ludhiana, Guragaon, Kalimpong, Pune, Bangalore and Delhi. Carnation has tremendous potential for export market.

4.2.1. Wilt, *Fusarium oxysporum* f.sp. *dianthi*

i) Symptoms: Fusarial wilt is widely prevalent in carnation all over the world. The initial symptoms are foliar yellowing (Fig. 4.8) and production of 'crookneck' or bent shoots. The leaves and shoots wither and become brownish. Initially the symptoms are confined to a few branches or a part of the plant but ultimately the entire plant shows the symptoms. The symptoms are first noticed during mid-day when the temperature is quite high inside the polyhouses, but they are not visible when the temperature goes down. Stems when cut open show brown discolouration at the vascular region. Sometimes pith and cortex also get discoloured. Finally the stems show shredding of the internal tissue. The disease spreads through contaminated planting material as young rooted cuttings carry the pathogen without expressing the symptoms. Spread of the disease within the greenhouse occurs through root contacts, contaminated soil carried by workers and implements.

Fig. 4.8. Fusarium wilt on carnation.

ii) Management: The best way to control wilt of carnation is by integrated approach which includes use of clean stock, soil treatment and phytosanitation. Soil treatment with benzimidazoles (benomyl, carbendazim, thiophanate methyl at 0.1%), pthalamides (captaf, captafol at 0.2%) and sterol biosynthesis inhibitors (prochloraz) before planting is reported to give effective control of wilt. Soil fumigation with methyl bromide and treating the beds with benomyl before planting are effective. After transplanting, drenching with 0.1% carbendazim is quite effective.

Biocontrol agents like *Trichoderma harzianum, Bacillus subtilis, Streptomyces* spp. and non-pathogenic isolates of *Fusarium oxysporum* are also reported to give good control of wilt disease.

The substrate pH should be maintained above 6.5 as acidic soils favour the growth of *Fusarium*. During all the stages of crop development, calcium deficiency in the plants is known to increase their proneness to disease.

4.2.2. Foot Rot, *Phytophthora nicotianae* var *nicotianae*, *P. nicotianae* var *parasitica*

i) Symptoms: *Phytophthora* spp. attacks the roots and crown portion of carnation especially under high moisture conditions. The infection results in rotting of the stem at soil level resulting in wilt. In case of root infections, the leaves gradually get discoloured and start drying from bottom upwards. High soil moisture is the predisposing factor. The pathogen spreads through water and contaminated soil.

ii) Management : The reduction in soil moisture is an important step in the control strategy. The fungicides useful are benomyl (Benlate at 0.1%) or iprodion (Rovral at 0.1%) or combination product (Quintal which contains both Bavistin and Rovral in equal proportions) or fosetyl-AL (Aliette at 0.3%) or metalaxyl (Ridomil MZ at 0.2%).

4.2.3. Basal Rot, *Sclorotium rolfsii*

i) Symptoms: Rotting of stem at soil level extending to leaves, cottony growth of the fungus mycelium is commonly seen on infected portions. White to dark brown spherical sclerotial bodies are clearly visible at advanced stage of infection.

ii) Management: Regulation of soil moisture and avoiding excessive nitrogen to the soil helps to reduce the disease incidence. Soil drenching with chlotrothalanil/iprodion/dithionon at 2 g/litre is quite effective in controlling and further spread of the disease.

4.2.4. Leaf and Blossom Blight, *Alternaria dianthi, A. dianthicola*

i) Symptoms: Dark brown circular spots occur on the margins of leaves and also on flower buds. The infected parts get blighted. The spores of the fungus are spread through water emitted by misting and fogging devices.

ii) Management : The disease can be managed through application of pthalamides (Captafol at 0.2%) or chlorothalanil (Kavach at 0.2%) or iprodion (Rovral at 0.1%). Carnation cv. "Forca" is free from blight infection.

4.3. GERBERA

Gerbera (*Gerbera jamesonii*), commonly known as Transvaal daisy, Barberton daisy or African daisy, is an important flower grown throughout the world for beds, borders, pots and rock gardens. The flowers are of various colours and suit very well for floral arrangements. The cut blooms also have long vase life. In India, it is distributed in the temperate Himalayas from Kashmir to Nepal at altitudes of 1300 to 3200 meters above sea level. Gerbera belongs to the family Asteraceae and is native to South African and Asiatic regions. To meet the quality standards of export market, it has to be grown under naturally ventilated low cost polyhouses.

4.3.1. Powdery Mildew, *Erysiphe cichoracearum, Oidium erysiphoides* f. sp. *gerbera*

i) Symptoms: Powdery mildew is a major production problem in gerbera and accounts for considerable economic losses. The infection occurs both on leaves and flowers. The symptoms are seen on leaves and flower stalks in the form of grayish to white patches. The white powdery patches on the leaves reduce the effective leaf area available for photosynthesis and ultimately leading to reduced productivity and poor quality of flowers. The infection also occurs on petals and flower stalks resulting in malformed flowers and brittle stalks. The lesions on leaves and flowers produce fungal spores called conidia in abundance. These are produced under dry and humid conditions that are commonly prevalent in polyhouses. The dry spores are readily spread by the movement of workers and by wind.

ii) Management: Foliar application of sulphur compounds and carbendazim, dinocap, triadimefon, fenerimol, penconazole and triforine offer good control.

Foliar sprays of K_2HPO_4 at 4.3 g/l were found to be effective in reducing conidial production and thus controlled the disease (disease incidence of 5% as compared to 62% in control) which was on par with fungicidal spray (Hexaconazole at 0.1%).

4.3.2. Root Rot, *Phytophthora cryptogea*

i) Symptoms: Flaccidity of lower leaves is the initial symptom noticed generally. As the disease progresses, the infected plants wilt suddenly. In some cases the leaves turn purple. The infected plants show severe rotting of roots, which retain only the stellar portion because the cortical portion gets sloughed-off. The tissues of the crown and root become soft and brownish. Rotting can be seen extending upward through the petioles and occasionally the flower stalks. Gerbera does not tolerate excess moisture in the soil. The plants under such conditions succumb to infections caused by root rot fungi such as *Phytophthora* and *Pythium*. The pathogen proliferates well under wet weather conditions and propagules spread through contaminated soil and water.

ii) Management: Adopting strict phytosanitation and soil treatment with metalaxyl-MZ (Ridomil-MZ at 0.2%) or fosetyl-AL (Aliette at 0.2%) help in managing the disease. Regulation of soil moisture becomes decisive in containing the disease. Precautionary steps like preparation of raised beds (25 cm above soil level) and a friable substrate help in reducing the disease incidence. Pre-planting fumigation of beds and preventive application of broad spectrum fungicides help in preventing the infection.

4.3.3. Wilt, *Fusarium oxysporum* f. sp. *dianthi*

i) Symptoms: Discolouration and loss of turgidity of leaves are the initial symptoms noticed on infected plants. Cutting open of the crown portion shows clear discolouration of the vascular region. Severe wilting occurs under anaerobic conditions of the substrate caused due to excessive soil moisture (Fig. 4.9). The fungus is soil-borne and is also occasionally transmitted through the plantlets which act as symptomless carriers.

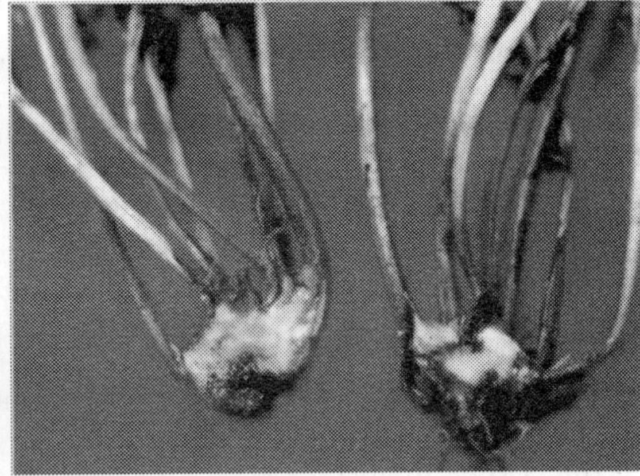

Fig. 4.9. Fusarium wilt on gerbera.

ii) Management: Planting disease-free rooted cuttings is the most important step in managing the disease. The substrate should be sterilized with chemicals or solarization before planting. Rouging and soil application of benzimidazoles (carbendazim, benomyl at 0.1%) combined with pthalamides (captaf at 0.2%) help in managing the disease.

4.3.4. Foot Rot, *Pythium* spp., *Sclerotium rolfsii*, *Rhizoctonia solani*

i) Symptoms: The plants show reduced growth, smaller leaves and flowers with weak stems. Leaves show purple discolouration, with rotting of both feeder and main roots.

ii) Management: Regular application of copper oxychloride or iprodion or dithionon or captaf to soil may help to control the disease. Soil application of FYM along with *Trichoderma harzianum* is helpful for initial suppression of the disease.

4.4. CHRYSANTHEMUM

Garden chrysanthemum (*Dandrathema grandiflora*), also known as 'Queen of the East', is a highly versatile and accommodating ornamental with a wide range of type, size and colour in flowers. In India, large flowered varieties are grown for cut flower, making garland, wreaths and 'veni', religious offerings and for bedding and potting purposes. Its commercial cultivation is being done in Maharashtra, Rajasthan, Madhya Pradesh and Bihar. Small flowered varieties are commercially grown for loose flowers in Tamil Nadu, Karnataka and Maharashtra. In India, chrysanthemum is grown in about 4,000 ha.

Although the list of diseases affecting Chrysanthemums is long, they are relatively trouble-free given full sun, fertile well-drained soil, and adequate watering. Maximizing plant vigour by appropriate cultural practices is an important disease management strategy.

4.4.1. *Septoria* Leaf Spot, *Septoria chrysanthemi*

i) Symptoms: The lower leaves are infected first. The spots are at first yellowish then become dark brown and black. Severe infection may result in the infection of entire leaves and they prematurely wither and drop off from the plant. The disease develops more rapidly in cool-wet weather. Spores of the fungus are spread by splashing water.

ii) Management: Regularly clean up and destroy infected plant debris and hand pick symptomatic leaves from lightly infested plants. Avoid splashing water onto plant foliage and water early in the day to allow foliage to dry quickly. In severe cases, fungicides such as chlorothalonil, mancozeb, myclobutanil, propiconazole, or thiophanate methyl may be applied.

The disease is effectively checked by fortnightly spraying of 0.1% carbendazim. Other fungicides such as captan, mancozeb or zineb also provide satisfactory control of the disease.

4.4.2. *Alternaria* Leaf Spot, *Alternaria* sp.

i) Symptoms: The pathogen cause small dark brown spots with yellow surroundings. The spots merge and leaves begin to rot. Severe infection results in withering of leaves, with dead leaves hanging to the stem for some time.

Symptoms first appear as yellow spots which turn brown to black. Spots often occur on lower leaves first and can coalesce into large necrotic areas and finally death of the entire leaf.

ii) Management: Spraying mancozeb or copper oxychloride each at 2 g/litre of water gives effective control of the disease.

4.4.3. Wilt, *Verticillium albo-atrum*

i) Symptoms: Wilt is the most serious disease of chrysanthemum. Symptoms of Verticillium wilt often appear only after blossom buds have formed; young vigorous plants may be symptomless. Foliage becomes yellow and wilted, sometimes only along leaf margins and on one side of the plant. Leaves begin to die from the base of the plant upward and often remain attached. Stems may exhibit dark streaks in the vascular system. Infected plant remains stunted and fails to produce flowers. This disease is favored when cool weather is followed by hot.

ii) Management: Start with pasteurized growing media and pathogen-free cuttings. Most cultivars are resistant. Avoid susceptible cultivars including 'Bright Golden Ann', 'Echo', 'Glowing Mandalay', Mountain Peak', 'Paragon', 'Pert', 'Puritan', and 'Wedgewood'.

The pathogen survives in the soil and hence continuous cropping in sick soil should be avoided. Drenching soil around plant base with 0.2% carbendazim, benomyl or pentachloro nitrobenzene (PCNB) helps in reducing the disease incidence.

4.4.4. *Fusarium* Wilt, *Fusarium oxysporum* var. *trachaephilum*

i) Symptoms: The first signs of this disease are yellowing of foliage, stunting, and wilting often along one side of plant. Plants may appear water stressed and foliage may become brown and die. Stems show a reddish brown discolouration of the vascular system. Fusarium is spread in contaminated soil and from infected cuttings and is favoured by warm temperatures, high relative humidity, over watering, and poor drainage.

ii) Management : Start with pathogen free cuttings or plants and pasteurized growing media. Adjust pH to 6.5 to 7.0 and use nitrate nitrogen fertilization. Avoid highly susceptible cultivars such as 'Bravo', 'Cirbronze', 'Illini Trophy', 'Orange Bowl', 'Royal Trophy', and 'Yellow Delaware'.

4.4.5. Gray mold, *Botrytis cinerea*

i) Symptoms: Gray mold may occur on petals, leaves, or as a stem canker as brown, water-soaked spots. Infected plant parts may be covered with gray to brown, powdery masses of spores. Senescing tissues are most susceptible. Gray mold is favored by extended periods of cloudy, humid, and wet weather.

ii) Management: Practice good sanitation including removing senescing flowers and leaves. Avoid wetting flowers when watering and do not over crowd the plants. Provide good air circulation and keep humidity low by a combination of heating and venting. Apply preventive fungicides as soon as disease is detected. Fungicides such as chlorothalonil, mancozeb, copper sulfate pentahydrate, fenhexamid, azoxystrobin, and thiophanate methyl are registered for Botrytis control. Be sure to rotate applications among chemical classes as fungicide resistant strains of Botrytis have been reported.

4.4.6. Flower Blight, *Stemphylium lycopersici*

i) Symptoms: The infection is noticed on calyx and petals in the form of necrotic brown spots. Early infection on buds may prevent blooming. Infected flowers wither and die. These fungi become serious when temperature and humidity are high.

Stemphylium lycopersici causes small, necrotic, light brown to dark brown lesions which may coalesce and cause blossom death. The pathogen is favoured by wet conditions and temperatures between 60° to 85° F.

ii) Management: Rogue and dispose of infected plants. Provide good air circulation and do not over crowd plants. Avoid overhead irrigation and keep plants dry.

The disease can be controlled by spraying 0.1-0.2% mancozeb or zineb at 15 days interval. Benomyl at 0.1% sprays have also been reported to be quite effective.

4.4.7. Powdery Mildew, *Erysiphe cichoracearm*

i) Symptoms: As the name implies, it is characterized by a white to ash-gray powdery growth on leaves and occasionally stems. Foliage may become puckered or distorted; severely infected leaves will shrivel and die. The disease is most serious during hot, humid weather. Unlike most fungal diseases, free water is not required for powdery mildew infection; high humidity encourages disease development.

ii) Management: Powdery mildew can be avoided by proper plant spacing, good air circulation, low relative humidity, and adequate light levels. Apply preventive fungicides at the first sign of disease with myclobutanil, triadimefon, propiconazole, sulphur, potassium bicarbonate, or thiophanate methyl.

4.4.8. Rust, *Puccinia chrysanthemi, P. horiana*

i) Symptoms: Two species of *Puccinia* causes rust on chrysanthemums *P. chrysanthemi* and *P. horiana. P. chrysanthemi* is most common in late summer and is characterized by dirty-brown pustules and yellowish-green spots on upper surface of leaves. *P. chrysanthemi* causes minor damage in the field and is uncommon on greenhouse plants. Severe infestation may damage large areas of leaves and lead to defoliation and reduced flower production.

P. horiana causes chrysanthemum white rust. Symptoms include white, pinkish or brownish pustules produced on underside of leaves with white, yellow, to pale-green lesions on upper leaf surface. Chrysanthemum white rust results in leaf distortion, discolouration, defoliation, and plant death. White rust is primarily a disease of greenhouse crops; when it occurs outside, direct sunlight and low humidity kill the spores.

ii) Management: Chrysanthemum varieties resistant to rust include 'Achievement', 'Copper Bowl', 'Escapade', 'Helen Castle', 'Mandalay', 'Matador', 'Miss Atlanta', 'Orange Bowl', and 'Powder Puff'. Infected plants to be destroyed to prevent disease establishment

4.4.9. Root Rot, *Pythium* sp., *Phytophthora* sp.

i) Symptoms: Infected plant parts (root, stem and leaves) show sudden wilting.

ii) Management: Provision of good drainage conditions to prevent water logging, soil solarization by covering with black polythene mulch during summer and soil drenching with mancozeb at 2g/litre of water is effective for the management of the disease.

4.5. GLADIOLUS

Gladiolus (*Gladiolus* spp.) is very much liked for its majestic spikes containing attractive, elegant and delicate florets. These florets open in sequence over a longer duration and hence have a good keeping quality of cut spikes. It is cultivated for cut flowers, garden and interior decoration, making bouquets, bedding, rockeries and pots. West Bengal, Maharashtra, Uttar Pradesh, Punjab, Haryana and Andhra Pradesh are the major gladiolus-growing states.

4.5.1. Wilt, *Fusarium oxysporum* f. sp. *gladioli*

i) Symptoms: This disease is caused by the fungus *Fusarium oxysporum* var. *gladioli* and is often referred to as Fusarium yellows or Fusarium Wilt. The symptoms are yellowing of the leaves starting with the older outside ones first and this is accompanied by the apparent stunting of newer leaves. In addition, the spike itself will often be stunted and faded in colour. The plant may also have blackened areas at the base which spread onto the corm (Fig. 4.10). Eventually, if left, the whole plant will wilt and when lifted for destruction, (for this is the only course of treatment), the interior of the corm when opened will appear marbled with a brown colour.

ii) Management: Where *Fusarium* wilt occurs, fertilize sparingly: nitrogen should be applied in the nitrate form rather than as ammonia or urea. A soil pH 6.6 - 7.0 is preferred. Harvest corms in dry weather if possible and cure them rapidly using forced air at 80 - 100°F if available. Inspect stored corms and rogue out suspects.

Planting healthy corms in clean soil, crop rotation, soil solarization and hot water treatment of corms at 50-55°C for 10 min were effective in controlling wilt disease. As a prophylactic measure, corms should be treated in 0.1% benomyl, carbendazim and thiabendazole solution before sowing. Captan at 0.2% can be used in combination with the above fungicides to minimize the chances of development of resistant strains of the pathogen. Solarization of field soil by mulching with transparent polythene sheets (100 gauze) raised the soil temperature which was lethal to wilt pathogen. *T. harzianum*/*T. viride* either alone or in combination with compost (mushroom, vermin, paddy straw) reduced wilt effectively (58 to 92%). Soil application of *T. harzianum, T. viride, T. koningii, T. virens* increased germination (95.6 to 100.0% compared to 79.0% in control), decreased the plant mortality (2.33 to 8.70% compared to 28.67% in control), increased plant height (76 to 85 cm compared to 51 cm in control), increased spike length (65 to 72 cm compared to 46 cm in control) and increased the number of cormels/plant (15 to 22 compared to 13 cm in control). Under field conditions, the disease can be effectively controlled by application of neem cake at 1.25 tonnes/ha along with *T. harzianum* before planting, followed by soil drenching with 0.1% carbendazim one month after planting.

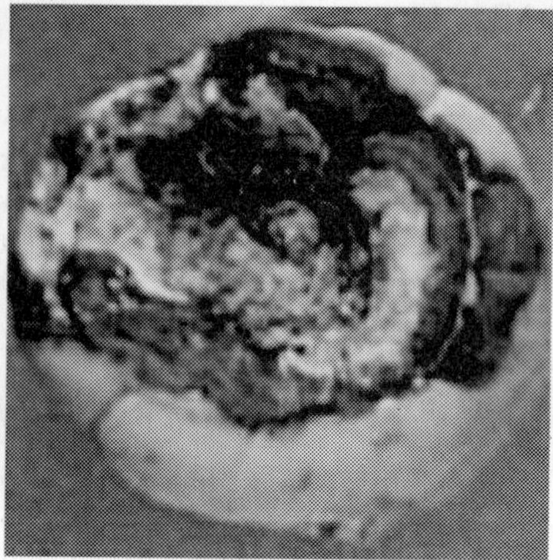

Fig. 4.10. Fusarium infection on gladiolus.

Gladiolus cvs. IARI Selection-1, *Psittacinus* hybrid and Suchitra were found resistant to wilt under Pune conditions.

4.5.2. Corm Rot, *Pencillium gladioli, Rhizopus arrhizus, Fusarium* spp., *Curvularia* spp., *Stromatinia* spp., *Botrytis* spp.

i) Symptoms: The fungi cause heavy loss in field as well as in storage by causing dry rot, hard rot, soft rot, etc. Most of these rots develop as a result of mechanical injury to corms during harvesting and storage. Generally, corms harbour more than one pathogen and symptoms are difficult to separate out pathogen-wise.

Corms that have been damaged in some way during the harvesting operation, including damage caused by the removal of the mother corm, or have not been cured properly whilst drying can often attract a *Penicillium* based rot. This rot manifests itself by the appearance of a fluffy blue or green mould which grows around and over the area of the wound.

ii) Management: Most of these rots can be prevented by avoiding injury to corms during harvesting and storage. They should be stored in cool and dry containers. Badly infected corms should be discarded.

The disease can be controlled effectively by hot water treatment and by corm dipping in carbendazim or captan or chlorothalanil all at 0.2% solution. The corms should be dried after treatment.

4.5.3. Stromatinia Rot, *Stromatinia gladioli*

i) Symptoms: This corm disease is caused by a fungus called *Stromatinia gladioli* and is most prevalent during periods of cool wet weather and as such can be a particular problem in northern European climates. The first signs of disease are usually on the growing plant and are evident by the yellowing of leaves which eventually shrivel and die. Evidence of the disease on the corms is the appearance of reddish brown markings which are sunken and if the corm is dissected, dark markings are seen as a marbling emanating from the centre (Fig. 4.11).

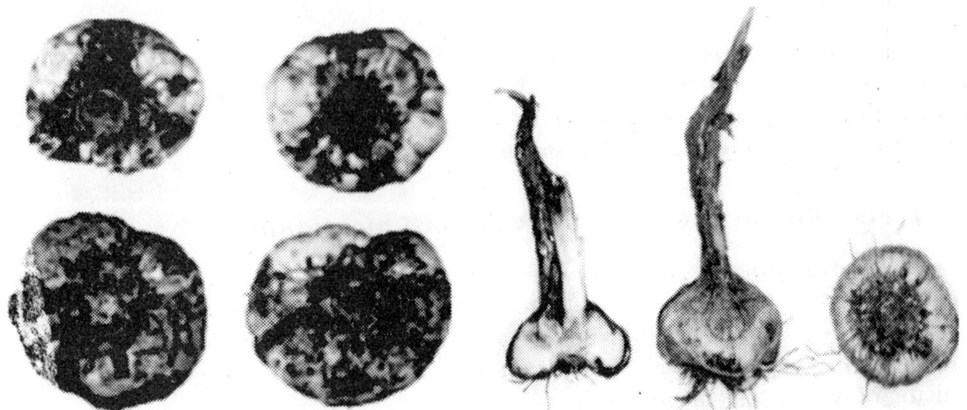

Fig. 4.11. *Stromatinia* rot of gladiolus corms.

ii) Management : A particular problem with this disease is that the fungus produces an over wintering particles called sclerotia. As a result diseased corms should never be planted and as a precaution corms should always be harvested in dry conditions. If *Stromatinia* is suspected in growing plants that soil should not be replanted in subsequent years as the disease is capable of passing from soil to plant.

4.5.4. Leaf and Flower Blight, *Curvularia trifoli, C. eragrostidis*

i) Symptoms: Oval to elongated brownish spots are formed on leaves. On young flowers symptoms are seen on the calyx and then on petals as water-soaked patches which later turn brown to black. The infected flower buds shrivel and rot.

ii) Management: The disease can be controlled by spraying mancozeb at weekly or 10 days interval.

4.6. TUBEROSE

Tuberose (*Polianthes tuberosa*) is one of the most important bulbous ornamentals grown both for cut and loose flowers, and also for the extraction of its highly valued natural flower oil. It is also grown for garden decoration in pots, beds and borders. It is commercially grown in Karnataka, West Bengal, Maharashtra, Punjab, Andhra Pradesh and Tamil Nadu. At present, the total area under tuberose cultivation in the country is estimated at about 3,000 ha.

4.6.1. Leaf Spot/Blight, *Alternaria polyantha*

i) Management : Spraying of 1% Bordeaux mixture is recommended to control the disease. Tuber treatment with *Trichoderma harzianum* and *Pseudomonas fluorescens* at 4 g/kg tuber and seedling dip in *T. harzianum* suspension before planting resulted in effective control of blight and increased the plant vigour.

4.7. CROSSANDRA

Crossandra (*Crossandra undulaefolia*) is mainly cultivated commercially for its loose flowers in Tamil Nadu, Karnataka and Andhra Pradesh. It is also grown in Sri Lanka, Tropical Africa and Madagascar. It is a perennial herb and belongs to the family Acanthaceae.

4.7.1. Foot Rot or Root Rot, *Phytophthora nicotianae*

i) Symptoms: This disease leads to sudden death of plants. In young seedlings, symptoms appear as lesions on rootlets followed by rotting of the entire rootlet. On the collar region, peculiar berry root can be seen. Leaves show pink discolouration and drooping. In advanced stage of infection, wilting of whole plants can be noticed.

ii) Management: Growing seedlings in raised beds drenched with captan, application of neem cake along with *Trichoderma harzianum* helps to control the disease. Soil drenching with AL-Fosetyl for 2-3 times at monthly intervals during monsoon is effective.

4.7.2. Wilt, *Fusarium* sp., *Rhizoctonia solani*

i) Symptoms: The characteristic symptoms of the disease are drooping of leaves, wilting of plants and rotting of roots (Fig. 4.12).

ii) Management: Drenching the plants with captaf at 2 g/litre of water is effective in controlling the disease.

4.8. CHINA ASTER

China aster (*Callistephus chinensis*) is one of the important annual flower crops. It ranks next to chrysanthemum and marigold among traditional flowers. In recent years, it has gained more popularity due to its multifarious uses including as cut flower. The flowers fetches very good price and returns to the farmer when their production coincides with their demand during particular season/festival. It is more popular among small and marginal farmers due to its easy cultivation. It is grown for cut flowers, herbaceous borders, bedding, bouquet and pots and cultivated in most parts of the country. It is commercially grown in Karnataka, Tamil Nadu, Andhra Pradesh, Maharashtra (Pune) and West Bengal. In Bangalore and Pune alone it is being grown in an area of 500 and 400 ha, respectively.

Fig. 4.12. Fusarium wilt on crossandra.

4.8.1. Wilt, *Fusarium oxysporum* var. *callistephi, Sclerotium rolfsii*

i) Symptoms: Both the pathogens are widely distributed and gross symptoms are same. The leaves wilt and roots are rotted. Generally the lower leaves show wilting symptoms and gradually wilting progresses upwards. A close examination of the stem base in Fusarial wilt shows a thin coating of pale pink or rose coloured fungal growth which is more prominent especially under humid conditions. In Sclerotial wilt, there is buff to dark brown round bodies with infected network attached to the stem base near the soil level.

ii) Management: Use of clean soil, avoiding bringing in soil on tools from beds previously infected with wilt fungi and discouraging continuous cropping in infested soil may help in reducing the disease incidence. Soil solarization should be done during summer months. Planting may be done on ridges.

Soil drenching with 0.2% benomyl or carbendazim gave good control of fusarial wilt, while soil drenching with 0.2% PCNB (Brassicol) is effective for sclerotial wilt.

Cvs. Heart of France, Crego, Ostrich, Feather, Stardust and Roment are reported to be resistant to Fusarial wilt.

4.8.2. Collar/Root Rot, *Phytophthora cryptogea*

i) Symptoms: The symptoms include sudden wilting of plants and rotting of stem at ground level.

ii) Management : The control measures include avoiding excess moisture by growing the plants on ridges and use of fungicides like captan, mancozeb, metalaxyl-MZ and fosetyl-AL.

4.8.3. Foot Rot, *Pythium* sp.

i) Symptoms: The main symptoms are yellowing of leaves, stems and roots of affected plants. The infected plants appear to be water-soaked and black.

ii) Management : Irrigation should be regulated and proper drainage should be provided. The seedling should be transplanted on ridges.

4.9. MARIGOLD

4.9.1. Damping-Off, *Pythium* sp.

i) Symptoms: Brown necrotic spots appear on the germinating seedling first on the root followed by stem. Brown ring forms on the stem at ground level. The roots begin to rot.

ii) Management: Drenching the nursery beds with mancozeb at 2 g/litre of water gives effective control of the disease.

4.9.2. Powdery Mildew

i) Symptoms: The characteristic symptom of the disease is covering of leaves with white powdery mass.

ii) Management : Spraying with wettable sulphur at 3 g/litre of water is effective in managing the disease.

4.9.3. Leaf, Spot, *Alternaria tagetica, Septoria* sp., *Cercospora tagetis, C. tageticola*

i) Symptoms: The main symptom of the disease is appearance of brown, brownish-grey and blackish spots on leaves.

ii) Management: Spraying with mancozeb at 2 g/litre of water is effective in controlling the disease.

4.9.4. Wilt/Stem Rot, *Phytophthora cryptogea*

i) Symptoms: The fungus attacks roots and collar portions of the plants. In nurseries, the infection results in damping-off and is aggravated by high soil moisture. In the field, the infected plants show wilting.

ii) Management: The disease can be controlled by soil treatment with captan, mancozeb, metalaxyl-MZ and fosetyl-AL.

4.9.5. Collar/Root Rot, *Pellicularia filamentosa, P. rolfsii, Pythium ultimum, Sclerotinia sclerotiorum*

i) Symptoms: These fungi are reported to infect root and collar portions of marigold.

ii) Management: Control measures such as soil sterilization and planting disease-free seedlings are useful.

4.9.6. Flower Bud Rot, *Alternaria dianthi*

i) Symptoms: The pathogen infects the young flower buds. The infected buds shrivel and become dark brown in colour. The initial symptoms are brown necrotic spots on sepals and peduncle which leads to dry rotting of flower buds. Mature buds though less susceptible, do not open and the infection is rarely seen on fully opened flowers. The pathogen also infects leaves causing blight. The infection is visible in the form of brown necrotic spots on margins and tips of older leaves.

ii) Management: The disease can be controlled by spraying mancozeb.

4.10. NARCISSUS

4.10.1. Basal Rot, *Fusarium oxysporum* f. sp. *narcissii*

i) Symptoms: Infected plants are stunted with distorted and yellow leaves which usually die. The basal plate is decayed and reddish brown in colour. The white to pinkish fungus is sometimes seen at the base between the scales. Eventually the bulb rots.

ii) Survival of Pathogen and Epidemiology: Fungus survives as chlamydospores in soil for long periods. Short rotations are not effective. Disease is favoured by warm soils and is limited below 55°F; disease is favoured by high levels of nitrogen fertilization. Potassium fertilizer may decrease disease sensitivity.

iii) Management: Avoid excessive fertilization. Dig up bulbs in diseased areas as soon as possible. Store bulbs below 64°F. Dip bulbs in thiabendazole as soon as possible after digging. Hot water treatment (see stem and bulb nematode) is also effective in reducing the disease. Rotate out of narcissus for at least 3 years. Plant when soil is cool. Cvs. Malvern City, Rijinveld's Early Sensation, White Lion, 'Soleil d'Or', 'Dulcimer' are moderately resistant; cv. St. Keverne is resistant.

4.10.2. Crown Rot, *Sclerotium rolfsii*

i) Symptoms: A bulb rot that is at first wet and later becomes dry and woody. A white fungus mat and small (0.03 to 0.125 inch), round, tan to brown sclerotia occur on decaying bulbs and in surrounding soil.

ii) Survival of Pathogen and Epidemiology: Disease is favoured by warm weather. Fungus survives in soil for long periods (10 years). Fungus has a wide host range. Can be spread by infected bulbs.

iii) Management: Avoid infested fields for 4 years or longer. Deep ploughing is sometimes practiced to bury sclerotia. Treat bulbs in hot water (See stem and bulb nematode). PCNB applied in the furrow at planting can be effective.

4.10.3. Fire, *Botryotinia polyblastis*

i) Symptoms: Water-soaked areas on petals that become brown and wither. Flowers are attacked first, and later the foliage. Foliage spots are small, elliptical, tan and usually near the tips. Yellow streaking follows the leaf spots.

ii) Survival of Pathogen and Epidemiology: Fungus survives as sclerotia in leaf debris. Sclerotia germinate in spring to produce fruiting bodies (apothecia), which produce airborne spores (ascospores). Conidia, which are also airborne, cause secondary spread. The disease is favoured by mild, humid conditions.

iii) Management: Remove flowers before fungus produces apothecia in March/April. Protect flowers and foliage with a fungicide.

4.10.4. Scorch, *Stagonospora curtsii*

i) Symptoms: Yellow to brown lesions at leaf tips. Elongate reddish brown leaf spots. Small black pycnidia form in necrotic areas.

ii) Survival of Pathogen and Epidemiology: The fungus survives in the neck and between scales. Disease is favoured by mild, moist conditions. Spores are spread in splashing water. The fungus infects members of the Amaryllidaceae.

iii) Management: Treat bulbs with hot water (See stem and bulb nematode). Protect plants in the field with fungicides. Make first application as leaves emerge.

4.10.5. Smoulder, *Botrytis narcissicola*

i) Symptoms: A dark brown lesion first appears on leaf tips. Infected leaves may curl when infection occurs on inner edge. Masses of woolly gray spores (conidia) and small black sclerotia form on diseased tissues, especially near soil.

ii) Survival of Pathogen and Epidemiology: Fungus survives in infected bulbs and as sclerotia in soil. Conidia are airborne. Favoured by cool, wet weather.

iii) Management: Rotate with other crops. Dig bulbs early. Dip bulbs in suitable fungicide such as thiabendazole. If disease is severe in field, spray with iprodione, fenhexamid, or thiophanate-methyl.

4.10.6. White Mold, *Ramularia vallisumbrosae*

i) Symptoms: Small, sunken, grey or yellow spots on leaves and green parts. Spots enlarge and darken to a yellow- brown with yellow margins. Masses of white, powdery spores (conidia) occur on leaves under moist conditions. Small, dark sclerotia are produced in older infected tissues.

ii) Survival of Pathogen and Epidemiology: Fungus survives as sclerotia in dead leaves and on bulbs. Sclerotia germinate to produce conidia as leaves emerge. Disease is favoured by warm, moist conditions.

iii) Management: Do not replant for 1 year in fields where the disease has occurred. Protect foliage with mancozeb. Copper fungicides are also helpful.

4.11. ORCHIDS

4.11.1. Basal Rot

i) Symptoms: Basal rot is also commonly known as 'Southern Blight'. It is directly related to watering methods and practices. Either over watering, or water left standing in the crown of plants combined with cool humid situations can bring on this condition.

On soft and broad leaved orchids there is little warning. Basal rot, as the name suggests, starts at the bottom or the crown of the plant and proceeds upward. Therefore, by the time you notice a problem, it is already too late.

ii) Management : If the condition is discovered on a plant with multiple growths, and only a small portion is afflicted, you can attempt to save the plant by cutting away the diseased portion with a sharp knife and treating the wound with sulphur. Keep on the dry side after the "operation" and hope the rot does not spread.

4.11.2. Flower Blight, *Botrytis cinerea*

i) Symptoms: This is a fungal disease which affects primarily the flowers of phalaenopsis, cattleya, cymbidium and dendrobium. This condition occurs most frequently in cool, moist conditions where there is little air movement. Symptoms are tiny black spots on the petals which soon grow larger and in extreme circumstances are covered with a grey fuzzy mold. If caught early, take measures to increase night time temperatures to no less than 65° F and provide maximum air circulation.

ii) Management: Remove and destroy badly infected material such as dead flowers which have fallen to the floor and are covered with mold. Spray flooring with a strong solution of fungicide. Spray plants with a mild solution of physan. This will not reverse the damage already manifest on your bloom, but should dry up the affected areas and halt the further spread of spores.

4.11.3. Fusarium Wilt, *Fusarium proliferatum*

The most serious disease caused by *F. proliferatum* on orchids is damping-off and rots of young plants. Hundreds of young seedlings are lost in community pots due to this devastating disease. Rots of these plants kill many and stunt the growth of survivors. As surviving plants mature they are plagued by the recurrence of fusarial diseases during wet periods.

i) Symptoms: Various parts of orchid plants are attacked by this fungus. On dendrobium plants at least 9 months old, new shoots are blackened by *Fusarium* infections and slow rot develops especially at the leaf tip. Leaves are lost and the entire apical tip of new shoots are killed. *Fusarium proliferatum* causes small (2-5 mm) circular to oval spots with irregular edges. Spots expand very slowly. As the leaf matures, some of the spots develop a black edge, while the centre of the spots becomes sunken. The leaf tissue surrounding the spot is slightly chlorotic or pale green.

With mature dendrobium plants, only the young leaves or wounded leaves are susceptible, while mature unwounded leaves are resistant. The occurrence of leaf spots on mature leaves is the result of infection when leaves were young. The distribution pattern of leaf spots reflect the leaf position of the unopened, young leaves of a new shoot. A characteristic pattern of spots for this disease is a row of spots across the width of the leaf (Fig. 4.13). Sunken brown spots also occur.

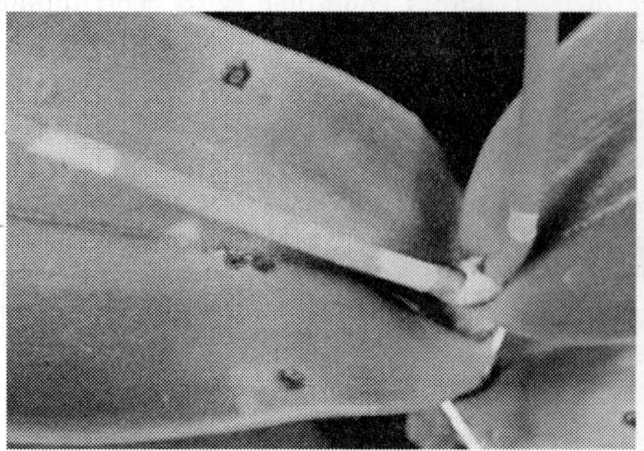

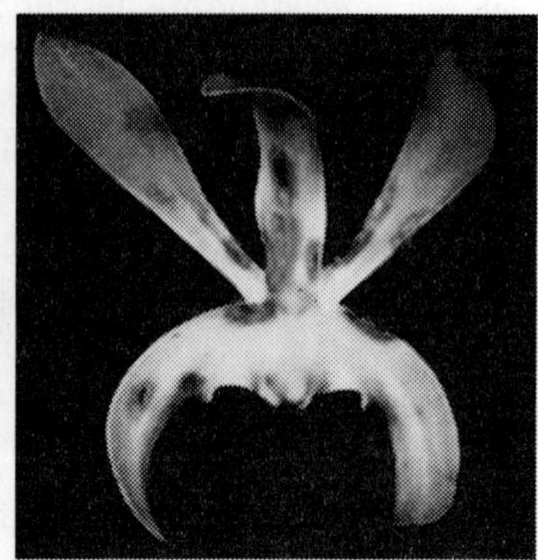

Fig. 4.13. Fusarium infection on leaf and flower of orchid.

The fungus infects the leaf sheath causing black to dark brown rots with irregular edges. Rot expands very slowly and after a few months, the centre of sheath blight or rot can be white to tan. Circular to elongate spots also develop on flowers. Spots are tan to deep brown with a dark border (Fig. 2.14).

ii) Biology: *Fusarium proliferatum* produces microscopic, long, canoe-shaped spores called conidia. These asexual conidia have 3 to 7 cells and are produced on specialized hyphae called conidiophores. A second, smaller type of conidia with 1 or 2 cells, called microconidia are also produced. Both spore forms are produced in very large numbers (millions) and germinate on suitable substrates, such as the host and grow into a colony which produces more spores. Growth of *Fusarium* is favoured by moisture and temperatures in the range of 25-31°C.

iii) Epidemiology: *Fusarium proliferatum* produces many spores on the surface of infected host tissue. Some microconidia are also produced within rotting tissue. White cottony masses on the diseased tissue are spores and mycelia of the fungus. These spores are blown or splashed to nearby plants or other healthy tissue. For many orchid plants such as dendrobium, new shoots are formed below the canopy of the older leaves. Thus fungal spores formed on older shoots easily splash or drip on young shoots emerging from the base of the plant.

Young shoots are very susceptible to infection and rapid disease development. Likewise, fragile young plants in community pots are extremely susceptible to this fungus and many are killed. Since growers commonly salvage all plants that survive, the *Fusarium* will persist on mature plants and cause leaf, sheath, shoot, and flower rots. These plants serve as sources of inoculum, and disease spread within a greenhouse is insured by overhead watering.

Fusarium spores are distributed by wind, splashing water or the movement of spores by contact (hands, clothing, gloves, etc). Insects, snails, and slugs also move spores from diseased to healthy plants. *Fusarium* will survive in diseased plant tissue for many months and spores will also survive in potting media and in the immediate environment of the diseased plants. Greenhouse walls, benches, floors, etc. will be contaminated as long as diseased plants are present.

iv) Management

Great emphasis must be placed on keeping young, disease-free seedlings healthy. The production of new seedlings should be in a greenhouse separated from the production of mature plants. This seedling-house should not have any mature plants of any kind, especially samples of mature blooming plants.

All diseased leaves, dead shoots and infected flowers need to be gathered and discarded from the nursery. Do not drop infected material to the ground as this provides the fungus a good opportunity to infest the ground for long periods.

Spores of *Fusarium* need water to germinate and penetrate the host. Thus controlling moisture will reduce disease levels. Avoid watering in the early evening as this will expose leaves for a long period of wetness during the night. Water during the day and keep leaves as dry as possible. Increase air movement to reduce humidity levels. Remove weeds and other shrubs that prevent adequate air movement.

Fungicides such as mancozeb will also reduce infection levels.

4.11.4. Leaf Spot, *Phyllosticta capitalensis*

Phyllosticta leaf spots and blights are one of the most serious problems in the commercial orchid industry. The quality and quantity of exports shall be affected due to leaf loss in dendrobium fields grown for cut sprays, and leaf rots of potted plants that are shipped.

i) Symptoms: The fungus causes leaf spots on orchids. On dendrobium, young leaf spots are typically yellow or black, circular, and small (Fig.4.14). These spots remain unchanged for many months. However, the fungal growth within the leaf tissue eventually increases and tan to brown rot develop. Within the next week or two, the entire leaf is killed and the characteristic, black web-like pattern formed by this fungus on diseased leaves.

On dendrobium, flower spots are small, faint, and commonly blue or lavender on purple cultivars (Fig. 4.14). When the flowers are harvested or drop from age, a black rot rapidly covers the flowers in 1 or 2 days.

Fig. 4.14. *Phyllosticta* leaf and flower spot on orchid.

ii) Biology: *Phyllosticta* produce small, black fruiting bodies (pycnidia) that contain conidia on the surface of infected tissue. These conidia are released when moisture is present and also require moisture to germinate. In addition, black fruiting bodies called perithecia, are also formed, shortly after pycnidia are produced. Perithecia produce sexual spores or ascospores. Ascospores are produced in sacs and there are 8, single-celled spores per sac.

iii) Epidemiology: In the field or greenhouse, disease spread is due to movement of both conidia and ascospores. Conidia are splashed from older diseased leaves to other leaves nearby. Ascospores are discharged into the air and are carried to long distances. When either spore type lands on young dendrobium tissue, it germinates and forms a small structure called an appressorium. Moisture is needed for this process. The appressorium adheres to the surface of the host tissue. The fungus then grows from the appressorium and penetrates the host. Without moisture the spores fail to germinate and are unable to infect the host.

After *Phyllosticta* penetrates the leaf, for sometime there are no symptoms or signs of the pathogen on the infected leaves. It takes 3 to 6 weeks before a small faint spot can be observed on the infected leaf. Many more weeks are required before a clearly defined yellow spot appears. Because of this, many growers are surprised to see spots developing many months following plant purchase.

In the field or in commercial greenhouses, there is sufficient sunlight for the plant to inhibit the expansion of *Phyllosticta*, confining the fungus to the yellow spot stage. However, when light levels are reduced the plant loses its ability to inhibit *Phyllosticta*. The fungus starts to grow rapidly and large rots develop. Leaves are lost and only the cane or pseudo stem remains. This is a major problem for growers who ship infected potted plants. Within shipping boxes, darkness and high humidity favor disease development and plants arrive with severely diseased leaves. Similarly, light levels in garden shops are also too low to prevent disease development.

iv) Management

Since many orchids are propagated by seeds or cloning, it is possible to start new plants without the fungus. These plants should be protected by: a) isolating them and keeping them away from mature plants that could have pathogens; growing new plants in a small, separate greenhouse or glasshouse would be ideal; b) preventing the movement of ascospores to new plants by keeping them in a separate greenhouse or in a section separated by solid plastic; c) reducing the number of ascospores by removing all rotted leaves and especially those in pots, and on the bench or ground.

For larger plants and for those already infected: a) remove all rotted or diseased leaves; clean benches, pots, and walkways by removing all dried sheaths, leaves, old stems etc., since the fungus survives in diseased plant material; b) trim old dead stems and flower spikes; c) apply fungicides such as mancozeb to kill fungal spores.

Since moisture is needed for disease spread by asexual spores (conidia), for germination of the spores, and for host penetration, moisture control is very important. Potted plants grown under solid cover will be nearly free of this disease. Water plants early in the day to allow leaves to dry as opposed to just before dark, when foliage would remain wet for many hours. Remove tall bushes around fields to encourage air movement in the field and reduce long periods of high humidity.

Application of fungicides such as mancozeb will also reduce infection levels. However, these fungicides are not effective, once the fungus has penetrated the host. Thus, disease prevention through cultural practices and fungicide application is crucial.

4.11.5. Heart/Black Rot, *Phytophthora palmivora*

i) Symptoms: Black Rot is a particularly aggressive infection of Cattleyas caused by the fungus, *Phytophthora*. A new shoot suddenly turns black: the rot moves rapidly, killing the rest of the plant. A whitish 'bloom' of fungus spores may be seen on the diseased tissue. Heat-stressed orchids are more susceptible. This disease is more prevalent during wet weather: it is spread primarily by splashing water.

ii) Management: Disease spread primarily by splashing water. Shelter plants from rain. Isolate infected plants. Removal of infected plants followed by treatment with systemic fungicides such as metalaxyl-MZ and fosetyl-AL is effective in controlling the disease.

4.11.6. Black Leg / Dry Rot, *Fusarium* sp., *Rhizoctonia* sp.

i) Symptoms: Dendrobiums and Vandas growing in waterlogged potting or bedding materials are susceptible to a slow but inexorable rot of the roots and stem from below. The causative fungi are *Fusarium* and *Rhizoctonia*. Pseudo bulbs become spongy and discoloured. The leaves, especially in Vanda, will turn yellow and drop off, one by one, until none are left and the plant dies. Plants can be salvaged. Uninfected pseudo bulbs of Dendrobiums can be removed to allow keikis to form. If a Vanda has many aerial roots, sever the stem above the line of infection. The disease is caused primarily by poor culture.

ii) Management : This disease is caused primarily by poor culture. Fungicides are not recommended. Use potting/bedding materials appropriate to the plant type. Replace potting materials before they become stale and waterlogged. Burn infected material.

4.11.7. *Pythium* Rot, *Pythium ultimum*

i) Symptoms: Pythium causes infection of seedlings as they are very susceptible. If the seedlings are placed closer in seed pans the infection spreads very easily under wet conditions.

ii) Management : Fungicides such as mancozeb, metalaxyl-MZ and fosetyl-AL should be applied to control the disease.

4.12. ANTHURIUM

4.12.1. Anthracnose, *Glomerella cingulata* (*Colletotrichum gloeosporioides*)

i) Symptoms: The infection occurs on young plants soon after planting. Symptoms start as tiny brownish spots on the flower spadix. During high humidity these spots enlarge, appear water-soaked and turn necrotic. Sometimes the entire spadix will turn black as lesions coalesce. The shape of most lesions is, however, angular due to the shape of the spadix tissue. As the disease becomes more severe, masses of orange coloured spores form on necrotic areas. Leaves and spathes are rarely if ever infected. Warm wet weather favours the outbreak of the disease which spreads fast in densely planted beds.

ii) Management: Sanitation and use of common fungicides (0.1% carbendazim + 0.2% copper oxychloride or 0.1% thiophanate methyl or 0.2% mancozeb or 0.15% kitazin) at fortnightly intervals helps in management of the disease.

4.12.2. Black Rot and Leaf Blight, *Phytophthora nicotianae* var. *parasitica, P. citrophthora*

i) Symptoms: Symptoms appear as black spots with water soaked margins on spathe, spadix, leaves, stalks and petioles. The rotting spreads faster on leaves and spathe through water droplets. The infection spreads rapidly during wet weather with moisture on leaf surface.

ii) Management: Aliette at 0.3%, dimethomorph at 0.2% and other anti-oomycete fungicides are effective for the management of the disease. Root rot caused by Pythium and Phytophthora can be controlled by the application of potassium phosphonate 0.3%.

Table 4.2. Management of fungal diseases of ornamental crops

Crop	Disease	Symptoms	Chemical Dose/litre	Time of application
Crossandra	Fusarium corm rot		Soil drenching with carbendazim at 2 g/l of water and foliar spray at the same concentration.	
	Leaf spot		Foliar spray with mancozeb at 2 g/l of water.	
Rose	Leaf spot	White mealy growth on leaves/flower buds	Wettable sulphur-2 g Triadimefon/Myclobutanil/ Tridemorph-0.5 ml	September to February or cloudy weather
	Black spot *Diplocarpum rosae*	Tan coloured spots and defoliation	Captan-2 g Iprodion/Bitertanol-1 g	September to December
Chrysanthemum	Septoria leaf spot, *Septoria chrysanthemi*	Reddish brown spots on leaves	Mancozeb-2 g Carbendazim-1 g	Rainy season
Gladiolus	Fusarium wilt *Fusarium oxysporum* pv *gladioli*	Wilting of plants with loss of turgidity	Carbendazim+Captan 2 gm each /lit for corm treatment after lifting and before planting	Before planting & drenching after germination

Jasmine	Cercospora leaf spot, *Cercospora jasminicola*	As in brinjal	Thiophanate Methyl/ Bitertanol-1 g Carbendazim-1 g	Rainy season
Zinnia	Powdery mildew, *Oidium zinni*	As in other crops	Wettable sulphur-2 g Tridemorph-0.5 ml	As in other crops
	Alternaria leaf spot, *Alternaria alternata*	As in other crops	Mancozeb/Iprodione-2 g	Rainy season
Marigold	Powdery mildew, *Oidium tageti*	As in carrot	Wettable sulphur-2 g	Cloudy weather
	Alternaria leaf spot, *Alternaria tageti*	As in carrot	Mancozeb/Iprodione-2 g Triadimefon-1 g	November-December

Table 4.3. *Trichoderma* spp. and other antagonistic fungi for the management of diseases of ornamental, medicinal and aromatic crops.

Crop	Disease / Pathogen/s	Biocontrol agent	Mode of application
Rose	Grey mold, *Botrytis cineria*	*T. viride*, *T. harzianum*	Cutting treatment
Gladiolus	Yellows & corm rot, *Fusarium oxysporum* f. sp. *gladioli*	*T. virens*, *T. harzianum*	Corm & soil treatment
Chrysanthemum	Wilt, *Fusarium oxysporum*	*T. harzianum*	Soil appln.-160 kg/ha

Chapter 5

MEDICINAL PLANTS

In medicinal plant sector, the WHO has estimated that about 80% of the population in developing countries rely on traditional medicines mostly plant drugs for primary health-care needs. Even modern medicines contain about 25% drugs derived from plants. A large number of people are earning their livelihood from different activities of medicinal plants such as collection, cultivation, marketing, processing, etc. The area under cultivation of the above crops is low, but the cash returns are high because of intensive cropping.

In recent times, due to increasing realization of health hazards and toxicity caused by synthetic drugs (side effects), there has been a renewal of interest in the use of plant based drugs throughout the world. A campaign is going on worldwide to utilize more and more plant derived chemicals in the human health care/personal care system. As a result, the demand for plant-based raw materials has increased enormously in both national and international markets. Herbal plants are finding diverse uses as raw materials, not only for medicines but also as biopesticides.

Medicinal crops form one of the important groups due to their demand in various pharmaceutical industries and also to earn foreign exchange by way of export. The potential for earning foreign exchange by India from the exports of medicinal and aromatic plants is estimated over US$ 3,000 million per annum. The export of these plants and their products has a tremendous potential in advanced countries like Europe, USA and Japan. The international market for medicinal plants related trade is estimated at US$ 60 billion/year having a growth rate of 7% per annum. Destination of foreign tourists is now aimed at India for health treatment. Kerala and Gujarat have become their favourite destinations since Ayurvedic treatment is very popular in these two states.

Unless we start cultivating the important medicinal plants, it will not be possible to meet the increased demands. However, cultivation of medicinal plants has to be taken up in polyculture models rather than in monoculture models. There is a huge task to be carried out at every level to make the Indian system of medicine a viable, sustainable and modern. For this task, urgency is felt which can only be established through regular and systematic scientific research on collection, conservation, evaluation and development of good agricultural practices to compete in the world market.

5.1. SOLANUM

Solanum (*Solanum viarum*), also called as Khasi Kateri, is a steroid-bearing perennial, tall bush distributed all over Assam, Sikkim and Manipur between 1,600-2,000 m elevations. Its berry pulp is rich in Solasodine alkaloid, which is a starting chemical for production of steroids. About 22 species including *Solanum viarum* contain an alkaloid Solasodine in their berries. Solasodine serves as an excellent raw material for the manufacture of Cortisone and Cortisonal hormones. It is used in the production of contraceptive pills, corticosteroids and sex hormones. It is cultivated in 3,000-5,000 ha, mainly in Maharashtra and Andhra Pradesh. It is also cultivated in West Bengal, Assam, Madhya Pradesh and Gujarat on a small scale.

5.1.1. Wilt, *Fusarium oxysporum* f. sp. *lacimiti*

i) Symptoms: The disease generally appears after first showers and is characterized by wilting and sudden death of plants. The disease spreads very fast and with in a fortnight majority of the plants in a field show wilting symptoms and many of them ultimately die irrespective of the age of the plant.

ii) Management

- Direct sowing of seeds instead of transplanting of seedlings can reduce the disease incidence.
- Early planting during October and November helps to escape the disease.
- Application of PCNB to the field at 300 g/100 m^2.
- Drenching of the basal portion of infected plants and surrounding soil with 0.1% benomyl or captan.
- Mix 2.5 kg *Trichoderma viride* in 50 kg of FYM and apply along plant lines.

5.1.2. Powdery mildew

i) Symptoms: Powdery white mass appear on leaves.

ii) Management: Spray wettable sulfur at 3 g/litre of water.

5.2. PERIWINKLE

Periwinkle (*Catharanthus roseus*), commonly known as vinca, is a common bedding plant. It is popular in container gardens and new flower colours produced on compact cultivars, have increased its use in many public and home gardens.

It contains several alkaloids found to be effective in the treatment of leukemia and other diseases such as diabetes, hypertension, etc. The more important alkaloids obtained from Vinca are Leurosine, Vinblastin, Vincristin, etc.

5.2.1. Wilt/Root Rot, *Phytophthora nicotianae*

i) Symptoms: Wilting of plants is very common (Fig. 5.1). The wilt begins from infected leaves or as a result of root infections. Leaf spots are initially small, irregular water-soaked areas easily visible on the under surface of the leaf. These spots expand to rot the entire leaf.

Fig. 5.1. Wilt/root rot of periwinkle.

Diseased leaves are gray-olive green to brownish. The rot progress from leaves to the stem, and sections or branches of the plant are killed. Yellowing of leaves and black leaf rot also occur near the stem. Movement of the fungus into the primary stem or into the larger stems causes wilting of the entire plant or large sections of the plant.

Root rot is not initially noticeable. Small feeder roots are infected and larger roots become diseased with time as the fungus feeds and destroys the root system. Infected plants may be stunted, lack vigour, or appear to be suffering from nutritional deficiency. On smaller plants, movement of the fungus from the roots to the collar causes rot at the soil line. These plants wilt rapidly since upward movement of water is inhibited. On larger plants, collar rots may progress into a lower branch and gradually the plants are killed. Infected branches commonly have no leaves and only dead twigs remain.

ii) Biology: *Phytophthora nicotianae* produces microscopic, hyaline (clear) spores called sporangia. These asexual spores are papillate (have a pointed tip), nearly spherical (round), and release zoospores when water is available. These zoospores are able to swim for many hours. They are attracted to host tissue or organic matter. Zoospores eventually stop swimming, encyst (round-up and form a wall) and germinate. If the environment is not wet enough, sporangia germinate by producing fungal threads or germ tubes.

The fungus also produces a spherical spore with thickened walls. These are chlamydospores and do not release zoospores. They germinate by producing germ tubes. These spores are able to survive in soil for many months in the absence of the host.

Phytophthora nicotianae is heterothallic and has two mating types, A1 and A2. When both mating types are grown together, numerous sexual spores, called oospores are produced. Oospores are also thick-walled. The fungus is favoured by warmer temperatures (25 to 28°C) and moisture.

iii) Epidemiology: This pathogen produces sporangia on the surface of the diseased tissue. The sporangia release zoospores when water is present and zoospores or sporangia are splashed to healthy leaves or other plants. On the host, the germ tube produced by the zoospores will penetrate the host. Penetration of the host and growth of the fungus in the host begins the infection process. Continuous feeding and growth of the fungus results in rot and production of disease symptoms. Sporangia also produce germ tubes that penetrate the host epidermis and initiate the disease process.

Chlamydospores are produced inside the diseased leaves and stems, which fall to the ground. The diseased plant tissue decomposes and the chlamydospores remain in soil for months, accounting for the soil-borne nature of this fungus. Movement of contaminated soil or infected plants spreads the pathogen to long distances.

iv) Management: Use clean seed and monitor seedlings for any sign of disease. Remove any diseased seedling immediately. Control the moisture levels in the nursery, landscaped areas, or home garden by using covered greenhouses or benches, increasing air movement to reduce leaf wetness, preparing the soil by adding compost and insuring good drainage. Removal of diseased plants and plant parts reduces the amount of spores that spread the disease.

Fungicides are available that reduce disease levels. Mancozeb is a good contact fungicide and can reduce the number of leaf spots. Metalaxyl is an effective systemic fungicide that will protect leaves and new shoots.

5.2.2. Dieback, *Colletotrichum dematium*

i) Symptoms: The dieback has been found to affect the crop during the monsoon. The characteristic symptoms of the disease are withering of the terminal bud of the young plants followed by wilting and drooping of leaves. In later stages, the entire branches dry up and wither tip downwards. The fruiting bodies develop on the buds of the withered branches which become grayish in colour.

ii) Management: Mulching between the rows at seedling stage with dry spent straw of citronella or palmarosa reduces the incidence of dieback to a considerable degree. Cultivation of a new strain WH-40 developed at CIMAP Centre, Bangalore, which has high level of field resistance, minimizes the loss due to dieback.

Spraying of carbendazim or benomyl each at 0.1% at fortnightly intervals is effective for the disease management. Spraying 0.2% mancozeb or 0.3% captafol at fortnightly intervals is also effective.

5.3. ISABGOL

Isabgol (*Plantago ovata*) is an important cash crop of Gujarat. In Gujarat, it is cultivated in 33,600 ha area producing 30.742 tonnes of seed. Isabgol is also grown in Bihar (Sasaram), Gujarat (Ahmedabad, Banaskantha, Bhavnagar, Jamnagar, Junagadh, Khaira, Kutch, Mehsana, Rajkot), Haryana (Rewari), Madhya Pradesh, Rajasthan (Sirohi) and Uttar Pradesh.

The husk contains xylose, arabinose, galacturonic acid, rhamnose and galactose. Husk is a safe laxative, beneficial in habitual constipation, chronic dysentery and diarrhea. The seed and husk are used to cure inflammation of the nervous membrane of gastrointestinal and genito-urinary tracts, duodenal ulcer, gonorrhea and piles. It can also be used as cervical dilator for termination of pregnancy. Mucilaginous husk absorbs toxins from the gut and helps in excreting them from the body. It acts as a demulcent and anti-irritant in soothing the intestinal ulcers and reduces the serum cholesterol. Seed contains protein (17-19%) and yellow semi-drying oil (5%) containing linoleinic acid (0.2%), linoleic acid (47.9%), oleic acid (36.7%), palmitic acid (3-7%), stearic acid (6.9%), liguoceric acid (0.8%).

5.3.1. Damping-Off, *Pythium ultimum, Rhizoctonia solani*

The crop suffers from both the phases of damping-off i.e. pre-emergence and post-emergence. Older plants are less susceptible than the young seedlings.

i) Symptoms: In pre-emergence phase, the young seedlings are killed before they reach the surface of the soil. Germinating seeds undergo complete rotting. In post-emergence phase, new emerging plants are affected first. The initial infection appears as a slightly darkened spot on collar region of the seedlings. The infection extends rapidly and the invaded cells collapse and the seedling is over-run by the fungus and dies shortly. The disease normally appears if the prevailing temperature is high with high relative humidity at the time of sowing. Unseasonal rain after seeding the crop also helps in development of this disease.

ii) Management: Treatment of seeds with metalaxyl or fenaminosulf or captafol each at 5 g/kg of seeds protect the seedlings. Seed treatment with *Trichoderma harzianum* at 5 g/kg of seeds protected the seedlings from damping-off disease.

5.3.2. Downy Mildew, *Peronospora plantaginis, P. alta*

The downy mildew disease causes enormous quantitative and qualitative losses. Many times, it becomes a limiting factor in successful cultivation of isabgol. The disease is responsible for 30% loss in seed yield.

i) Symptoms: *Peronospora plantaginis* develops chlorotic areas on the upper surface of the leaves and ashy-white frost-like mycelial growth on the lower surface of the leaves. Later, the infected leaves become chlorotic and brownish in colour followed by curling, crinkling and drying of the plants (Fig. 5.2). On the other hand, the fungus *P. alta* develops chlorotic streaks, extending along with midrib of the leaves. Profound downy growth is observed on the lower leaf surface. Ultimately the whole leaf turns necrotic and in severe cases, the entire plant gives blighted appearance. The abundant oospores formation causes the thickening of the affected leaves.

Fig. 5.2. Downy mildew on isabgol.

The disease is both seed and soil-borne. The sporangia upon germination, produces germ tubes and penetrates through stomata. Sporangia and sporangiophores emerged through stomatal opening before death of the tissues. Studies on sporangial germination under field conditions revealed that the temperature of around 20°C and relative humidity of around 89% favour sporangial germination between 2.00 a.m. and 8.00 a.m. and also at 10.00 p.m. Maximum disease developed during the 75 to 95 days age of the crop.

ii) Management

a) Cultural Methods: Seed rate higher than 2.5 kg/ha increased the disease severity in November sown crop and reduced the yield significantly. However, December sown crop generally escapes from the disease.

b) Chemical Methods: Overnight seed soaking before sowing in Aureofungin at 0.25 g/10 litres of water followed by 3 sprays of Aureofungin at 15 g/ha controlled the disease effectively. The first spray is to be given immediately after appearance of the disease and should be repeated twice at an interval of 8-10 days.

Seed treatment with Apron SD at 5 g/kg of seed coupled with 3 sprays of 0.2% captafol or 0.05% metalaxyl or 0.3% mancozeb effectively controlled the disease. The sprays are to be initiated just after the appearance of the disease and should be repeated at 10 days interval. Seed treatment with metalaxyl at 4 g/kg seed and 3 sprays of metalaxyl at 2 g/l of water at 15 days interval starting from first appearance is best for disease control. Similarly seed treatment with 0.3% mancozeb + three sprays of 0.2% ridomil resulted in maximum seed yield.

Bordeaux mixture (6-3-50) at 600 to 800 l/ha should be sprayed 30 DAS followed by 2 or more sprays at an interval of 15-20 days.

5.3.3. Powdery Mildew, *Erysiphe cichoracearum*

i) Symptoms: Powdery mildew normally appears at the time of flowering. The disease is characterized by small whitish or grayish spots on the leaves which gradually enlarge and cover the whole plant surface with talcum-like powdery mass.

ii) Management: The disease can be controlled by spraying with 0.2% karathane or wettable sulphur as soon as the disease appears.

5.3.4. Wilt, *Fusarium* sp.

i) Symptoms: The disease is manifested either as pre-emergence damping-off or as a late season wilting of plants as they start to produce seed scales. Wilting, which first occurs in the outer leaves, is preceded by a change in leaf colour from green to silver. Wilting of the whole plant is rapid; leaves may become prostrate within 36 to 48 hours after incipient wilting. Discolouration of the tap root (commencing at the tip) takes place accompanied by the presence of large quantities of hyphae in xylem elements, vascular discolouraion and cortical root rot.

ii) Management: As a prophylactic measure, spraying of 6-3-100 Bordeaux mixture or copper oxychloride or mancozeb or zineb each at 0.2% commencing from 30 DAS and repeated twice at 15-20 days interval is effective for the management of wilt.

5.4. ASWAGANDHA

Ashwagandha (*Withania somnifera*) is a plant of immense medicinal importance growing all over North-Western and Central India. Roots of this plant are the major source of alkaloids and steroidal lactones (withanoids), which are intensively used in various pharmaceutical industries. It has adaptogenic, immuno-modulator, aphrodisiac, anti-stress and mildly sedative properties. Its roots yield valuable drugs which are employed in rheumatic pain, inflammation of joints, nervous disorders, cough, cold, epilepsy, and cardio and nerve tonic. At present, 4,000-5,000 ha area is under its cultivation mainly in Madhya Pradesh (Manasa Neemuch and Jawad tehsil of Mandsaur district) and neighbouring districts of Rajasthan (Kota and Charu).

5.4.1. Seedling Blight

i) Symptoms: Seedling blight is known to be the major disease of ashwagandha. The disease reduced the plant population drastically by seedling mortality, which ultimately reduced the yield.

ii) Management: Seed treatment with 0.2% PCNB (Brassicol) (27.5% mortality), 0.3% mancozeb (31.5% mortality) and 0.2% delan (32.75% mortality) gave significant reduction in seedling mortality and increased the dry root yield.

Seed treatment with *Trichoderma viride* at 4 g/kg of seed along with metalaxyl at 6 g/kg of seed showed minimum seedling mortality (18.7%).

5.4.2. Leaf Blight

i) Symptoms: Leaf blight is a serious disease of ashwagandha at Malwa region of Madhya Pradesh. The symptoms appeared as small, light brown spots on leaves and flowers which enlarged, coalesced with each other and gave a blighted appearance.

ii) Management: Spraying of Cuman-4 and Rovral both at 0.3% were found effective against this disease.

5.4.3. Dieback

i) Symptoms: The pathogen caused necrosis of tender twigs from the tip backwards. The entire branch or the entire top of the plant withered away. The dead twigs were water-soaked, turn brown, become grayish or straw coloured in advanced stage of the disease.

ii) Management: Spraying of mancozeb at 3 g/litre of water gives effective control of the disease.

5.5. OPIUM POPPY

Opium poppy (*Papaver somniferum*) is grown in Banswara, Bara, Bhilwara, Bundi, Chittorgarh, Jhalawar, Kota and Udaipur districts of Rajasthan; Faizabad, Barabanki, Bareilly and Shahjehanpur in Uttar Pradesh and Neemuch, Mandsaur and Ratlam in Madhya Pradesh. It is a source of raw opium which is used in narcotic, veterinary and pharmaceutical industries. It contains alkaloids like Morphine, Codeine, Narcotine, Papaverine and Thebain. Poppy seed is used for culinary purpose - breads, sweets and confectioneries. Capsule husk is brewed with tea (Bonda chai).

5.5.1. Collar Rot, *Rhizoctonia solani*

i) Symptoms: Collar rot is one of the serious diseases of poppy. A black lesion develops at the joint of the root and shoot, resulting into falling-off of the plant and rapid loss of crop stand.

ii) Management: Crop rotation to provide at least 3 years gap for poppy cultivation in the same field is effective. Seed treatment with thiram and soil drench with 0.4% zineb are found effective. Foliar spray of 0.15 to 0.2% ridomil along with micronutrients and urea right from the seedling stage has been found most effective for the control of collar rot.

5.5.2. Downy Mildew, *Peronospora arborescens*

Downy mildew has been reported as an important disease from major opium poppy growing countries like India, China, Egypt, France, The Netherlands, Hungary, Greece, Spain, Portugal, Italy, Turkey, Australia, USSR and Yugoslavia. The epiphytotic form of the disease was observed in Argentina, Germany, Iran, Hungary, Austria and India. In India, the occurrence of the disease has been reported from Rajasthan, Madhya Pradesh, Uttar Pradesh, and Delhi.

i) Symptoms: The plants with systemic infection have two kinds of symptoms viz., lower portion is healthy and upper portion diseased. In the upper diseased portion of the plant, the leaves remain small, chlorotic, curling downward at the edges and are closely placed. Heavy sporulation occurs on the entire lower surface of the diseased leaves. The sporulation on leaves is persistent. Buds generally remain under developed and do not open. The capsule, if formed, shrinks, gets wrinkled and dries gradually after infection. Heavy sporulation occurs on peduncle, buds and capsules which is not persistent. Rudimentary brownish black structures develop in place of normal seeds. Such capsules do not yield opium as well. The infected plants die prematurely.

In the plants with leaf spot infection, the infected area becomes first chlorotic and then necrotic. The leaf spots enlarge by coalescing with each other. Sporulation may occur on the underside of spotted area but is not heavy and not persistent. In spite of excessive leaf spots, plants do not die prematurely.

The nitrogen fixing preceding crops like black gram and groundnut increased the disease incidence. The disease incidence increased with increase in soil moisture and supply of nitrogen. The disease stimulatory effect of nitrogenous fertilizers was mitigated by the application of FYM. The disease severity was more in late sown crops as prime growth period coincided with low temperature and high humidity. As a result, latex, morphine, seed and husk yields were significantly low in the late sown crops.

ii) Management: At Udaipur, it was observed that early sown (30th October) crop was less affected by downy mildew incidence. The occurrence and severity of the disease decreased with lower plant density.

Seed treatment with metalaxyl and four sprays of metalaxyl at 0.2% at 35, 55, 75 and 95 days after sowing found most effective in reducing primary and secondary disease incidence. It also increased latex yield (26.7%) and seed yield (24.7%) over control. Apron seed treatment (8-10 g/ha) and sprays of ridomil MZ, Cuman-L, and mancozeb also were highly effective in controlling the disease.

Opium poppy cv. Jawahar-540 is having some resistance against the disease has been recommended for Madhya Pradesh.

5.5.3. Powdery Mildew, *Erysiphe polygoni*

Powdery mildew has been found to occur in severe form on poppy in Rajasthan, Madhya Pradesh and Uttar Pradesh. Poppy cultivars suffer heavy losses on account of this disease.

i) Symptoms: Powdery mildew is observed in the fields at or after flowering stage i.e. nearly 14-16 weeks after sowing. Late sown crop (after 15th November) develops heavy powdery mildew disease. First appearance of the disease is in February or early March. The base of the stem is attacked first where it appears as small, circular patch consisting of radially arranged mycelium. This enlarges gradually and sporulation occurs soon. Several new colonies continue to appear covering large part of the stem. Seven to ten days after the appearance of the disease, blackening develops on the affected parts. When the powdery mildew develops to its maximum, ½ to ¾ of stem appears black interspersed with green patches. On leaves the mildew spreads all over. Growth of mildew is luxuriant in shade. Late forming branches are severely affected and their growth is checked. The buds on such branches remain rudimentary and wither.

ii) Epidemiology: The disease is very severe in Rajasthan during February to April when the minimum temperatures were 11.2 to 17.2°C and maximum temperatures were 26.6 to 34.8°C.

iii) Management: Crop sown before 15th October almost escape the disease. The disease can be effectively and economically controlled by only one spray of 0.5% wettable sulphur at the time of first appearance of the disease in the field or when the crop has reached the age of 80 days or when the maximum daily temperature has reached 27°C and above.

5.5.4. Root Rot, *Macrophomina phaseolina*

i) Symptoms: The disease appeared in the month of December when the plants were about 2-months old. The first visible symptom was in the form of small black spot on the upper part of the tap root near the soil surface. Gradually the leaves started withering and drying. The lesions on the tap root later become deep and the infected plants were broken down at the point of black necrotic lesion. In case of severe infection, about 20% plants died prematurely.

ii) Management: Soil application of *T. harzianum* reduced the percentage of infected plants (0 to 10% compared to 20 to 40% in control), increased plant height (94 to 100 cm compared to 85 to 88 cm in control) and increased the opium yield (114 to 191 mg compared to 92 to 94 mg in control).

Integrated Disease Management

Seed treatment with metalaxyl 35 SD at 8 g/kg of seed + sprays of 0.2% metalaxyl at 35, 55 and 75 DAS along with recommended agronomic practices was the best to reduce the disease infection by 54.05% at Faizabad, 55.83% at Mandsaur and 82.28% at Udaipur and increased latex yield by 122.2, 72.59 and 53.39% at these places, respectively. Similarly, seed yield at these locations were found to increase by 182.57, 37.32 and 79.72%, respectively.

5.6. SARPAGANDHA

Sarpagandha (*Rauvolfia serpentina*) is a perennial native Indian herb. Its roots are used for controlling high blood pressure and certain forms of insanity in Ayurvedic system of medicine since ancient times. It has received worldwide recognition after isolation of its bioactive reserpine alkaloid in allopathic medicines. In addition, its sedative property is utilized by Ayurvedic physicians in treatment of insomania, epilepsy and asthma. It is commercially cultivated in Madhya Pradesh, Uttar Pradesh, West Bengal, Assam and Orissa. The root is its economic part, containing 55 alkaloids of which reserpine, deserpidine, ajmalcine, serpentine and yohimbine being pharmacologically active and important. Of these, ajmalicine, reserpine and serpentine are highly potent in medicine. The commercial root crop contains 1.4-3.0% of total alkaloids.

5.6.1. Foliar Blight/Spot, *Alternaria tenuis*

i) Management

Six applications of mancozeb at 0.3% as foliar spray were found to be most effective giving 187.8 g fresh root/plant and 78.8% protection against foliar blight/spot (*Alternaria alternata, Cercospora rauvolfia, C. serpentina, Colletotrichum gloeosporioides, Corynespora cassiicola, Curvularia lunata, Macrophomina phaseolina* and *Rhizoctonia solani*) diseases. Application of two doses of neem cake at 50 g/plant at 30 days interval plus two sprays of *B. subtilis* in September and October along with three foliar sprays of mancozeb at 15 days interval plus two foliar sprays of carbendazim at 0.15% at 21 days interval gave 201.5 g fresh root/plant, and 75.5% protection against foliar blight pathogens.

Seed treatment with contact fungicides is recommended. Spraying of maneb rovral at 0.2% also gives effective control of the disease.

5.7. DIOSCOREA

Various species of Dioscorea (*Dioscorea floribunda*) are the source of diosgenin, a steroidal sapogenin, and is the most commonly used precursor for synthesis of certain drugs which include cortisones, sex hormones and oral contraceptives.

5.7.1. Leaf Spot, *Cercospora dioscorea*

i) Symptoms: The fungus causes many leaf spots scattered all over the leaves. They are irregularly angular and dark brown to black, with indefinite margins. There is heavy defoliation causing considerable reduction in the yield of tubers.

ii) Management: Spraying of 0.1% benomyl or carbendazim at fortnightly intervals gives effective control of the disease.

5.7.2. Leaf Fall/Dieback, *Colletotrichum* sp.

i) Symptoms: The first symptom observed is the drying of young leaves and the branches from the tip downwards. There is heavy defoliation and die-back. In severe cases, the entire plant may die. The acervuli are formed on the bark of the vines which are dead or dying.

ii) Management: Spraying of 0.1% carbendazim or 0.2% mancozeb at fortnightly intervals is effective in managing the disease.

5.8. PYRETHRUM

The genus *Chrysanthemum* consists of about 300 species of herbs and under-shrubs of which only a few such as *C. cinerariaefolium. C. coccineum* and C. *marschalki*, yield the commercial insecticide pyrethrum. The term pyrethrum is commonly applied to the dried flower heads of the above 3 species which is a source of Pyrethrin I, Pyrethrin II, Cinerin I, Cinerin II, Jasmoline I and Jasmoline II. Pyrethrum is one of the safe insecticides known which has a low mammalian toxicity and is rapidly metabolizes if accidentally swallowed. It acts as a contact poison with rapid knock down effect on insects. It affects the nervous system resulting in muscular convulsions, paralysis and death of insects. Major pyrethrum growing areas include South Indian hills above 1800 m above MSL and Kashmir valley.

5.8.1. Wilt, *Fusarium* sp.

i) Symptoms: This fungus is found to cause a severe wilt disease of *Chrysanthemum* spp. The chief symptoms are retarded growth, sudden wilting and death of plants followed by browning and rotting of roots and collar region. Another fungus, *Sclerotinia minor* is also usually found associated with the wilted plants.

ii) Management: Drenching the basal portion of the infected plants and the soil around with 0.1% benomyl or 0.2% captan is effective in management of the disease.

5.8.2. Root Rot, *Fusarium solani*

i) Symptoms: The fungus causes a root rot disease in pyrethrum. The typical symptoms such as yellowing, wilting and stunting of the aerial parts and reduction in number of tillers, start appearing in June-July. The wilting usually starts with tillers and proceeding inward progressively until most of the tillers and the main plant wilts. Infected root show black discolouration, extensive rotting and disintegration of cortical tissues.

ii) Management

- Seedlings should be grown in disease-free soil.
- Raised nursery beds should be prepared.
- Nursery beds should be drenched with 0.15% benomyl or 0.1% zineb.

5.9. SENNA

Senna (*Cassia angustifolia, C. acutifolia*) is a well known drug made from leaves and pods. It is grown principally under dry land conditions and usually as a semi-arid irrigated crop. Leaves and pods of senna contain two glycosides viz., Sennoside A and Sennoside B which are used as laxatives. It is also used in medicine as a cathartic. It is used as an ingredient of herbal tea in Europe. The major senna growing areas include Madurai, Tirunelveli and Ramanathapuram in Tamil Nadu; Pune in Maharashtra; Anand and Mehsana in Gujarat and Jodhpur in Rajasthan.

5.9.1. Leaf Spot, *Alternaria alternata*

Leaf spot is the most serious among the diseases causing lot of damage to the senna crop. The infested leaves drop off prematurely and affect the yield considerably.

i) Symptoms: The disease becomes visible first as minute pale yellow spots on the leaf blade and its margins. As the spot grow in size they become circular and irregular in shape with their coalescence and turn dark brown to black in colour. The lesions vary from a minute spot to about 8 mm diameter in size. With the advancement of the disease, leaf tips and margins die and the necrotic tissues increase in size.

The disease is known to adversely affect the sennosides content in leaves and it is inversely proportional to the disease intensity. Sennosides content in leaves of highest disease intensity has almost one-third to that of healthy leaves and about one-half when there is moderate infection. Loss of sennosides in severely affected leaves was over 60% as compared to that in disease-free leaves. Even a mild incidence of the disease had affected the recovery of sennosides considerably (36.63%). As the price of the produce is dependent upon its quality, even a mild incidence of leaf spot affects the revenue of the crop.

ii) Management: Carbendazim at 0.05% and macuprax at 0.2% gave best control of the disease, but the yield is higher in macuprax treatment. Spraying of 0.15% mancozeb at fortnightly interval for 3 times in a period of 5-6 weeks is also recommended. Spraying of Bordeaux mixture, benomyl, captafol and copper oxychloride is also recommended.

5.9.2. Damping-Off, *Rhizoctonia bataticola*

i) Symptoms: In the North-Western India, senna is affected by the damping-off disease at seedling stage. The damping-off in seedlings is the most devastating disease.

ii) Management: Seed treatment with thiram or captafol at 2.5 g/kg of seed gave good protection. The disease incidence under field conditions can be controlled by soil drenching with 0.1% carbendazim solution.

5.9.3. Dry Rot, *Macrophomina phaseolina*

i) Symptoms: The dry rot occurs at late stage of crop growth.

ii) Management: Soil drenching with thiram or carbendazim gives effective control of the disease.

5.9.4. Leaf Spot, *Cercospora* sp., *Phyllosticta* sp.

i) Symptoms: The leaf spots attack the plants mainly during cloudy weather.

ii) Management: The pathogens are controlled by spraying 0.15% mancozeb.

5.9.5. Leaf Blight, *Colletotrichum* sp.

i) Symptoms: In Senna, the disease was found to be caused by *Colletotrichum* sp. (new species), which leads to premature defoliation of the leaves.

ii) Management: The disease has been successfully managed by three applications of propiconazole at 0.1% at 15 days interval.

5.10. LIQUORICE

Liquorice (*Glycerrhiza glabra*) is grown in Gujarat, Haryana and Rajasthan. Dried peeled or unpeeled underground stem and roots contain glycyrrhizin, which is 5 times sweeter than sucrose. It is used as flavouring agent in pharmaceutical industry for manufacturing cough syrups.

5.10.1. Collar Rot, *Macrophomina phaseolina* (= *Rhizoctonia bataticola*)

i) Symptoms: The increased incidence of the disease was observed with increase in temperature from February to July. During high atmospheric day temperatures above 35°C during summer, the old and new sprouts were attacked near collar region. Leaves on the affected sprouts turned pale yellow and dried up, which subsequently fell down and only deep brown to black branches with sclerotia and pycnidia were seen. In majority of the cases, this disease was confined to collar portion. However, above-ground stem portions were also occasionally affected. High incidence of the disease was observed on young sprouts as well as emerging sprouts during high temperature.

ii) Management: Reduction in soil temperature during summer by: different crop mulches, increasing the frequency and levels of irrigation and use of soil amendments and organic manures help to reduce the incidence of the disease.

5.10.2. Wilt/Root Rot, *Fusarium oxysporum*

i) Symptoms: Affected plants show discolouration, followed by rotting of under-ground parts (stolons) which results in sudden wilting/drying of the plants.

ii) Management: Reduction in soil temperature during summer by: different crop mulches, increasing the frequency and levels of irrigation and use of soil amendments and organic manures help to reduce the incidence of the disease.

5.10.3. Leaf Spot, *Cercospora cavarae*

i) Symptoms: This disease is invariably observed in Gujarat starting from late monsoon and prevailing up to onset of summer. The diseased portion on the upper leaf surface turns light in colour which eventually turns brown at later stage. Exactly underneath the upper diseased portion, initially angular water-soaked spots were observed. These spots were restricted to interveinal areas. They were gradually covered with spores and they turned dark brown and became velvety. Under epiphytotic conditions, severe defoliation is also observed. This disease was also reported from North-West Himalayas.

ii) Management : Spraying maneb or zineb at 0.2% gives effective management of the disease.

5.11. CINCHONA

Cinchona (*Cinchona* sp.) is grown in Nilgiris, Anamalai, Naduvattam, Shevaroy and Palani hills of Tamil Nadu; some parts of West Bengal; South Kanara and Kodagu in Karnataka. Bark is source of Quinine and about 20 other alkaloids of which cinchonidine, quinidine and cinchonine are the most important. The other alkaloids of lesser importance are quinamine, quinicine, cinchonicine, hydroquinine, hydroquinidine, hydrocinchonine, quinamine, conquinamine and poricine.

5.11.1. Seedling Blight, *Phytophthora parasitica*; Root Rot, *Phytophthora cinnamomi*; Damping-off, *Pythium vexans, Sclerotium rolfsii*

i) Symptoms: The above pathogens are important which affect seedling stand in the nursery.

ii) Management: Soil drenching with copper oxychloride at 5 g/litre of water is effective in managing the above diseases.

5.12. BELLADONNA

Indian belladonna (*Atropa accuminata*) and European belladonna (*Atropa belladonna*) are the source for tropane alkaloids which include hyoscine, hyoscyamine and atropine (0.13-0.7%) which are used in the manufacture of tinctures, treatment of whooping cough, epilepsy, Parkinson's disease, renal and bilary colic.

Indian belladonna is grown in the farms of Kashmir State Forest Department which is a remunerative crop for North-Western hilly region of India.

5.12.1. Damping-Off, *Pythium* spp.

i) Symptoms: The pathogen is responsible for death of seedlings in nursery.

ii) Management : Drenching the soil with 3 g/litre of copper oxychloride is effective.

5.12.2. Leaf Spot, *Cercospora atropae*

i) Symptoms: The pathogen is responsible for angular spots on leaf surface.

ii) Management : Spraying mancozeb at 2 g /litre of water is effective.

5.13. GLORIOSA

Gloriosa (*Gloriosa superba*) tubers and seeds contain colchicine at 0.15-0.3% and 0.7-0.9%, respectively. Colchicine is used in treatment of gout.

5.13.1. Leaf Blight, *Curvularia lunata*

i) Symptoms: Leaves turn light yellow and later small black spots are seen which ultimately spread to the entire leaf.

ii) Management : Spraying mancozeb at 2.5 g/litre of water is effective.

5.13.2. Tuber Rot

i) Symptoms: Tubers become soft and discoloured. Leaves turn yellow and plants die.

ii) Management: Soil drenching around the base of plants with carbendazim at 2g/litre of water is found effective.

5.14. LONG PEPPER

Long pepper (*Piper longum*) is grown in Karnataka, Kerala, Madhya Pradesh, Maharashtra, Tamil Nadu, Eastern Uttar Pradesh and lower hills of West Bengal. Its spikes contain the alkaloid piperine (4-5%) and piplatin, piperolactum A, piperolactum B and piporadione. Roots contain alkaloids piper longuminine (0.2-0.25%) and piper longumine (0.02%). Further purification of piper longumine yields alkaloids like cepharadione A, cepharadione B, cepharanone B and aristolactum A11. Dried immature roots and fruits find extensive use in Ayurveda and Unani medicine in treatment of chronic bronchitis, cough and cold.

5.14.1. *Colletotrichum* Leaf Spot, *Colletotrichum* spp.

i) Symptoms: The pathogen is responsible for leaf and vine rotting.

ii) Management: Spraying of 1% Bordeaux mixture or copper oxychloride at 0.2% once during May and another 2-3 sprays during rainy season is effective in controlling the leaf spot disease.

5.14.2. *Cercospora* Leaf Spot, *Cercospora* spp.

i) Symptoms: Necrotic spots and blight appear on leaves during warmer months.

ii) Management : Spraying 1% Bordeaux mixture or copper oxychloride at 0.2% gave effective control of leaf spot disease.

5.15. ALOE VERA

Aloe vera (*Aloe indica*) is grown in Gujarat, Madhya Pradesh, Maharashtra and Rajasthan. Yellow bitter juice present in the inner parenchymal cells of aloe vera yields 'Drug aloe' (Aloin) and '*Aloe* gel' which are used in medical preparations, cosmetics and ointments.

5.15.1. Leaf Spot, *Alternaria alternata*

i) Symptoms: Leaf spot is the main disease which not only affects the growth and yield of the plant but also affects its quality (bad odour of the gel).

ii) Management : Spraying with 0.2% mancozeb or chlorothalanil gives effective control of leaf spot disease.

5.15.2. Leaf Blight, *Colletotrichum* sp.

i) Symptoms: This is one of the serious diseases of *Aloe vera*, caused by *Colletotrichum* sp.; the disease greatly reduced the yield of gel.

ii) Management : The disease has been successfully managed by three application of propiconazole at 0.1% at 15 days interval.

5.15.3. Black Spot

Among four selected bioinoculants namely, *Bacillus subtilis, Glomus aggregatum, G. intraradices* and *Trichoderma harzianum* evaluated against black spot disease either alone or in different combinations; maximum plant height attained was 56 cm in *G. intraradices* treated plants followed by 53 cm in *B. subtilis* and *B. subtilis* + *G. aggregatum* treatments. Maximum plantlets production (14) was recorded in the treatments of *B. subtilis* + *T. harzianum* + *G. aggregatum* followed by 12 in others. Maximum herb yield obtained was 3.65 kg/plant in *G. aggregatum* treated pots, followed by 3.35 kg/plant in *G. aggregatum* + *B. subtilis* and *G. intraradices* treated pots.

5.16. AMBRETTE

Ambrette (*Abelmoschus moschatus*), also called as Muskdana and Kasturi Bhendi, grows wild all over the Deccan plateau and Karnataka, in the hilly region and at the foot-hills of the Himalayas. It is raised as a kharif crop in Punjab, Nagaland, Arunachal Pradesh, Bihar, Terai area of Kumaon, adjoining areas of Lucknow, Mahmudabad and Sitapur (Uttar Pradesh). Major constituent of oil is Ambrettolic acid and its lactone - 'ambrettolid'. The volatile oil finds use in perfumes, cosmetics, talcum powder, toiletries and confectionaries. Various parts of the plant find extensive use in Ayurvedic medicine. It is also used in manufacture of 'zarda', a flavoured chewing tobacco.

5.16.1. Anthracnose, *Colletotrichum hibiscicum*

i) Symptoms: The fungal pathogen infects almost all parts of the plant. Infection at seedling stage has been observed.

ii) Management : The seed should be treated with Agrosan GN or Ceresan to save the crop from anthracnose before sowing. Seed treatment with contact fungicides and spray of Bordeaux mixture are recommended.

5.16.2. Fusarium Wilt

i) Symptoms: Fusarium wilt occurs in seedling stage which also precedes its death. The disease prolongs over 2-3 weeks.

ii) Management : The crop may be sprayed with 1% Bordeaux mixture to control the infection.

5.16.3. Leaf Blight

i) Symptoms: Leaf blight disease causes damage to ambrette by way of premature leaf drying which results in death of leaves and pods. During high humidity, a white colony growth appears on diseased surface of leaves. Dark brown patch covers the whole leaf causing defoliation and death of plants.

ii) Management : Seed treatment with ceresan makes plants less susceptible to the disease. Repeated spray of 0.5% zineb at monthly interval controls the leaf blight effectively.

5.17. ROSEMARY

Major growing areas of rosemary (*Rosmarinus officinalis*) include Nilgiris and Bangalore. Leaves and flowers yield "Rosemary oil" used in perfumery, cosmetics and culinary purpose.

5.17.1. Leaf Blight, *Rhizoctonia solani*

i) Symptoms: Rosemary plantation is adversely affected by this soil-borne fungus causing severe leaf blight disease. The incidence of the disease is more after rains.

ii) Management: A fortnightly spray of 0.1% benomyl or 0.3% mancozeb is effective for management of the disease.

5.18. KALMEGH

5.18.1. Leaf Blight, *Colletotrichum* sp., *Rhizoctonia* sp.

i) Symptoms: Two distinct leaf blight diseases of kalmegh (*Andrographis paniculata*) have been investigated. One type is caused by *Colletotrichum sp.* and another by *Rhizoctonia sp. Rhizoctonia* sp. was found more destructive than the *Colletotrichum* sp. However, both caused premature defoliation of leaves.

ii) Management

The disease has been successfully managed by three applications of propiconazole at 0.1% at 15 days interval.

Table 5.1. Disease management in medicinal plants

Crop	Disease	Etiology	Control measures
Opium poppy (*Papaver somniferum*)	Downy mildew	*Peronospora arborescens*	Seed treatment with metalaxyl and four sprays of metalaxyl at 0.2% at 35, 55, 75 and 95 days after sowing found most effective in reducing primary and secondary disease incidence. It also increased latex yield (26.7%) and seed yield (24.7%) over control. Opium poppy cv. Jawahar-540 is having some resistance against the disease and recommended for Madhya Pradesh.

Senna (*Cassia angustifolia*)	Leaf spot	*Alternaria alternata*	Spraying of Bordeaux mixture, benomyl, captafol and copper oxy chloride is recommended.
Isabgol (*Plantago ovata*)	Downy mildew	*Peronospora alta, P. Plantaginis, Pseudoperonospora plantaginis*	Seed treatment with metalaxyl at 4g/kg seed and 3 sprays of metalaxyl (0.2%) at 15 days interval starting from first appearance is best for disease control. December sown crops are mostly disease free.
Sarpagandha (*Rauvolfia serpentina*)	Leaf blight and Bud rot	*Alternaria tenuis*	Spraying of maneb (0.3%) or rovral (0.2%). Seed treatment with contact fungicides.
Liquorice (*Glycyrrhiza glabra*)	Leaf spot	*Cercospora cavarae*	Spraying maneb or zineb at 0.2%.
Long pepper (*Piper longum*)	Leaf spot and rotting	*Colletotrichum* spp., *Cercospora* spp.	Spraying of Bordeaux mixture - once during May and 2-3 sprays during rainy season.
Khasi kateri (*Solanum viarum*)	Wilt	*Fusarium oxysporum*	Direct sowing of seeds instead of transplanting of seedlings can reduce the disease incidence

Table 5.2. *Trichoderma* spp. for the management of diseases of aromatic crops

Crop	Disease / Pathogen/s	Biocontrol agent	Mode of application
Opium poppy	Sclerotinia rot & blight, *Sclerotinia sclerotiorum*	*T. harzianum, T. viride, T. koningii, T. virens*	Soil treatment
	Downy mildew, *Peronospora arborescens*	*Trichoderma* spp.	Seed treatment

Chapter 6

AROMATIC CROPS

In recent years, there has been an increased interest in the cultivation of aromatic plants to meet the requirements of cosmetic, flavouring and perfumery industries. The aromatic plants provide raw material for the production of flavours, herbal cosmetics, perfumery etc. Aromatic plants and their products, particularly the essential oils, are now becoming one of the more important export items from many developing countries of Asia. Chemically essential oils are terpenes, which act as carriers of the aromatic substances. Most essential oils also contain camphors, and the more odiferous compounds present in them consist of oxygen derivatives of terpenes, alcohols, esters, aldehydes and ketones. India has enjoyed a prominent position in the manufacture of superior perfumes and aromatics by using essential oils. The important aromatic plants are lemon grass, vetiver, patchouli, palmarosa, citronella, mints, geranium, lavender, basil, jasmine etc.

In the world market, India stands in second position. Most of the essential oils produced though marketed within the country, but a sizeable amount has also been exported. Among the essential oils exported from India are Japanese mint oil, ambrette seed essence, sandal wood oil, citronella oil, lemon grass oil, palmarosa oil, ginger oil, clove oil, eucalyptus oil, vetiver oil, pepper oil, etc. The annual exports of the derivatives of aromatic plants are to the tune of Rs. 600-700 million.

6.1. JASMINE

Jasmine (*Jasminum* spp.) is one of the leading traditional flowers of India. Its flowers are used for making garlands, adorning hairs of women, in religious and ceremonial functions, and for producing perfumery oil. It is commercially cultivated in Karnataka, Tamil Nadu, Andhra Pradesh, Maharashtra, Gujarat, Haryana, Punjab, West Bengal, Rajasthan, Assam and Uttar Pradesh. Apart from internal trade, fresh flowers of jasmine are exported to Malaysia, Singapore and Sri Lanka.

6.1.1. Wilt, *Sclerotium rolfsii*

i) Symptoms: The young leaves turn yellow and the twigs start drying from tip downwards. The plants wilt in patches.

ii) Management

- Soil drenching with 1% Bordeaux mixture or 0.3% copper oxychloride reduced damage due to wilt.

6.1.2. Leaf Spot, *Cercospora jasminicola*

i) Symptoms: Many species of *Jasminum* are susceptible to this disease. The symptoms are noticed mainly on leaves as circular to oval necrotic spots which show dark brown to reddish borders. The under side of the spots show dark grey coloured growth of the pathogen. Severe infection often leads to heavy defoliation. Symptoms are sometimes observed on stems.

ii) Management

- The disease can be controlled by spraying 1% Bordeaux mixture or 0.1% benomyl or carbendazim at 15 to 20 days interval. The frequency of application may be increased during rainy season.

6.1.3. Rust, *Uromyces hobsoni*

i) Symptoms: This disease is prevalent in some species of *Jasminum* and is more common in South India. Symptoms are noticed on all the aerial parts of the plant including flowers. The earliest symptoms are yellowish to orange coloured spots on leaves, young twigs and floral buds. These spots enlarge and swell considerably. The lower side of the swollen region shows several orange yellow cup-like projections which contain orange yellow spores of the fungus. The infected leaves and young stems become greatly hypertrophied and distorted. Towards the end of the season, dark brown to black swollen structures (called telial galls) develop on areas which earlier showed orange yellow cup-like projections.

ii) Management

- The infected leaves and other plant parts should be collected and destroyed to reduce the inoculum for the next season.
- The disease can be controlled to some extent by spraying 0.2% mancozeb or zineb.

6.2. MINTS

Mints (*Mentha* spp.) are a group of aromatic herbs belonging to the family Lamiaceae, which are considered to be the most important cash crops in Indo-Gangetic plains. Cultivation of mints has been done in a large scale in many tropical and sub-tropical countries of the world including India, China, Brazil, Japan and USA. Indian farmer grow it as a bonus crop on an area of more than 1,50,000 ha in Central and Northern parts of Indo-Gangetic plains. The crop fit well in the traditional food based cropping system with other crops like paddy, wheat, potato, mustard, maize, okra, carrot, onion, spinach, pigeon pea, cowpea, etc. Similarly it can be grown as an intercrop with sugarcane and some legumes. In India, the highest area under commercial cultivation of mint has been in Uttar Pradesh (Terai region), but during last few years it has spread through out the country, especially in North and North-Western states including Madhya Pradesh.

Among different oil yielding species of mint, Japanese mint (*M. arvensis*) occupies the highest area under mint cultivation followed by Peppermint (*Mentha spicata*), Spear mint (*M. piperata*), Bergamot mint (*M. citrata*) and Scotch spearmint (*M. cardiaca*).

Japanese mint essential oil is used as a source of menthol, menthyl acetate, menthone and terpenes which finds wide use in medicine, perfumes, food and cosmetics. The oil of peppermint, which contains less menthol but has a sweeter aroma and taste, is used in tooth paste, chewing gums, candies, high grade liquors, medicines and other pharmaceutical preparations. Spearmint oil having carvone and limonene, is mainly used for flavouring; whereas, Bergamot mint oil is mainly used in perfumery and flavouring due to higher content of linayl acetate and linalool.

6.2.1. Powdery Mildew, *Erysiphe cichoracearum*

This disease was first reported by Ganguli and Pandotra (1962) on Japanese mint. It causes heavy toll in April-May leading to reduction in yield by 20% due to defoliation.

i) Symptoms: The important symptoms of the disease are; small chlorotic spots appearing on the surface and brownish discolouration on the corresponding lower surface prior to the appearance of powdery phase. Later, circular white powdery patches appear mostly on the upper portions of the surviving leaves. Heavy defoliation occurs in severe cases having a banded appearance on stem and branches. Later, these white patches turn to purplish black and produce large number of perithecia.

ii) Management: To control powdery mildew, remove infected plants and do not water from above.

6.2.2. Wilt, *Verticillium dahliae*

i) Symptoms: Another fungal wilt disease *Verticillium dahliae* is a major problem on mint crops. The typical symptoms of the disease are dwarfing, unilateral development of the branches, etiolation of leaves, wilting and ultimate death of the plants. The pathogen is soil-borne and spread through soil, roots and stolons used as seed material. None of the commercial varieties are resistant.

ii) Management : The disease can be controlled using crop rotation and fungicides.

6.2.3. Stolon Rot, *Macrophomina phaseolina (Rhizoctonia bataticola)*

The disease was reported on *Mentha cordiaca* by Green (1961) and on *M. arvensis* and *M. spicata* by Hussain and Janardhan (1965).

i) Symptoms: The important primary symptoms are yellowing of leaves and stunted growth of the plants. The wilting of the plants occurs in advance stages leading to the death of above-ground portion. Under-ground stolons exhibit pinkish brown lesions in the earlier stages and gradually turn into dark brown to black patches. These patches increase in size and finally result in the soft decay of a portion or the entire affected stolon.

ii) Management: Avoid excess irrigation. Uproot and burn the affected plants. Apply *Trichoderma* sp. to soil.

6.2.4. Sclorotium Rot, *Sclorotium rolfsii*

Ganguly and Pandotra (1964) reported this disease from India.

i) Symptoms: The characteristic symptoms of the disease are appearance of white mycelial strands around infected regions of the stem, which gradually become pale and the plant starts wilting. In later stages, sclerotial bodies develop and start girdling the stem which results in toppling of the plant at the collar region and finally the death of the infected plant.

ii) Management: Pandotra and Shastri (1968) reported that *Mentha spicata* var. *luciniata* and *M. arvensis* var. *piperascens* were able to tolerate the disease even after infection.

6.2.5. Rust, *Puccinia menthae*

The most significant disease problem encountered with mint is a rust fungus which, if left unchecked, will totally defoliate the plants. Scotch spearmint appears to be more sensitive to rust attack than the native spearmint, but both are prone to significant oil loss if the rust is left unchecked. Ganguly and Pandotra (1962) reported this disease from India. Although this disease is known to cause severe losses in other countries where this crop is grown extensively, but under Jammu conditions, this disease appears almost at the end of the crop season and causes comparatively very little or no damage to the standing crop. This disease mainly causes defoliation which ultimately reduces the yield of mint oil.

i) Symptoms: Brown, circular pustules appear on the undersides of leaves followed by leaf yellowing and defoliation. A yellow spore stage (uredospores) is in reddish blisters on young shoots in spring, and a black spore stage (teliospores) is on stems and regrowth in fall (Fig. 6.1). Telial stage is characterized by swelling on the upper portion of the stem. Swollen shoots with elongated, chlorotic internodes are associated with systemic infection.

Fig. 6.1. Lesions of peppermint rust on the underside of mint leaf.

Puccinia menthae over winters on mint stubble and on wild and escaped mint. It completes its complex life cycle on one host (autoecious) and is macrocyclic (five spore stages), of which the aecial stage can be systemic in plants.

ii) Management

- Rabbing (flaming) in April 1 to May 15 is absolutely necessary for rust control.
- Early harvest (before leaves drop due to rust) may limit rust damage.
- Spray chlorothalonil or pycroclostrobin or propiconazole. Begin applications when plants are 5 to 10 cm high or when conditions become favourable for disease. Make second application 10 to 14 days after the first application. Do not apply within 90 days of harvest.
- The disease can be controlled using crop rotation.
- Removal and destruction of disease affected plants.

6.2.6. Leaf Blight, *Alternaria* sp.

i) Symptoms: The main symptoms of the disease are appearance of the round to oval or slightly irregular dark brown spots on the upper surface of the leaves with concentric zones surrounding the pale yellow margin. Later, these spots coalesce, enlarge and occupy the entire leaf or a portion of the leaf which leads to the dropping of leaves.

ii) Management: To control leaf blight, remove infected plants and try reducing soil moisture.

6.2.7. Leaf Spot, *Curvularia lunata*

Thakur *et al.* (1974) reported this disease on Japanese mint.

i) Symptoms: The typical symptoms appear in the form of small dirty brownish spots, scattered all over the leaves in initial stages. These spots are very minute, which increase in their size later on and form big spherical or irregular patches. In the advance stage, the spots coalesce and form large dark brown patches of irregular size. The symptoms of the disease appear at all stages of the crop growth of leaves, but the bigger spots generally appear on the older and lower leaves. The remaining portion of the infected leaves become chlorotic and finally wither away.

ii) Management: To control leaf spot, remove infected plants and try reducing soil moisture.

6.3. BASIL

Among several species of basils, sweet basil (*Ocimum basilicum*) is considered to be the most important crop for its high quality of essential oil. Major components of its oil are linalool (43-50%), menthyl chavicol (18-33%), eugenol and isoeugenol (5-6%) and the minor constituents are alpha and beta pinene, camphor, geraniol, etc. The oil of sweet basil owes its importance to its extensive use in condimentary products, cosmetics, toiletry, perfumery and confectionary industries, particularly in European countries.

6.3.1. *Fusarium* Wilt, *Fusarium oxysporum* f. sp. *basilicum*

i) Symptoms: A devastating basil wilt disease is caused by a soil-borne pathogenic fungus, *Fusarium oxysporum* f.sp. *basilicum*. *Fusarium* sp. causes a major production-limiting disease in commercial farms. Plants infected with this disease usually grow normally until they are 15 to 30 cm tall, then they become stunted and suddenly wilt. Early signs of this fungal disease include slow growth and yellowing of the young shoots. Advanced symptoms include wilting, die-back, and discoloured stems. Infected plants may show a darkened discolouration of the stem beneath the bark. Longitudinal sections of the stem will reveal this symptom. Under moist conditions, a light white to pink, cottony, growth can be observed along the infected stem areas. The pathogen is seed-borne and survives in the soil for many years. Once a field has become infested with the *Fusarium* wilt pathogen, infective propagules may persist in the soil for 8-12 years. Interestingly, only sweet basil is affected.

ii) Management : Soil fumigation with methyl bromide before planting is effective in controlling the disease.

Control can be achieved by using clean seed and fresh potting soil. Seed can be disinfested by soaking in 10% household bleach (1 part bleach to 9 parts water) for 1 minute, thoroughly rinsing with tap water, and planting immediately. The seed should be soaked in cold water for four hours followed by a heat treatment of 20 minutes in 133-136° F.

Outdoor plantings should be rotated out of basil. Growers should avoid growing sweet basil or members of the mint family. Mints will not exhibit symptoms of the disease but may carry over the inoculum from year to year.

The spread of *Fusarium* can be reduced by quick removal of all diseased plants and leaves. Avoid planting in wet, poorly drained areas.

The basil cultivar "Nufar" has shown some resistance. Some of the specialty basils, such as lemon basil and purple basil, show some resistance to the disease.

6.3.2. *Colletotrichum* Leaf Spot, *Colletotrichum* sp.

i) Symptoms: Dark spots form on leaves and the dead tissue within leaf spots may drop out, causing a shot-hole symptom. The disease can cause defoliation, tip dieback, stem lesions, and sometimes loss of entire plants. Spores are water-splashed from diseased tissue.

ii) Management: Sow seeds in sterile containers in sterile soil. Since wet conditions favour disease development, reduce leaf wetness periods by reducing humidity and increase plant spacing to increase air movement. Avoid overhead irrigation. Remove diseased plants to reduce inoculum levels.

6.3.3. Damping-Off, *Rhizoctonia solani*, *Pythium* spp.

i) Symptoms: Seedlings either rot and do not germinate or germinate and collapse. The disease frequently attacks the base of the seedling and leaves a shriveled dried-up stem at the base of the seedling.

Symptoms appear as wilting and a slow or rapid collapse of the plant. The roots can appear brown and water-soaked instead of white. A water-soaked lesion can often appear at the base of the stem.

ii) Management: Control can be achieved by using sterile potting media, clean pots, and fungicide-treated seed. If seeds in a seed tray become symptomatic, discard all of the seeds since asymptomatic seeds may be infected. Affected plants should be rouged out and discarded.

6.3.4. *Botrytis* Blight, *Botrytis cinerea*

i) Symptoms: *Botrytis* blight occurs on many different kinds of herbs. The disease appears when plants are grown under conditions of high humidity, such as in a terrarium. The disease appears as tan to brown spots on leaves that can be associated with a grayish mold colonizing the damaged area. Senescing flowers are particularly susceptible and frequently have papery spots on them.

ii) Management: *Botrytis* blight can be suppressed by avoiding overhead irrigation and watering in the morning to allow time for leaves and flowers to dry. Proper plant spacing can improve air drainage and rapid leaf drying. Control can also be achieved with the use of fungicide sprays (neem oil and thiophanate-methyl) applied as soon as symptoms are visible.

6.3.5. *Cercospora* Leaf Spot, *Cercospora canescens*

i) Symptoms: The disease causes characteristic leaf spot on both surfaces which are brownish black on the upper surface and light grey on the lower surface. A large number of spots may form on a single leaf which often coalesce and become necrotic. The infected leaves dry up and there is heavy defoliation.

ii) Management: Spraying of 0.1% benomyl or carbendazim at fortnightly intervals is effective in controlling the disease.

6.3.6. *Glomerella* Leaf Spot, *Glomerella cingulata*

i) Symptoms: The infection starts from the upper surface of leaves usually from the leaf margins and tips. At first, the spots are small and irregular. Gradually infection spreads to cover the entire leaf surface. The leaf spots coalesce and become irregular large patches of dark brown colour. This results in heavy defoliation.

ii) Management: Spraying of 0.1% benomyl or carbendazim at fortnightly intervals is effective in controlling the disease.

6.4. LEMON GRASS

Lemon grass (*Cymbopogon flexuosus*), West Indian lemon grass (*Cymbopogon citratus*) and Jammu lemon grass (*Cymbopogon pendulus*) are grown in Assam, Kerala, Maharashtra, Tamil Nadu and parts of Uttar Pradesh. The chief constituent of palmarosa oil is Citral. Citral is the starting material for preparation of α-Ionone (used in flavours, cosmetics and perfumes) and β-Ionone (used in the manufacture of synthetic vitamin A).

6.4.1. Leaf Blotch, *Curvularia* spp.

Barua and Bordoloi (1983) reported the presence of *Curvularia verruciformis* to occur on lemon grass. *Curvularia andropogonis* causing severe leaf blight disease has been reported on lemon grass in different parts of India during rainy season (Alam and Hussain, 1976; Alam *et al.*, 1986). A leaf spot caused by *Curvularia trifolii* was found to be severe on some lemon grass varieties at Lucknow. The pathogen causing leaf blotch (*Curvularia* sp.) can also infect *Cymbopogan martini* var. *motia, C. pendulus* and *C. winterianus* (Gupta *et al.*, 1977).

i) Symptoms: The disease is characterized by the appearance of minute, circular, reddish-brown spots, mostly on the leaf margin and tips. These spots later enlarge and coalesce to form longitudinal, elongated, reddish-brown, necrotic lesions resulting in premature drying of leaves. Older leaves are more susceptible for infection. Lenticular spots have a distinct dark brown border. The size of the spots varied from 1-8 mm in length and 0.5 to 3.0 mm in width. The spots often coalesce and form blotches.

ii) Management: Spraying of 0.2-0.3% mancozeb at 15 days interval is effective in managing the disease.

6.4.2. *Helminthosporium* Leaf Spot, *Drechslera sacchari* (*Helminthosporium sacchari*)

i) Symptoms: The disease is characterized by the appearance of oval-shaped brown spots with pale centres and dark purple borders which in advanced stages, coalesce to form necrotic blight areas.

ii) Management: Spraying of 0.2-0.3% mancozeb is effective in managing the disease.

6.4.3. Red Leaf Spot, *Colletotrichum graminicola*

i) Symptoms: The disease appears on lower leaves in the rainy season and forms small brownish spots, scattered all over the ventral surface of the leaves. Under favourable conditions, the disease also infects the leaf sheaths and midrib. As the disease advances, many such spots may coalesce to form bigger patch leading to drying of the leaves. Later, the spots show concentric rings in the centre. Under warm and humid conditions, enormous black fruiting bodies of the pathogen are present on older lesions.

The pathogen perpetuates in the remaining portions of the plants (stumps) on the field from one season to another.

ii) Management: Two sprays of carbendazim at 0.1% or 3 sprays of maneb at 0.3% starting from first appearance of disease at 20 days or 10-12 days intervals, respectively gives effective control of the disease.

6.4.4. Rust, *Puccinia nakanshikii*

i) Symptoms: Rust disease on lemon grass was reported from Sri Lanka (Bandare, 1981) and Hawaii (Gardener, 1985). Mildly and moderately infected leaves show discrete, linear, cinnamon brown uredinia on the abaxial surfaces, associated with chlorotic streaks. Severe infection causes extensive necrosis corresponding with coalesced uredinia. Telia also occur on severely infected leaves. No aecial host is known.

Varma (1987) reported the presence of rust caused by *Puccinia purpurea* from Kerala.

ii) Management: Uproot and destroy the disease affected plants.

6.4.5. Little Leaf/Grassy Shoot, *Balansia sclerotica*

This disease is now becoming more serious on lemon grass and palmarosa as reported from Madhya Pradesh (Indore), Kerala and Uttar Pradesh (Lucknow). It is being a floral disease; the oil yield is adversely affected because the inflorescence consists of higher percentage of oil. This disease also affects seed production.

i) Symptoms: The characteristic symptoms are little leaf formation and stunting of the plants. In place of normal inflorescence, proliferated, small, leafy structures are produced. The symptoms appear like viral/mycoplasma infection.

ii) Management: Uproot and destroy the disease affected plants.

6.4.6. Smut, *Tolyposporium christensenii*

The disease was reported on lemon grass from India by Raghunath (1965). It is occurring in severe form in Kerala on lemon grass and on palmarosa in Madhya Pradesh (Indore). This is also an inflorescence disease which adversely affects the oil yield and seed production.

i) Symptoms: The symptoms are characterized by the transformation of seed into the smut sori which are slightly long and covered by dull yellow membrane.

ii) Management: Uproot and destroy the disease affected plants.

6.5. PATCHOULI

Patchouli (*Pogostemon patchouli*) is a highly aromatic bushy under-shrub. It is cultivated for its highly fragrant leaves which contain a very sweet smelling oil of lasting sticky odour. The crop is grown in small pockets in Karnataka, Kerala and Tamil Nadu. It has a very characteristic aroma and blends well with other essential oils. The oil is used in very low concentration (2 ppm) in scenting soaps, cosmetics, after-shave lotions, detergents and many fancy products. The oil is also used to flavour foods, beverages, candy and baked products. In combination with sandal wood oil, it is used in blending tobacco and making incense sticks.

6.5.1. Wilt, *Rhizoctonia* spp.

i) Symptoms: The disease is characterized by sudden wilting of plants. Infected plants are found to have a brown discolouration at the collar region and the main roots below. The disease is very severe under wet or moist conditions of soil.

ii) Management: Soil drenching with 0.2% PCNB (Brassicol) at the base of the infected plants helps to reduce the disease incidence.

6.6. SCENTED GERANIUM

Scented geranium (*Pelargonium graveolens*) is an aromatic perennial herb. It is grown in cooler, sub-tropical climate of Mysore and Bangalore (Karnataka), and extended to Hyderabad (Andhra Pradesh). Its oil has refreshingly delicate long-lasting rosy odour with fruity under-note, containing 66-78% primary alcohols (rhodinol). It is used in manufacturing of perfumes, creams, talcum powder and body lotion. It is stable in alkaline medium and is therefore used for scenting soaps.

6.6.1. Botrytis Blight, *Botrytis cinerea*

i) Symptoms: Botrytis gray mold is caused by the fungus *Botrytis cinerea*. Symptoms can develop at any stage and on any plant part (Fig. 6.2). Under conditions of high moisture and humidity, infected plant tissues and debris develop the diagnostic gray, fuzzy growth characteristic of the fungus *Botrytis cinerea*. The symptoms on leaves vary from distinct spots or lesions to large dead areas, often with concentric rings. V-shaped lesions can also develop. Lesions often occur when spent flower petals fall onto the leaves. Symptoms on stems appear at the base of plants as light- to dark-brown lesions. These lesions grow until they rot the entire stem. Plants can develop brown lesions after cuttings are taken. Flowers can turn brown and drop prematurely during periods of high moisture and relative humidity. Senescing flowers are covered with a gray, fuzzy mass of the fungus. Botrytis on the flower of a Geranium will greatly reduce the appearance of the flower and will result in lower sales for the grower.

The fungus can survive in the field year round as conidia, mycelium, or sclerotia on living or dead plant material. The fungus is very active and rapidly reproduces when relative humidity is at or above 85%, with little or no air circulation. The optimum air temperature for germination is between 72 and 77°F. The fungus penetrates the living tissue through wounds in the plants from broken stems, damaged leaves from transport, or from cuttings that were harvested.

Fig. 6.2. *Botrytis* blight symptoms on geranium leaves and flowers.

ii) Management

- Practice good sanitation by pulling weeds and removing plant debris.
- Use soil amendments to develop a well-drained soil.
- Use disease-free plants.
- Avoid overhead irrigation and water early in the day and not later than 11 a.m.
- Use adequate plant spacing to promote good air circulation.
- Remove diseased plants and plant parts. Remove old flowers and senescing leaves.
- Fungicides are available for chemical control.

The easiest way to control this disease is through cultural practices. These practices include cleaning up waste around the growing areas including leaves, soil, and removal of excess water that adds to relative humidity. By improving air circulation with Horizontal Air Flow (HAF) fans, the leaves of the plants will dry faster and not be as susceptible to infestation. Chemical treatments of Daconil are effective but should be used only when needed. Over use of these chemicals can possibly lead to resistance by the fungus.

6.6.2. Black Leg/Root Rot, *Pythium* sp.

The Black Leg disease is caused by the fungal pathogen *Pythium* sp. *Pythium* is very commonly found in greenhouses where bedding plants are grown. *Pythium* is normally associated with various root rots of poinsettias, petunias, impatiens, and geraniums. In geraniums, *Pythium* can infect the roots when the media conditions are extremely saturated. The fungus initially attacks the root tips and works its way to the crown of the plant through the vascular system. Black Leg can attack at any stage of development. When the symptoms appear it is usually too late and many healthy roots are dead by the time the problem is noticed.

i) Symptoms: Root and stem rots cause yellowing, stunting and wilting of the above-ground portions of the plant. Plants often have a noticeable brown to black lesion at the soil line, since the fungi attack and girdle the plant. Under wet, humid conditions, a gray or off-white webby growth can be seen on the infected plants (Fig. 6.3).

Fig. 6.3. Black leg on geranium.

ii) Management : Improve soil drainage, use soil amendments and avoid standing water. Water management is essential to control root rots.

- Avoid heavy use of nitrogen fertilizer.
- Use disease-free plants.
- Remove diseased plants.
- Apply fungicides as soil drenches.

6.6.3. Leaf Rust, *Puccinia pelargonii-zonalis*

i) Symptoms: Leaf rust, caused by the fungal pathogen *Puccinia pelargonii-zonalis,* is characterized by small yellow spots on the upper surface of the leaf and rust-colored pustules that develop in the spots on the underside of the leaf. The pustules break open and release spores. Spores are often formed in a circular fashion around the original pustule. When infection is severe, leaves become yellow and drop prematurely. Pustules containing the rust-colored spores can be seen on stems, petioles and stipules. The disease is more common in greenhouse production, but it is a potentially important disease in the landscape. (Fig. 6.4)

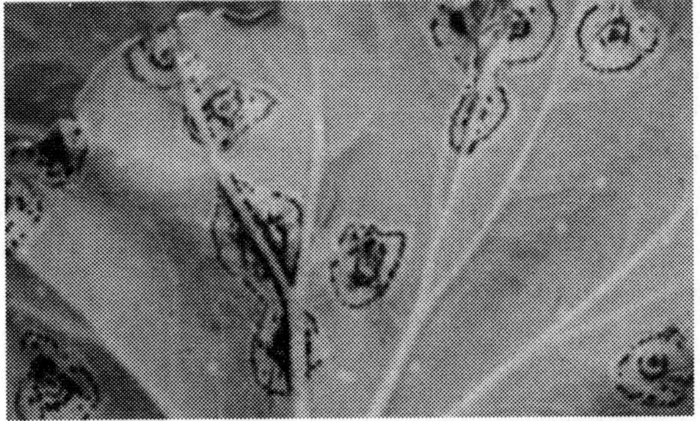

Fig. 4.7. Leaf rust symptoms on Geranium.

ii) Management

- Practice good sanitation by pulling weeds and removing plant debris; use amendments to create a well-drained soil.
- Use disease-free plants.
- Avoid overhead irrigation.
- Remove diseased plants.
- Apply fungicides as soil drenches.

6.6.4. *Alternaria* and *Cercospora* Leaf Spots

i) Symptoms: *Alternaria* leaf spots are small and blister-like, and appear on the undersides of older leaves. Mature spots become sunken and brown with diffuse yellow halos. If conditions remain suitable for disease development, the spots are apparent on upper leaf surfaces as well, and lesions show concentric rings. Chlorosis and defoliation occur when infection is severe.

Cercospora leaf spots appear as small, pale, sunken lesions that later turn gray and produce many spores. This gives an appearance of dark, raised centres. Cholorosis may be seen close to the lesions, and defoliation also occurs when infection is severe.

ii) Management

- Remove and discard infected leaves and plant debris.
- Do not let the foliage stay wet very long.
- Water before 11 a.m.
- Thin annuals and prune perennials selectively to improve air circulation and increase light penetration.
- Fungicides are available for chemical control.

- Three to four sprays with 0.2% captafol or 0.3% chlorothalanil or 0.1% ridomil at 12-15 days interval gives effective control of the disease.

6.6.5. Wilt, *Fusarium oxysporum* f. sp. *redolens*

i) Symptoms: A slow wilting caused by *Fusarium* is a very severe disease of scented geranium. The infection seems to take place through the cracks on the lower part of the stem or branches at or near the soil level. The disease is characterized by wilting of side branches of the plant which slowly proceed inwards finally resulting in the death of the entire plant. The leaves and stem of the infected plants become brown and there is rotting of plant at collar region which proceeds downwards involving the roots. The disease is found to be more severe during rainy months.

ii) Management: Avoid water logging and planting in areas where this disease already exists. Treat the rooted cuttings with 0.3% mancozeb before planting. Drenching of the soil at the base of the affected plant with 0.1% benomyl, carbendazim or captan helps in reducing the disease incidence.

6.7. CITRONELLA

Citronella (*Cymbopogon winterianus*) is a large, perennial, stemless, aromatic grass. Its oil from leaves has a strong characteristic lemonic odour extensively used in scenting soaps, cosmetics, producing deodorants and mosquito repellent products. The oil is used as starting material to produce pure citronellol, geraniol and related upstream high-value aroma chemicals used in perfumery industry. It is cultivated in 6,000-10,000 ha land in high rainfall tracts of Assam, Arunachal Pradesh, Meghalaya, Nagaland and Manipur. It is also cultivated on small scale in Karnataka, coastal Andhra Pradesh and Orissa.

6.7.1. Leaf Blight/Blotch, *Curvularia andropogonis*

The pathogen attacks citronella at Arunachal Pradesh, Manipur, Meghalaya, Nagaland and other citronella growing areas (Gupta *et al.*, 1977).

The pathogen which causes severe leaf blight on Jawa citronella was also found on lemon grass in different parts of India during rainy season (Alam and Husain, 1976; Alam *et al.*, 1986).

i) Symptoms: The fungus affects the crop in the beginning of monsoon. The disease appears as minute, circular, reddish-brown spots, mostly on the leaf margin and tips during the month of July. These spots later enlarge and coalesce to form longitudinal, elongated, reddish-brown, necrotic lesions resulting in premature drying of leaves. Older leaves are more susceptible for infection. In case of serious infections, the entire leaves dry up resulting in considerable decrease in leaf and oil production. Black dots in the form of sporulation of the fungus are observed on dead spots of the leaves.

Alam *et al.* (1986) reported that two strains of *Curvularia andropogonis* probably appear in nature, one causing leaf spot (strain CLS) and another causing leaf blight (strain CLB). Strain CLS infects all the four species of *Cymbopogan* severely, whereas strain CLB was severe on *C. winterianus* and *C. flexuosus* and do not infect *C. citratus* and *C. martini* var. *motia.*

ii) Management: The incidence of the disease can be reduced significantly in the field by prophylactic spraying of 0.3% mancozeb or zineb or 0.1% ridomil at an interval of 15 days.

6.7.2. Lethal Yellowing, *Pythium aphanidermatum*

i) Management: The disease could be easily managed under field conditions by slip treatment and three foliar sprays either with 0.1% ridomil-mancozeb or previcur-N.

6.7.3. Collar Rot and Wilt, *Fusarium moniliforme*

i) Management: Foliar spraying with carbendazim or Topsin-M each at 0.2% gives highly effective control of the disease.

6.8. PALMAROSA

Palmarosa (*Cymbopogon martinni* var. *motia*), also called as Rusa, Rusha and Rosha, is grown in Andhra Pradesh, Madhya Pradesh, Maharashtra, Karnataka, Tamil Nadu and Uttar Pradesh. The major trading centres include Amravati, Akola and Elichpur in Maharashtra; Khandwa and Betul in Madhya Pradesh; Zaheerabad and Mehboobnagar in Andhra Pradesh. Oil of palmarosa is used for fine perfumery and valued for its geraniol content. Oil imparts rose like aroma to wide variety of soaps and tobacco products. Oil contains 70-80% geraniol.

6.8.1. Leaf Spot, *Colletotrichum graminicola*

This disease was first recorded on palmarosa in India by Patwardhan (1963) from Pune.

i) Symptoms: The disease appears on lower leaves in the rainy season and forms small brownish spots, scattered all over the ventral surface of the leaves. Under favourable conditions, the disease also infests the leaf sheaths and midrib. As the disease advances, many such spots may coalesce to form bigger patch leading to drying of the leaves. Later, the spots show concentric rings in the centre. Under warm and humid conditions, enormous black fruiting bodies of the pathogen are present on older lesions.

The pathogen perpetuates in the remaining portions of the plants (stumps) on the field from one season to another.

ii) Management: Two sprays of carbendazim at 0.1% or 3 sprays of maneb at 0.3% starting from first appearance of disease at 20 days or 10-12 days intervals, respectively gives effective control of the disease.

6.8.2. Leaf Blotch, *Curvularia andropogonis, C. trifolii*

This disease was first reported by Upadhyaya *et al.* (1975) from Jorhat, Assam. It was reported to occur very severely under favourable climatic conditions of North-Eastern region of India. This pathogen not only infects palmarosa, but also *Cymbopogan winterianus* (citronella), *C. flexuosus* (lemon grass) and *C. pendulus* in Arunachal Pradesh, Manipur, Meghalaya, Nagaland, Uttar Pradesh, Madhya Pradesh, Bihar, Karnataka and Jammu and Kashmir (Gupta *et al.*, 1977). Janardhan *et al.* (1980) reported that the disease causes more than 31% total oil and 17.8% geraniol loss in palmarosa. The disease occurs in epiphytotic form during August and October.

i) Symptoms: The disease appears as minute, circular, reddish-brown spots, mostly on the leaf margin and tips. These spots later enlarge and coalesce to form longitudinal, elongated, reddish-brown, necrotic lesions resulting in premature drying of leaves. Older leaves are more susceptible for infection.

ii) Management: The disease can be effectively controlled by foliar spraying of 0.3% mancozeb at 15 days interval before the onset of the disease or even during initial stages of infection.

6.8.3. Leaf Blight, *Ellisiella caudata*

i) Symptoms: Leaf blight is one of the serious diseases of palmarosa. The disease appears in epiphytotic form during rainy season and causes considerable loss in the production of herbage and essential oil yield. Initial symptoms appeared in the form of small grey coloured necrotic spots all over the surface of infected leaves. In severe stage of infection, the necrotic lesions increase in size and coalesce to form big patches leading to premature drying of infected leaves. The fungus produces masses of spores on the dried necrotic lesions.

ii) Management: The disease can be effectively controlled by foliar spraying with captafol or chlorothalanil each at 0.3% at 15 days interval.

6.9. VETIVER

Vetiver (*Vetiveria zizanoides, V. lawsoni*) is grown in Madurai, Thanjavur, Tirunelveli and Trichy districts of Tamil Nadu; Kurnool and East Godavari districts of Andhra Pradesh; Palghat, Kozikode, Thrissur and Wayanad districts of Kerala. Cultivation is picking up in Madhya Pradesh, Uttar Pradesh and Haryana. Oil of vetiver is used in soap making. It is also used for making aromatic mats, brooms, screens, pillows and mattresses. Vetiver is used in soil conservation. It has deep, dense and strong fibrous root system. Grass is unpalatable to cattle. It is planted on the edges of field bunds and banks of major irrigation canals to prevent soil erosion.

Vetiver is subjected to a few diseases only due to its sturdiness.

6.9.1. Leaf Blight, *Curvularia trifolii*

i) Symptoms: A leaf blight disease was reported from India (Sarwar and Shrinath, 1965). Tan to dark, oval spots appear on the leaves later turning black as the leaves turn pale yellow. The spots become necrotic and coalesce to form blighted appearance. Leaves become yellow and gradually dry out.

ii) Management: Spray copper oxy chloride at 3 g /litre of water twice or thrice depending on the intensity of disease.

Table 6.1. Management of fungal diseases of aromatic crops

Crop	Disease/ Causal organism	Symptoms	Control
Ocimum	Cercospora leaf spot, *Cercospora oscemi*	Chocolate coloured brown spots on leaves, petioles and flower buds	Spraying with thiophanate methyl/bitertanol/carbendazim-1 g/l during rainy season and in Nov-Dec.
Lemon grass	Red leaf spot, *Colletotrichum graminicola*	Brown spots with concentric rings in the centre appearing on the lower surface of the leaves. The spots may be formed on leaf sheaths and midrib. Later the spots merge to form bigger patches and the affected leaves dry away.	Two sprays of 0.1% carbendazim just after the appearance of the disease at an interval of 20 days. Three sprays of 0.2% mancozeb at an interval of 10-12 days.
	Leaf blight, *Curvularia andropogonis*	Minute, circular, reddish brown spots mostly on the margins and tip of the leaves which merge to form elongated reddish brown necrotic lesions resulting in premature drying of leaves, older leaves are more susceptible to infection.	Spraying 0.2% zineb or 0.3% copper oxychloride at an interval of 15 days.
	Little leaf or grassy shoot, *Balansia sclerotica*	Stunted growth. Little leaf formation in place of normal inflorescence.	Spraying 0.3% zineb just before flowering stage at an interval of 10-12 days. Use of fresh seedlings for plantation and crop rotation.
Scented geranium	Alternaria Leaf Spot, *Alternaria tenuis*	Water-soaked spots on underside of leaf enlarge to 1/4-1/2 inch. Concentric rings of dark tissue form in spots.	Avoid overhead watering. Remove and destroy crop debris from benches, pots, and walkways. Apply azoxystrobin, copper hydroxide, chlorothalonil, fludioxonil, or mancozeb.

Blackleg, *Pythium* sp.	Cuttings first develop a brown water-soaked rot at the base. The coal black rot proceeds 3-4 inches up stem and kills the cutting rapidly.	Use sterile propagation media. Discard infected cuttings since affected rooted cuttings later develop root rot (see Pythium Root Rot). Keep hose ends off the ground.
Botrytis Blight, *Botrytis cinerea*	Flowers turn dark and fall prematurely. Where infected petals land on leaves, an irregular leaf spot forms. Cutting stubs have a brown rot that may extend down stem. Cuttings develop a dark brown rot near the base.	Avoid damaging plants in any way. Do not leave large cutting stubs. Remove and destroy fading flowers and leaves. Heat, ventilate, and space plants to ensure good air circulation and low humidity. Apply a mixture of chlorothalonil and mancozeb. Use copper hydroxide, fenhexamid, fludioxonil, or iprodione alone only once as a final clean up of the crop and only if disease pressure is low.
Pythium Root Rot, *Pythium* sp.	Root tips appear translucent and water-soaked and contain microscopic spores. The outer layers of root tissue strip off when pulled from soil leaving the central core of vascular tissue bare. The fungus may invade the lower stem.	Pot in sterile media. Keep hose ends off the ground. Discard blackleg-infected cuttings even if well rooted. Apply fosetyl-AL, mefenoxam, etridiazole, etridiazole + thiophanate methyl, metalaxyl, or propamocarb.
Rust, *Puccinia pelargonii-zonalis*	Chlorotic specks on upper leaf surface appear directly opposite pustules of rust-colored spores on underside of leaf. Spores erupt in concentric rings forming a "target" spot.	Purchase rust-free cuttings. Do not irrigate overhead. Discard unwanted geraniums at season's end. Apply chlorothalonil, myclobutanil, flutolanil, oxycarboxin, or triadimefon to protect plants.
Verticillium Wilt, *Verticillium albo-atrum* or *V. dahliae*	Middle and upper leaves collapse, dry, and fall. Vascular tissue of affected stems is browned. Symptoms are readily confused with those of bacterial blight.	Purchase culture-indexed cuttings. Use sterile potting mix. Destroy infected plants.

Palmarosa, (*Cymbopogon* spp.)	Red leaf spot	*Collectorichum graminicola*	Two sprays of carbendazim (0.1%) or 3 sprays of maneb (0.3%) starting from first appearance of disease at 20 days or 10-12 days intervals, respectively.
Scented rose (*Rosa damascena*)	Die-back	*Diplodia rosarum*	Pruned ends of the shoots should be treated with copper fungicides.
Ambrette seeds (*Abelmoschus moschatus*)	Anthracnose	*Collectotrichum hibiscicum*	Seed treatment with contact fungicides and spray of Bordeaux mixture are recommended.

Table 6.2. *Trichoderma* spp. and other antagonistic fungi for the management of diseases of aromatic crops

Crop	Disease / Pathogen/s	Biocontrol agent	Mode of application
Jasmine	Root rot, *Macrophomina phaseolina*	*T. viride, T. harzianum*	Cutting treatment
Chinese Rose	Wilt, *Fusarium oxysporum*	*Aspergillus niger*	Soil treatment
Menthol Mint	Stolon decay, *Sclerotinia sclerotiorum*	*T. harzianum, T. virens*	Sucker treatment

Chapter 7

TUBER CROPS

Tuber crops form an important part of the diet in many areas of the tropics and subtropics. They are most important after cereals and grain legumes. They are mostly grown for the carbohydrates stored in underground storage organs viz., roots or rhizomes. In addition, the leaves or leaf stalks of some of the tuber crops are rich in protein, vitamins and minerals and are used for either human consumption or cattle feed. Tuber crops are a major source of food and calories in many tropical countries. The annual tropical root crop production (sweet potato and cassava) is in the range of 8.921 million metric tonnes in 0.367 million ha with productivity of 20.1 tonnes/ha (Table 7.1). Root and tuber crops are predominantly grown in the upland or rain fed areas, mainly by resource poor small and marginal farmers. The major tuber crops include cassava (tapioca), sweet potato, colocasia, amorphophyllus, yams and coleus. Cassava and sweet potato are grown to a larger extent in southern and eastern India. Their cultivation has spread to Maharashtra, Gujarat and North-Eastern states as well. The Salem belt in Tamil Nadu and the Samalkot belt in Andhra Pradesh are known for cassava as an industrial crop. Yams and Colocasia are popular in Andhra Pradesh, Tamil Nadu, West Bengal, Uttar Pradesh and Orissa states.

Table 7.1. Area, production and productivity of tuber crops in India (2005-06)

Tuber crop	Area ('000 ha)	Production ('000 tonnes)	Productivity (tonnes/ha)
Sweet potato	122.8	1066.5	8.9
Cassava (Tapioca)	244.6	7854.9	31.4
Total	367.4	8921.4	20.1

(Source: Indian Horticulture Database 2006)

Tuber crops, especially cassava can be successfully substituted as alternative source of feed for livestock in both tropical and subtropical countries. Cassava has also been used as a substrate for the production of ethyl alcohol. The tropical root crops, in general, have a great potential in meeting basic food and energy needs of the developing world. Tuber crops are very efficient in solar energy transfer, having a clear superiority over cereals in biological efficiency and are comparatively free from severe pests and diseases. Cassava and sweet potatoes account for about 30% of the total production of root crops from developing countries. The potential of root crops as nutritionally rich sources of ß-carotene, anti-oxidants, dietary fibre and minerals like calcium have begun to be recognized as a result of the multifarious research programmes worldwide.

7.1. YAM

Yam (*Dioscorea* spp.) belongs to the family Dioscoreaceae, which is wide-spread throughout the tropics and subtropics. Normally yams are consumed as boiled, baked or fried vegetables. They contain 18-20% starch with mucilaginous substance. Yam also contains alkaloids, tannins and steroids which have pharmaceutical value. *Dioscorea alata* (greater yam), *D. rotundata* (white or white Guinea yam), *D. cayensis* (yellow Guinea yam) and *D. esculenta* (lesser yam) produce edible tubers. *D. alata* was first cultivated in S.E. Asia and is widely grown in Asia and the Pacific, is less favoured in Africa, but much grown in the Caribbean. *D. rotundata* is the most important species in W. Africa, where it is originated. *D. cayenensis* is a native of W. Africa, where it is extensively cultivated. *D. esculenta* originated in Indo-China and is extensively grown in S.E. Asia and elsewhere in the tropics.

In India, greater yam and lesser yam are under cultivation. Madhya Pradesh, North-Eastern states, West Bengal, Bihar, Orissa, Uttar Pradesh, Kerala, Tamil Nadu, Gujarat and Maharashtra are major yam-growing states.

7.1.1. Anthracnose, *Glomerella cingulata*

i) Distribution: The disease is distributed in Caribbean, East Africa, Far East, India and West Africa.

ii) Symptoms: In *Dioscorea alata*, the first signs of infection are small brown spots (2-5 mm in diameter), sometimes with a yellow edge, which appear mainly on the lower leaves (Fig. 7.1) and stems. The spots enlarge and spread giving the stem a blackened appearance and causing leaves to wither. Brown to black microscopic acervuli (fruiting bodies) may be observed on the lesions of the leaves and stem, particularly during the rains, giving the appearance of dark concentric rings on the lesions. Sometimes secondary infection occurs and the leaf is destroyed.

Stems are also attacked, but less frequently. If the infection reaches stele, death of the plant can result. Attacks on the stem generally occur as a result of direct penetration near the terminal bud which leads to rapid death of the plant. Direct infection can also occur at ground level.

The reduction in yield caused by anthracnose can be particularly serious in the more humid regions. *D. alata* is more susceptible but *D. esculenta* is more resistant. *D. cayenensis* and *D. rotundata* are only slightly susceptible, even in humid forest regions, their leaves only develop a few brown spots and stems are not attacked.

Fig. 7.1. Anthracnose on yam leaves.

iii) Development and Spread: The pathogen over seasons on crop debris as acervuli. Infection is favoured by wet and humid conditions. Many acervuli of the *Gloeosporium* type which become erumpent later are produced on lesions. Acervuli remain devoid of setae when humidity is over 95% but at lower humidity sterile setae mixed with conidiophores convert the acervuli to the *Colletotrichum* type. In very dry conditions, only setae are formed. Spores are mainly dispersed by rain splash and the disease is therefore most severe when there are heavy rains.

iv) Management: Crop debris should be removed as it is a major source of inoculum. Spraying maneb at 3.32 g a.i./litre at 10 days interval is effective. Benomyl alone or applied alternately with mancozeb on a weekly cycle has also been shown to give good control and increase yields. Benomyl + propineb are also effective. Zineb, ferbam and ziram sprays were shown to give 50%, 41% and 30% yield increases, respectively.

7.1.2. Leaf Spots, *Cercospora brasiliensis, C. carbonacea, C. pachyderma, Phaeoramularia dioscoreae, Pseudocercospora contraria*

i) Distribution: The disease is worldwide in distribution on all species of yams.

ii) Symptoms: Several *Cercospora* spp. and some species which were formerly classified as *Cercospora* are recorded as causal agents of leaf spots on yams. Generally the symptoms are brownish spots which darken with age and coalesce as the disease progresses. *Phaeoramularia dioscoreae* causes small brown spots which sometimes coalesce. *C. carbonacea* is the most commonly recorded causal agent and causes large blackish-brown to brown spots which may spread over the whole leaf.

iii) Development and Spread: The fungus over seasons on crop debris. Conidia are produced on both surfaces of infected leaves and are spread by rain. The disease is more common during warm and wet weather.

iv) Management: Spraying of maneb or captan at 2 g/litre of water every 10 days was found effective in the management of the disease.

Two to three sprays of 0.05% carbendazim at 15 days interval starting after the first appearance of the disease was found effective in decreasing the disease incidence (24.61% as compared to 62.64% in control) and increasing the tuber yield (19.75 MT/ha as compared to 7.72 MT/ha).

7.1.3. Tuber Rot, *Botryodiplodia theobromae, Rhizopus nodosus, Fusarium oxysporum, F. solani*

i) Distribution: The disease is distributed in China, India, Ivory Coast, Japan, Nigeria and Puerto Rico.

ii) Symptoms: Nematode infestation facilitates the entry of *Fusarium* spp. into the tubers. Infected tissues develop a brown coloured dry rot. Often the infection is not noticed until after harvest or until the tubers are cut opened. If the tubers are cut transversely, zonations of infection can be seen. Brown dry rot caused by *Fusarium* occurs centrally and there may be a blackish zone round this where there is considerable damage and secondary bacterial infection.

iii) Management: Cut tubers used as planting material should be allowed to dry in order to prevent the entry of the pathogen. The incidence of infection can be greatly reduced by treating the cut surfaces with Bordeaux mixture or coating with lime, chalk or wood ashes. Crop rotation helps to reduce the disease incidence. Also avoid planting yams in nematode infested soil.

Dipping the tubers in systemic fungicides such as benomyl and thiabendazole has also been found effective. A 5 minute dip in 3000 ppm benomyl followed by dusting the cut ends with 0.3% benomyl in talcum powder effectively controlled tuber rot. A 30 minute dip in 1000 ppm benomyl or thiabendazole gave 80 to 90% control of pre-harvest rotting in the field.

The set treatment (30 g/litre for 15 min.) and soil application (1 kg/ha) of *Trichoderma viride* recorded the lowest tuber rot incidence (7.33%) and highest harvest index (66.9%).

7.1.4. Dieback, *Colletrotrichum* sp.

i) Symptoms: This disease starts when water from heavy rains splashes the small spores of the pathogen from the soil up onto the young leaves. The fungus does not grow so well on the older leaves. The first signs of this disease are small round brown dead spots on the leaves. There is a yellow colour on the outside of the spots. Later the dead spots grow in size until the whole leaf has died. Then the whole branch with its stem and leaves may dieback. The tubers will not grow properly.

ii) Management

- Do not use the tubers from diseased plants as planting material.

- Do not plant yams in the infested soil, but use crop rotation with non-host crop.
- Spray the plants when they are very young with 0.1% benomyl.
- Plant the yams at a time when there are no rains. The danger of attack is less once the yams have grown higher, because the older leaves do not get the disease so easily.

7.1.5. Collar Rot, *Rhizoctonia* sp.

i) Symptoms: This fungus gets into the roots of tubers of the yams and rots them. There may be signs of the disease low down on the stem near the soil. It makes brown marks low down on the stems. Dark coloured spots on this part of the stem are called "collar rot". This fungus is carried by the tuber. If there are any marks of this disease on the leaves, they are on one side of the leaf only.

ii) Management: The most important way of stopping this disease is not to use tubers from diseased plants as planting material. The big tubers are probably good to use for planting.

7.2. SWEET POTATO

Sweet potato (*Ipomea batatas*) is an economically important member of the family Convolvulaceae and probably originated in Central America, where it has long been cultivated. It is now grown extensively throughout the tropics and subtropics. It is one of the life sustaining crops rich in carbohydrates that stand between man and starvation. Sweet potato ranks fourth and sixth on the list of dry matter production per hectare and edible energy production per hectare, respectively. It is also used for making starch, syrup and alcohol. It occupies a prime place in terms of calories production/ unit area and time. It is grown as a starchy food crop throughout the tropical, sub-tropical and frost-free temperate climatic zones in the world. It is being cultivated in 0.129 million hectares producing 1.148 million tonnes of sweet potatoes with an average yield of 8.87 tonnes per hectare (Table 7.2). India ranks sixth in area but the productivity is very low (8.9 t/ha), lower than the world productivity level.

Table 7.2. Area, production and productivity of leading sweet potato growing states in India (2005-06)

State	Area (in '000 ha)	Production (in '000 MT)	Productivity (in MT/ha)
Orissa	47.1	394.3	8.4
Uttar Pradesh	20.0	240.8	12.1
West Bengal	25.8	208.0	8.1
Maharashtra	5.0	70.0	14.0
Karnataka	3.6	42.1	11.6

Chhattisgarh	3.0	38.2	12.6
Tamil Nadu	1.4	29.6	21.1
Assam	8.5	29.5	3.5
Madhya Pradesh	4.7	28.2	6.0
Meghalaya	5.4	18.9	3.5
Others	4.9	49.1	10.0
Total	129.4	1148.6	8.87

(Source: Indian Horticulture Database 2006)

Sweet potatoes are affected by several different diseases in the field, some of which also affect the storage life of the tubers. Virus diseases are probably responsible for the greatest field losses and some fungus diseases can also cause economic losses. The use of clean planting material is the most important disease control measure and crop rotation aids control of many fungal diseases.

7.2.1. Cercospora Leaf Spot, *Cercospora ipomoeae, Psudocercospora timorensis, Phaeoisariopsis bataticola*

i) Distribution: *Cercospora ipomoeae* is very common throughout the tropics. *Psudocercospora timorensis* is distributed in Brunei, Fiji, Formosa, Ghana, Guinea, Hong Kong, India, Malaysia, Mauritius, Nepal, New Guinea, New Hebrides, Sabah, St. Lucia, Sterra Leone, Solomon Islands, Sudan, Tanzania and Uganda. *Phaeoisariopsis bataticola* is present in India, USA and Venezuela.

ii) Symptoms: Leaf spots are small (about 6 mm in diameter) and angular with brownish colouration. Leaves may drop from the lower parts of the stem in severe attacks. Damage to leaves is most severe during wet weather conditions.

iii) Development and Spread: Conidia are produced abundantly on the leaf spots and are spread by wind and rain.

iv) Management: Three sprays of 0.25% Dithane M-45 at fortnightly intervals was found effective in reducing the disease intensity (19.89% as compared to 50.78% in control) and in increasing the yield of tubers (21.76 MT/ha as compared to 12.91 MT/ha in control).

7.2.2. Wilt/Stem Rot, *Fusarium oxysporum* f. sp. *batatas*

i) Distribution: This disease appears to occur in temperate regions and found in USA, Japan and Northern India. Stem rot is widespread in USA and 10-50% losses in susceptible varieties may occur. Losses up to 95% have been recorded. In India, the disease is of common occurrence in North and North-East India.

ii) Symptoms: The fungus causes vascular wilt. The first symptom of the disease is yellowing of the youngest leaves. The vascular system becomes blackened causing stems to darken internally. This darkening may extend several feet among stems. Infected plants wilt, collapse and die. Infected tubers have a blackened ring about 6 mm under the skin and develop a surface rot with shallow, sunken, circular lesions in storage. Plants produced from these tubers are likely to be diseased.

iii) Development and Spread: The main sources of inoculum are infected tubers and plants. The fungus can live for several years on plant remains in the soil and can spread locally by water, wind, animals, machinery etc. Diseased plants develop as a result of either diseased planting material or healthy plants becoming infected from the soil. Conidia, which are readily dispersed, are produced on the dead leaves and stems.

The incidence of the disease varies with the soil type, which may influence disease severity more than the inoculum density.

iv) Management

a) Cultural Methods: The most important means of controlling stem rot is to use healthy planting material and crop rotation. Crop rotation will not eradicate the fungus but it will help to reduce the losses. Tobacco should be avoided in the rotation as some varieties are attacked by the stem rot fungus. Sweet potato should not be grown in the same field more often than once in 3-4 years.

b) Chemical Methods: The systemic fungicide thiobendazole has shown some promise as a dip treatment of tubers.

c) Host Resistance: Sweet potato cvs. Pusa Safed and Kalmegh have been reported resistant to wilt/stem rot disease.

7.2.3. Black Rot, *Ceratocystis fimbriata*

i) Distribution: The fungus occurs in North and Central America, parts of South America, West, Central and Southern Africa, India, Southeast Asia, Japan and the Pacific.

ii) Symptoms: Black rot is an important field and storage disease of sweet potato. The fungus infects underground parts of the plant. Infection may extend a short way up the stems causing blackening above the surface of the soil giving rise to the common name 'black shank'. The first symptoms are small, slightly sunken black spots on the underground parts of the plant. The spots enlarge and may cover almost the whole tuber. Fungal spores are often produced in the centre of these areas of infection. If the infected tubers are stored, the rot spreads rapidly causing heavy losses. Infected sweet potatoes have an unpleasant taste when cooked. Plants produced from infected tubers often have the disease.

iii) Development and Spread: The fungus persists in the soil between sweet potato crops. The sources of inoculum are soil and infected tubers. Plants may become infected in the seedbed or in the field. The disease spreads particularly quickly in stored sweet potatoes.

iv) Management

a) Cultural Methods: The best method of control is to use disease-free planting stock. Healthy seed sweet potatoes can be grown from cuttings, which will be free from the disease as only underground parts are affected. Cuttings should be planted in the soil free from black rot pathogen to ensure the production of disease-free roots for field planting. As the fungus persists in soil, sweet potatoes should only be grown once in every 2-3 years.

Sweet potatoes should not be manured with dung from animals that have eaten black rot infected tubers as it may contain spores of the fungus.

b) Chemical Methods: Tubers to be used for planting can be surface sterilized with fungicides. Thiabendazole has shown promise in trials. When inoculated sweet potatoes were dipped in a suspension of 3.32 g a.i./litre, development of the disease was prevented. The control given by thiabendazole and benomyl dips was greatly improved by raising the temperature of the dips to 44-54°C. Good control of black rot was also obtained with ferbam dips at 44-54°C.

7.3. CASSAVA (TAPIOCA)

Cassava (*Manihot esculenta*), popularly known as tapioca, is a member of the family Euphorbiaceae, which has long been cultivated in Central America and Brazil and was later carried to Africa (where it is now an important subsistence crop), Asia and the Pacific Islands. An economic crop can be obtained even under marginal conditions or on exhausted soils. The crop requires very little attention after planting. It has draught tolerance (mainly due to in-built mechanism to shed or drop the leaves under adverse moisture conditions) and produces more calories/unit area.. The tubers contain 25-30% starch, 1.5% protein and 2.1% minerals. It is being cultivated in 0.252 million hectares producing 7.925 million tonnes of cassava with an average yield of 31.36 tonnes per hectare (Table 7.3). The Salem belt of Tamil Nadu and Samalkot belt of Andhra Pradesh are known for cassava as an industrial crop.

Table 7.3. Area, production and productivity of leading cassava growing states in India (2005-06)

State	Area (in '000 ha)	Production (in '000 MT)	Productivity (in MT/ha)
Tamil Nadu	127.1	4857.4	38.2
Kerala	95.0	2585.1	27.2
Andhra Pradesh	19.0	388.7	20.5
Meghalaya	4.2	23.1	5.5
Pondicherry	0.7	19.5	27.9
Nagaland	1.3	15.9	12.2
Assam	2.9	14.0	4.8

Karnataka	1.1	12.0	10.7
Others	1.4	9.2	6.6
Total	252.7	7924.9	31.26

(Source: Indian Horticulture Database 2006)

Even though cassava is a hardy crop, it has been shown that diseases can cause serious losses. As cassava is propagated vegetatively, the use of disease-free planting material is the most important single factor in disease control. Under traditional farming systems, yields have been more than doubled by using disease-free planting material and improving cultural practices. Cassava bacterial blight is the most important disease of cassava and virus diseases are also of considerable threat to the crop.

7.3.1. Tuber Rot, *Phytophthora drechsleri*

i) Symptoms: *Phytophthora* root rot is prevalent in water-logged soils. The disease causes yield losses up to 80%. The pathogen attacks both mature and young plants causing sudden wilting and severe soft rotting of the swollen roots. Diseased roots develop a brown discolouration and disintegrate rapidly. Roots exude a fowl smelling liquid when partially rotted. In severe cases, plants may die.

ii) Management: The disease can be controlled by cultural practices such as providing good drainage, use of lighter soils, early harvest, crop rotation with cereals and fallowing. Wide spacing tends to discourage the spread of root pathogen. Crop residues should be removed and burnt and healthy planting material should be used for planting.

The set treatment (30 g/litre for 15 min.) and soil application (1 kg/ha) of *Trichoderma viride* recorded the lowest tuber rot incidence (7.33%) and highest harvest index (66.9%).

7.3.2. Brown Leaf Spot, *Cercosporidium henningsii*

Brown leaf spot (Fig. 7.2) is a major fungal disease in heavy rain fall areas. It causes considerable premature defoliation and loss in tuber yield up to 30% in susceptible variety like Malayan-4.

Fig. 7.2. Brown leaf spot on tapioca leaves.

i) Distribution: Brown leaf spot disease is widespread fungal leaf disease in all cassava growing areas.

ii) Symptoms: Older leaves tend to be more susceptible than younger ones and the disease tend to be more widespread and severe in crops over 5 months old. The disease cause characteristic brown spots (3-12 mm in diameter) on both sides of the leaves. Spots on the upper surface are angular and uniformly brown with a distinct dark border. Spots on the lower surface have less distinct margins and appear grayish in the centre because of the presence of fruiting bodies of the fungus. The lesions expand and become irregular and angular in shape as they are limited by leaf margins and veins. Commonly an indefinite halo appears round the lesions. As the disease develops the infected leaves turn yellow, become dry and fall off. Susceptible varieties can be severely defoliated during the rainy season.

iii) Development and Spread: Warm, wet conditions favour the development and spread of *C. henningsii*. Conidia of the fungus on older or fallen leaves are spread by rain splash and wind to younger leaves and to new plantings. The conidia need moisture to germinate on the leaf surface. The fungus penetrates through stomata and grows trough the intercellular spaces. The fungus survives during the dry season in old lesions, often on fallen leaves.

iv) Management

a) Cultural Methods: Measures designed to reduce the humidity within the stand of cassava, such as wider spacing and the varieties with a low leaf area index are recommended. The severity of the disease can be reduced if the planting is timed so that the most susceptible growth stage (over 5 months old) does not coincide with the wet season.

b) Chemical Methods: Spraying of copper fungicides, thiophanate, benomyl and carbendazim at monthly intervals starting from third month was effective in controlling the disease.

c) Host Resistance: Cassava cvs. Sri Prakash and Sri Visakam released from Central Tuber Crops Research Institute, Thiruvanathapuram, Kerala, India, are resistant to brown leaf spot disease.

7.3.3. Cercospora Leaf Blight, *Cercospora viscosae*

i) Distribution; The disease occurs commonly as brown leaf spot in warm cassava growing areas of Brazil and Colombia and also Southeast Asia.

ii) Symptoms: The disease is characterized by large brown spots without definite borders. Individual spots may cover a fifth or more of the leaf lobe. The lesions are uniformly brown on the upper surface but on the lower surface the centres of the spots are grayish due to the presence of fruiting bodies of the fungus. The lesions usually have concentric rings on the lower surface. *Cercospora* leaf blight usually occurs during the rainy season and can cause heavy defoliation in susceptible cultivars with resulting yield loss of over 30%. Plants over 6 months are more severely affected.

iii) Management: The same management methods used for brown leaf spot also control *Cercospora* leaf blight.

7.3.4. Concentric-Ring Leaf Spot, *Phoma* sp., *Phyllosticta manihobae*, *P. manihoticola*

i) Distribution: The disease occurs in Africa, Brazil, Colombia, India and Philippines.

ii) Symptoms: The disease occurs in cooler cassava growing areas during the rainy season when temperatures are below 22°C. Susceptible varieties may be severely defoliated and occasionally dieback or death may occur. The disease is characterized by large brown spots with indefinite margins. Lesions are usually found at the tips or edges of the leaf lobes or along the main veins or midrib. At first the upper surface of the lesions consists of concentric rings formed by the brown pycnidia (fruiting bodies) of the fungus. These rings are often absent on the older lesions as the epidermis disintegrates allowing mature pycnidia to be washed off by rain so that the lesions resemble those caused by *Cercospora viscosae*. Few pycnidia are produced on the lower surface of the lesions, which appear uniformly brown.

When humidity is high grayish-brown hyphae can be observed on the surface of the lesions. On the underside of leaves, veins and vein lets become necrotic in the region of the lesions. The lesions expand causing leaf blight. Eventually the necrosis spreads to the whole leaf and petiole. Leaves then wilt and drop, sometimes resulting in extensive defoliation. Dieback of young shoots can also be caused and in extreme cases the plant may die. Diseased stems become brown and are often covered with pycnidia.

iii) Development and Spread: The young upper leaves are more susceptible to the disease. The disease is more prevalent at high altitudes or in low land areas during the rainy season, because conditions are best for spore germination (20-25°C).

iv) Management: Field resistant cultivars have been observed and these may be used in a breeding programme for resistance.

7.3.5. Powdery Mildew, *Oidium manihotis*

i) Distribution: The disease is distributed worldwide and endemic in almost all cassava plantations.

ii) Symptoms: The first sign of the disease is a white fungal mycelium growing over the surface of the leaf. Yellowish lesions are formed where the fungus penetrates the leaf. The lesions are indefinite and contain pale brown angular spots of necrotic tissue.

The disease occurs most commonly during the dry season in warmer cassava growing areas. Mature leaves appear to be the most susceptible.

iii) Management: Resistance has been observed in some varieties.

7.3.6. Anthracnose, *Glomerella manihotis*

i) Distribution: The disease is widespread in distribution and appears to be serious in Southeast Asia, Africa and Brazil.

ii) Symptoms: Anthracnose tends to appear after long periods of rain. Young shoots seem to be the most susceptible. Leaf border blight and spots appear on young leaves, which become distorted. Infected tissue becomes necrotic and whole leaves may die. The disease also attacks stems, causing a dieback of young stems and the production of cankers on older stems. In the centre of the cankers, pinkish areas of the fungal fruiting bodies can be observed. Anthracnose tends to disappear at the onset of the dry season.

iii) Management: Planting material taken from infected stems is of poor quality, does not germinate well and results in weak plants and reduced yields. The best preventive measure is to use healthy planting material and to avoid planting at the height of the rains.

7.4. COLOCASIA

Colocasia (*Colocasia esculenta*), also known as cocoyam, taro, dasheen and eddoe, is a member of the family Araceae and occurs wild in S.E. Asia. It is cultivated throughout the humid tropics and is of great importance in the Pacific Islands. It is mostly a staple food or subsistence crop grown commercially in some countries.

7.4.1. Leaf Blight

The leaf blight of colocasia is a major and widespread disease causing 50% loss in tuber yield.

i) Distribution: This disease occurs widely in India, Indonesia, Malaysia, Sarawak and the Pacific. It has also been recorded from parts of Africa and the Caribbean.

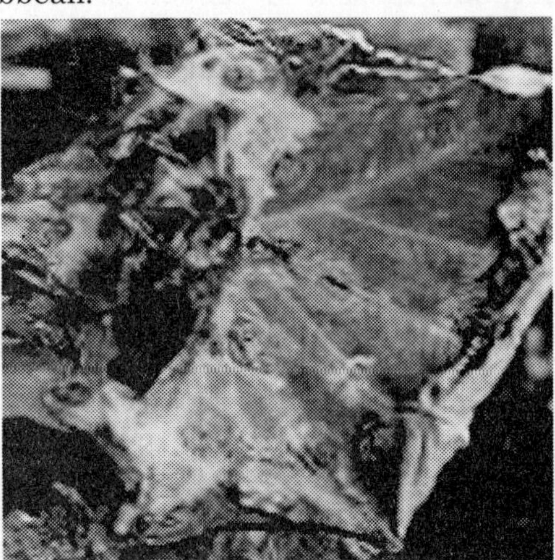

Fig. 7.3. Leaf blight damage on leaves of colocasia.

ii) Symptoms: Leaf blight is the most destructive leaf disease of colocasia. Lesions are initially small, dark and round but rapidly enlarge to 2.5-5.0 cm in diameter and become purplish to brownish in colour (Fig. 7.3). Drops of a clear liquid exude from the spots, and turn yellow, orange or purple when dry. There is usually chlorotic halo round the spots. As the disease progresses, the spots coalesce and have characteristic rings of yellow or brown colour. Eventually the whole leaf may be affected and may die.

iii) Development and Spread: Leaf blight is common in wetland and is favoured by humid, cloudy conditions and poor soil fertility. Losses up to 50% have been reported. Spores are produced on the leaf spots and are readily spread by rain.

iv) Management

a) Cultural Methods: Wide spacing minimizes spread of the disease due to rain splashing. Wide spacing together with interplanting and open sides to the colocasia field help to improve the air flow and thus keep the humidity down. Farmers should avoid walking through the field when it is wet as this encourages the spread of spores. Where possible the colocasia should be cultivated on higher land. Crop remains should be burnt after harvest.

In case of mild infection of young plants (up to 2-3 months), leaf rogueing can be effective in controlling or delaying the build-up of the disease. Rogued leaves should be burnt.

Strict hygiene should be observed at planting and infected leaves should be removed twice a week for 80-90 days after planting, or until a dense canopy is formed.

b) Host Resistance: Colocasia cvs. Jhankri and Muktakeshi are tolerant to leaf blight disease.

c) Chemical Methods: Tribasic copper sulphate and copper oxychloride have been used successfully at intervals of 7-15 days. Zineb and mancozeb have been shown to give effective control. The best control was obtained with sprays of mancozeb at 2.24-4.48 kg/ha at intervals of 5-7 days. It has been suggested that copper oxychloride sprays and phytosanitary measures could be used together in a practical control programme.

Four sprays of 0.25% zineb at 15 days interval was found effective in reducing the disease intensity (23.24% as compared to 56.15% in control) and increasing the tuber yield (20.91 t/ha as compared to 15.49 t/ha in control). Two sprays of ridomil at 0.25% was also found effective in management of the disease.

7.4.2. Brown Leaf Spot, *Cladosporium colocasiae*

i) Distribution: The disease is distributed in Borneo, Ethiopia, Fiji, Formosa, Guinea, Hawaii, Hong Kong, India, Nepal, New Caledonia, New Guinea, Nigeria, Pakistan, Sarawak, Salmon Islands and Tahiti.

ii) Symptoms: Superficial dark reddish-brown spots of 5-10 mm in diameter develop on both sides of the leaves. The spots are diffuse in appearance. The disease occurs in wetland and upland colocasia and is particularly severe on old leaves.

iii) Management: Fungicidal sprays applied for leaf blight control will also control brown leaf spot.

7.4.3. Corm Rot, *Pythium myriotylum, P. splendens, Phytophthora cinnamoni, P. nicotianae* var. *parasitica*

i) Distribution: The disease occurs in India and Pacific area.

ii) Symptoms: If young plants are attacked, they may be killed without producing leaves or they may become stunted. In plants over 3 months old, the first symptoms observed are wilting, chlorosis and stunting which may be followed by collapse of the whole plant. The rotting of the corms usually starts at the base and progresses upwards so that eventually the whole corm may be affected. The rotted tissue is firm and may be whitish-yellow to grey and blue or dark purple. Healthy and diseased tissues are clearly divided. In later stages, the diseased tissue crumbles like cheese and has an offensive smell.

Roots decay in association with the corms. In some cases there is little damage to the corm but many roots are killed leaving the plant loose in the ground. Rotting may also spread from the main corm to lateral suckers.

iii) Development and Spread: The fungi are encouraged by plentiful moisture, warm temperature and poor soil conditions. Any physical damage to colocasia plant predisposes them to infection. If colocasia is left in the field after maturity, it causes a build-up of the disease. These fungi survive between crops in crop residues. Damage may be confined to a few corms or may be widespread and cause 100% loss depending upon the conditions. Damage is particularly severe in wetland areas and poorly drained soil.

iv) Management

a) Cultural Methods: Colocasia should be harvested as soon as possible after it is mature and crop debris should be removed from the fields. Diseased plants should be rogued from the growing crop taking care to remove as much of the corm and decayed roots as possible. If a field is badly infected with the disease, colocasia should not be cultivated for 2 or 3 seasons and the soil should be allowed to dry out by deep cultivation. In irrigated wetland colocasia, water should be kept flowing and not be allowed to stagnate. Dry land colocasia should be planted in well drained soil. Healthy planting material should always be used for planting.

b) Chemical Methods: Dipping of planting material in copper oxychloride or captan has been recommended. Soil drenches with these fungicides have also been found to be effective.

7.4.4. Sclerotium Rot, *Corticium rolfsii*

i) Distribution: The disease is widespread in distribution.

ii) Symptoms: The pathogen causes a rot of dying petioles and a corm rot. The rotted tissue is brown and soft but not watery and tends to be stringy. Characteristic white mycelial growth and numerous dark brown sclerotia which resemble cabbage seed are observed. The disease may be more severe in colocasia that is over mature.

iii) Development and Spread: *Corticium rolfsii* is a common pathogen and attacks many crops; sources of inoculum are therefore readily available. The disease is more common in upland colocasia.

iv) Management

Prompt harvesting of colocasia and crop rotation help to avoid the disease. Infected plants should be removed and burnt. Soil drenches with quintozone or dicloran have been recommended for control.

7.5. ELEPHANT FOOT YAM/AMORPHOPHALLUS

Elephant-foot yam (*Amorphophallus paeoniifolius*) is a popular tuber crop of humid tropics. Among root and tuber crops, *Amorphophallus* is one of the most important tuber crops that occupy special significance as high value cash crop due to its high production potential and popularity as a vegetable. The genus *Amorphophallus* is indigenous to tropical Asia and Africa and *A. paeoniifolius* is found widely distributed in thickets and secondary growth forests at low and medium altitude in the Philippines, Malaysia, Indonesia, Sri Lanka and the South East Asia subcontinent. In India, it is widely grown in Andhra Pradesh, Bihar, West Bengal, Assam, Jharkhand, Gujarat, Tamil Nadu, Kerala, Uttar Pradesh, Maharashtra, Chhattisgarh, Meghalaya and Andaman and Nicobar Island. Chemical composition of Elephant-foot yam corms: moisture 60-80%, starch 25-28%, sugars 0.5-0.6%, crude protein 1.0-2.8%, fat 0.1-0.3%, ash 0.5-2.6%, Vitamin A-0.8 mg, Vitamin B1 - 0.08-0.09 mg, Vitamin B2 - 0.02-0.03 mg, Vitamin C - 5-12 mg.

7.5.1. Collar Rot, *Sclerotium rolfsii*

i) Management : *T. harzianum* was highly effective in suppressing *S. rolfsii* in elephant foot yam under field conditions. The conidial mixture of 2 bioagents *T. pseudokoningii* (effective in reducing the mycelial growth) and *T. virens* (good colonizer of sclerotia) was effective in reducing the incidence of storage rot. Application of *T. harzianum* as corm treatment + soil treatment after 30 DAP significantly managed the disease. Mulching with paddy straw/ polythene sheet reduced the collar rot disease.

Integrated Disease Management

Elephant-foot yam is affected by three major diseases namely, Collar rot caused by *Sclerotium rolfsii*, leaf blight caused by *Phytophthora colocasiae* and mosaic caused by virus.

Use of mosaic virus free healthy planting material (corm) + corm treatment with *Trichoderma viridae* at 5 g/litre or cow dung slurry mixed with mancozeb (0.2%) along with soil application of neem cake at 250 g/pit followed by one foliar spray with neem gold at 0.5% at 60 days after planting effectively controlled almost all the field diseases. Corm treatment with cow dung slurry mixed with mancozeb (0.2%) along with dimethoate (0.1%) has been found most effective against post harvest diseases and also enhancing sprouting. But care should be taken to dry the treated corms in shade before planting. Corm treatment was found most effective when cut pieces are used as planting material than whole corms. If diseases appeared in the field, then 2-3 foliar sprays of mancozeb (0.2%) at fortnightly interval has been recommended.

Use of mosaic virus free tubers + mulching after planting and two foliar sprays with Mancozeb (0.2%) and Monocrotophos (0.05%) at 60 and 90 days after planting has proved very effective in reducing the incidence of major diseases and pests of Elephant-foot yam.

Chapter 8

PLANTATION CROPS

Plantation crops serve a variety of human needs such as food, oil, industrial raw materials, beverages and confectionary items. They generate huge employment opportunities directly or indirectly to several million people in their production, processing, marketing and international trade sectors. Plantation crops are a large group of crops. These are grown over an area of 3.282 million ha with production of 11.262 million tonnes in India (Bijay Kumar, 2007) (Tables 8.1 and 8.2). The plantation crops contributed 6.09% of total horticulture production, in 2006-2007. India is the largest producer and consumer of areca nut, cashew nut, tea, third largest producer of coconut, fourth largest producer and consumer of rubber and sixth largest producer of coffee in the world. While the major plantation crops include coconut, areca nut, oil palm, cashew, coffee, tea and rubber, minor plantation crops are cocoa and betel vine. They play an important role in export as well as domestic requirement, employment generation and poverty alleviation particularly in rural sector. The annual income generated is more than Rs. 5,52,590 million and export earning is approximately Rs. 93,350 millions.

Table 8.1. Area, production and productivity of plantation crops in India (2005-06)

Plantation crops	Area ('000 ha)	Production ('000 MT)	Productivity (MT/ha)	% Share of India production
Coconut	2028.9	10190.1	5.0	90.5
Areca nut	381.1	483.1	1.3	4.3
Cocoa	29.5	10.2	0.3	0.1
Cashew	843.4	579.4	0.7	5.1
Total	3282.9	11262.8	3.4	

Table 8.2. State-wise area, production and productivity of plantation crops in India (2005-06)

State	Area in '000 ha	Production in '000 MT	Productivity in MT/ha
Kerala	1096.6	4545.1	4.1
Tamil Nadu	497.7	3415.0	6.8
Karnataka	735.1	1095.9	1.5
Andhra Pradesh	286.1	706.6	2.4
Maharashtra	180.0	376.1	2.0
Orissa	170.8	267.0	1.5
West Bengal	44.3	261.9	5.9
Assam	104.3	217.8	2.0
Goa	81.9	115.8	1.4
Gujarat	20.4	99.1	4.8
Andaman & Nicobar	30.0	64.7	2.1
Lakshadweep	2.7	36.5	13.5
Meghalaya	17.9	21.9	1.2
Pondicherry	5.2	20.3	3.9
Tripura	6.7	11.7	1.7
Mizoram	2.0	5.3	2.6
Nagaland	1.1	2.0	1.8
Total	3282.9	11262.8	3.4

There is a good scope for switching over the production of plantation crops viz., coffee, tea, coconut and areca nut to organic farming, as it is possible to realize higher returns from the unit quantity exported when grown organically. The plantation crops are generally grown in ecologically fragile hilly tracts, adopting organic farming methods would protect environment and also prevent contamination thereby producing products of high nutritional quality. Further, the plantation crops are amenable for organic farming as they produce huge amount of waste biomass for recycling, which can meet major portion of nutrients required.

Export of Plantation Produce

India is the largest producer in the world of cashew nuts, coconuts and tea. India is largest producer, importer, processor, exporter and second largest consumer of cashew. Indian exports during 2006-2007 were US$ 866 million in tea and coffee (Table 8.3).

Table 8.3. Export of plantation crops

Commodity	Apr-Mar 2006	Apr-Mar 2007	% Growth	% Share
Tea	390.92	432.00	10.51	0.34
Coffee	358.83	434.77	21.16	0.34
Cashew	585.76	553.41	-5.52	0.44

8.1. COFFEE

Coffee (*Coffea arabica, C. canephora*) is an important commercial plantation crop mostly grown on the hilly slopes of the traditional coffee tracts in South India (Karnataka, Kerala, Tamil Nadu and Andhra Pradesh) and non-traditional Orissa, Maharashtra, Madhya Pradesh and North-Eastern Council States. The coffee industry is well established and flourishing enterprise. The major coffee plantations in India are in the Southern States with Karnataka being the major coffee producing state. Recently some coffee plantations have come up in the North Eastern States of India as well. Besides being one of the major foreign exchange earner to a tune of around Rs. 2000 million annually, coffee industry supports more than four lakh people to earn their livelihood. The two main species of coffee in cultivation are *Coffea Arabica* (Arabica) and *Coffea canephora* (Robusta). *C. arabica* produces finer seeds which are more valued and grows best at higher elevations (1000 and 1500 m MSL). *C. canephora* is sturdier and better adapted to the warm climate prevailing in the lower elevations (500 and 1000 m MSL). Robusta, in general, is more tolerant to many diseases and pests including nematodes, as compared to Arabica. Coffee in India is grown under a canopy of shade trees of various species along with other plantation crops like black pepper, cardamom, banana, cocoa, etc. The present area under coffee in India is above 100,000 hectares, nearly three-fourth of which is under the fine quality Arabica species with most of the remaining area under Robusta.

India produced 290,000 tonnes of coffee accounting for approximately 4% of the global production. Over 80% of the coffee produced in India is exported. Although over 50 varieties are produced, the popular ones are the Indian Robusta which is used for blending and Arabica. There is a strong demand in the world market for production of high value, high quality, especially organic coffee. Today, the area under certified organic coffee is 2500 ha with an estimated production of 1500 MT of which 1000 MT is being exported currently.

With the current low price trends which are highly unremunerative to the growers, that most of the growers have shifted the cultivation practices to organic farming system. Today, more than 10,000 ha have been certified under organic tea cultivation.

8.1.1. Leaf Rust, *Haemeleia vastastrix*

i) Symptoms: Small pale-yellow spots appear on the lower surface of infected leaves. Later, orange-yellow spore mass appears resulting in defoliation and die-back (Fig. 8.1). The pest is responsible for serious *crop* loss and causes fluctuations in production.

ii) Management: Maintaining optimum shade, judicious pruning and spraying of 0.5% Bordeaux mixture as pre-monsoon, mid-monsoon and post-monsoon application can effectively control the disease in Arabica coffee.

Spray tridemefon at 2 g + Teepol 0.5 ml/litre of water at fortnightly intervals at times of intermittent rainfall.

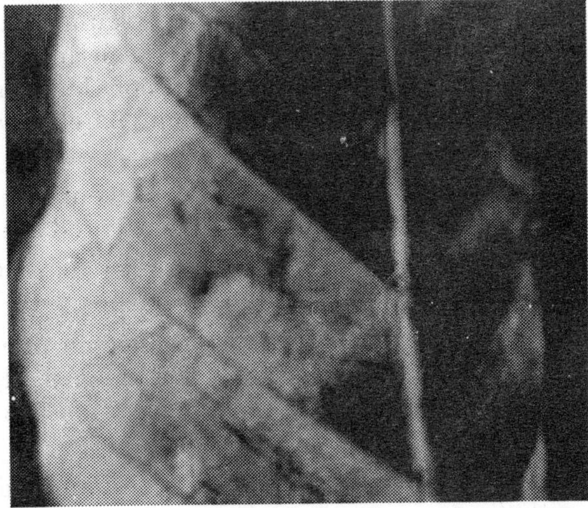

Fig. 8.1. Rust on coffee leaf.

8.1.2. Brown Root, *Fomes noxius*; Red Root, *Poria hypolaterita*; Black Root and Santavery Root Disease, *Corticium koleroga*

i) Symptoms: Stump or brown root-rot affected bushes show a gradual yellowing of leaves and defoliation and death. Affected roots are brittle and show dark brown wavy lines of the fungus. Stem near the ground level becomes soft and spongy. Brown fungal encrustation can be seen on the affected roots (Fig. 8. 2).

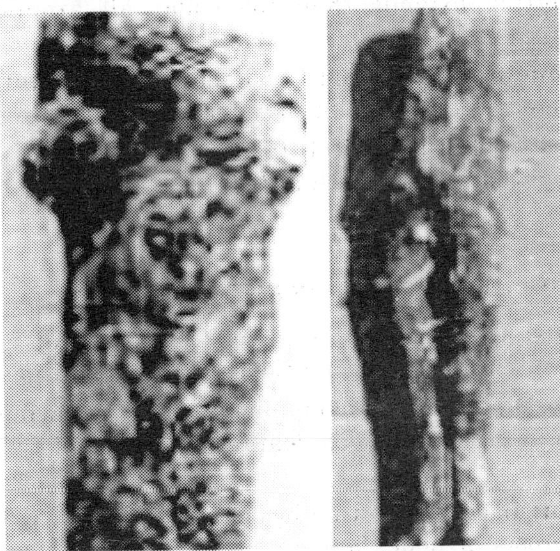

Fig. 8.2. Black root and red root on coffee.

The pest infestation causes blackening and rotting of affected leaves, young twigs and berries. Affected leaves get detached and hang down by means of slimy fungal strands. Defoliation and berry drop occur.

ii) Management: Thinning out the shade, regular pruning, handling, centering (opening up the centre of the bush), cutting and burning the infected plants and plant parts followed by prophylactic sprays with 1% Bordeaux mixture in the black rot prone endemic patches would effectively check the disease.

The following measures are suggested to prevent the spread of root diseases to neighbouring areas and to recover the infected areas:

- Uprooting of shade tree stumps after timber extraction followed by uprooting and burning of affected coffee bushes.
- Making an isolation trench around the affected plants by enclosing few healthy plants and application of 1-2 kg agricultural lime to the uprooted pit and exposing the spot for at least 6 months before replanting.
- Application of 5-10 kg/plant of well decomposed FYM/compost fortified with *Trichoderma* sp. to the surrounding healthy plants.
- Soil application of 2 kg *T. harzianum* or *T. viride*/plant during June and October gave good control of brown rot and red rot diseases. Mix 2.5 kg of *Trichoderma* spp. in 100 kg of FYM. Leave it overnight. Spread these equally around plant basins twice a year.

8.1.3. Collar Rot or Damping-Off

i) Symptoms: The pathogen is responsible for pre-emergence damping off in germinating beds and collar rot in primary or secondary beds. Seedlings show wilting, rotting extends from the collar all around and seedlings collapse.

ii) Management: Maintaining good shade and sanitation by removing and burning affected seedlings helps in reducing the disease incidence.

Application of PCNB (brassicol) 0.4% or terrachlor or rhizoctol or Flit 406 or ferbam or dithane M 22 each at 0.4% is effective for the management of the disease.

8.1.4. Anthracnose, *Colletotrichum gloeosporioides*

i) Symptoms: The pathogen is a weak parasite but becomes vigorous under favourable environmental conditions. It causes 3 diseases:

a) Radcliffe Disease: The symptoms include yellowing of one or more functional leaves, necrosis of nodes, leaf-fall, die-back of twigs and death of many branches.

b) Stalk Rot: It causes rotting of berries and leaves. The other symptoms include necrosis of nodes and internodes from the junction of brown wood and green wood towards apex followed by berry drop and defoliation due to decaying of stalk portion.

c) Brown Blight: Roundish spots which are first brown, then turn grey making the leaf look blighted.

ii) Management: Providing proper drainage helps to remove excess water from the root zone. Spraying with 0.5% Bordeaux mixture prevents defoliation. Maintaining adequate shade avoids sun scald.

8.1.5. Fusarial Root Rot or Wilt

i) Symptoms: The pathogen is responsible for sudden wilting and yellowing of leaves leading to defoliation and death of aerial parts. Roots turn brown to pinkish.

ii) Management: Burn the plants at site, isolate affected patches by digging deep trenches, ring the bark and poison the shade trees while thinning. Remove the affected plants and treat the soil at 1kg lime per plant to raise the soil pH. Treat the soil around the affected area with PCNB (brassicol) at 0.4%. Maintain the vigour of plants.

8.1.6. Berry Blotch

i) Symptoms: Necrotic spots on the exposed surface of green berries enlarge and cover the major portion. Fruit skin shrivels and sticks fast.

ii) Management: Spray 1% Bordeaux mixture during June and late august, maintain medium shade overhead.

8.1.7. Brown Eye Spot

i) Symptoms: Usually severe in the nursery but found on young coffee plants also. Small circular necrotic spots appear with dark brown margin and pale white centre. Leaves turn yellow and are shed.

ii) Management: Spray nursery seedlings with 0.4% flit 406 and young coffee plants with 1% Bordeaux mixture during April-May and September-October, avoid exposing the plants to direct sun.

8.1.8. Black Rot-Koleroga, *Koleroga noxia*

i) Symptoms: The symptoms include blackening and rotting of affected leaves, twigs and developing berries. Affected leaves get detached from branches and hang down by means of slimy fungal strands. Affected leaves and berries become dry, showing a white web consisting of closely interwoven mycelium. Defoliation and berry drop takes place from infected branches in advanced stage.

ii) Management: Spraying of 1% Bordeaux mixture gives effective control of black rot of coffee.

8.1.9. Pink Disease

i) Symptoms: The symptoms include pink encrustation on infected branches and development of longitudinal cracks through which pink encrustation bursts. Cobweb like mycelial branches develop on affected branches. Infected branches lose leaves and die.

ii) Management: Mix 2.5 kg of *Trichoderma viride* in 100 kg of FYM. Leave it overnight. Spread these equally around plant basins twice a year.

8.2. TEA

India is the largest producer and consumer of tea (*Camellia sinensis*) in the world. During 2006-2007, India produced approximately 950,000 tonnes of tea of which around 218,000 tonnes was exported. The total area under tea is 5,07,200 ha. It accounts for 20% of the total area under tea in the world, 28% of world production, 22% of global tea consumption and 15% of the total global tea exports. Tea accounts for approximately £1 billion in turnover in India. Tea is grown mainly in three major regions in India - Assam, Darjeeling in North East and the Nilgiris region in Southern India. The largest producers of tea are Assam (50.7%), West Bengal (22.1%), Tamil Nadu (15.9%) and Kerala (8.3%).

Other areas where tea is grown to a small extent are Karnataka, Tripura, Himachal Pradesh, Uttarakhand, Arunachal Pradesh, Manipur, Sikkim, Nagaland, Meghalaya, Mizoram, Bihar and Orissa. The Northern States of India account for around 75% of the tea grown in India and the Southern States accounting for around 25% of the production. The tea industry provides direct employment for one million workers of which sizeable number are women.

More than half of the one million hectares of tea cultivation is in India and Sri Lanka, on areas ranging from 225,000 to 350,000 hectares. Between them, these two major producers account for more than three-quarters of the world's export trade of 600,000 metric tonnes of this commodity. This apparently disproportionate share in the export trade is because in these two countries, domestic consumption is far outstripped by total production.

With the current low price trends which are highly unremunerative to the growers, that most of the growers have shifted the cultivation practices to organic farming system. Today, more than 10,000 ha have been certified under organic tea cultivation.

8.2.1. Dieback and Canker, *Glomerella cingulata*

i) Symptoms: This is one of the most serious of all tea diseases and is caused by the fungus *Glomerella cingulata*. Leaves on affected branches suddenly turn yellow and wilt. Branch tips usually die. Gray blotches appear on the bark and stem, and then sunken areas (cankers) develop, eventually girdling the stem. Parts of the plant above the stem canker lose vigour, wilt and die. Damaged plants show more symptoms during hot, dry weather.

ii) Management : Keep tea as healthy as possible. Plant in a well-drained acidic soil, avoid wounding and fertilize properly. Remove diseased twigs by pruning several inches below the cankered areas. Disinfect pruning tools between all cuts, using a solution of one part household bleach to nine parts water. Fungicides, such as thiophanate-methyl and copper salts of fatty acids can be applied during wet periods and normal leaf drop periods to protect fresh leaf scars from infection.

8.2.2. Flower Blight, *Ciborinia camelliae*

i) Symptoms: This serious disease of tea causes the flowers to turn brown. Flower blight appears in early spring when moisture is present and is caused by the fungus *Ciborinia camelliae*. Symptoms begin as small, brown, irregular-shaped spots on the flower petals. These spots quickly enlarge to cover most of the flower. The entire flower turns brown and usually drops within 24 to 48 hours. Only the flowers of the plant are affected.

ii) Management : Sanitation is the best control. Pull off and destroy all infected flowers. Rake up and remove all leaves, flowers and plant debris that have fallen to the ground. Replace the mulch under the plant. This fungus survives in the soil. Spores of the fungus can be wind-borne for up to 1.6 km. Therefore, control is best achieved when applied to other tea plants in the community.

Application of soil drenches, such as PCNB, around the plant in late December or early January is often helpful in reducing the intensity of disease. Fungicide sprays recommended for the flowers include mancozeb and triadimefon.

8.2.3. Root Rot, *Phytophthora cinnamomi*

i) Symptoms: This fungal disease is caused by *Phytophthora cinnamomi*. The first symptoms are a uniform leaf yellowing, poor growth and wilting of the entire plant. Infected root systems lack small feeder roots and appear discoloured. Infected roots are reddish-brown to dark-brown in colour (healthy roots are white). Death of the plant can occur rapidly, or the plant may remain in a state of decline for several years. All varieties of common tea are susceptible and all varieties of *Sasanqua camelliae* are resistant to this root rot.

ii) Management : This disease is difficult to control once plants are infected, so prevention is very important. In areas where this disease has been a problem, request tea cultivars grafted onto a sasanqua rootstock. Procure healthy plants that show no signs of wilting or yellowing of the leaves.

The fungus thrives in areas with poor drainage and warm soils. Always choose locations that have good drainage for planting. The drainage of existing areas can be improved by using raised beds. Fungicides can be effective on a preventative basis only, and repeat applications are required. Fungicides containing etridiazole and mefenoxam can be applied, but will not cure an infected plant.

8.2.4. Blister Blight, *Exobasidium vexans, E. camelliae*

i) Symptoms: The blister blight is the major disease affecting the tender harvestable shoots of tea resulting in enormous crop loss. Translucent spots appear on tender leaves. Tender stem also gets affected. Leaf galls are most often observed during the spring flush of growth. New shoots and leaves become enlarged, thickened and fleshy, and appear abnormal. The colour of the affected areas turns from light green to nearly white or pink. The galls later rupture on the undersides of the leaves revealing a whitish mass of spores (Fig. 8.3). The galls eventually harden and become brown. Plants are seldom severely damaged.

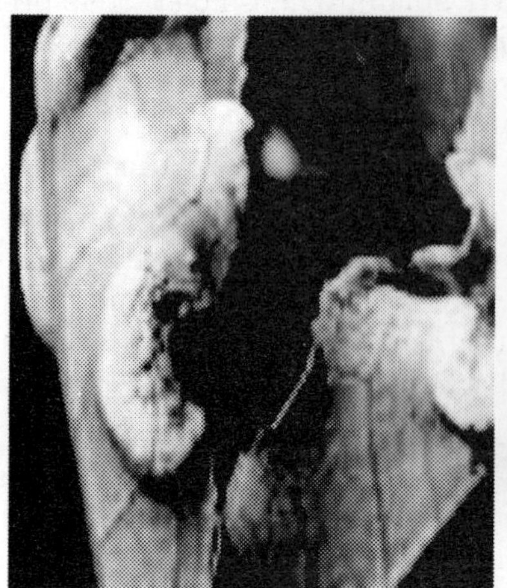
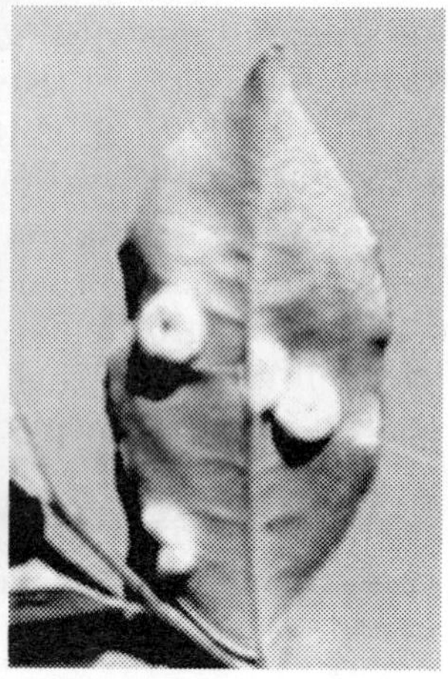

Fig. 8.3. Tea leaves infected by blister blight disease.

ii) Management : Remove and destroy young galls before the lower leaf surfaces turn white and spores are released, otherwise the disease will be worse during the next year. Rake up and remove fallen leaves. Avoid wetting the leaves when watering. Humid, moist, shady conditions favour gall formation. Chemical controls, such as mancozeb are limited in effectiveness and must be applied before infection occurs. Start sprays at bud break and continue through the first of June at 7- to 14-day intervals.

The incidence of the disease can be reduced by thinning the shade, timely weed control and pruning at right time. Need based application of 1% Bordeaux mixture also help in controlling the disease.

Hexaconazole, bitertanol, propiconazole and tebuconazole are the most effective systemic fungicides recommended against blister blight. A combination schedule of triazole fungicides (hexaconazole, propiconazole) and morpholine fungicide (tridemorph) has been recommended recently.

Spray hexaconazole (1 ml/l of water) or tridemorph (2 ml/l of water) at 15 days interval followed by copper oxychloride (3 ml/l of water) at 7 days interval to control blister blight of tea.

8.2.5. Grey Blight and Dieback, *Pestalotia theae/Pestalotiopsis theae*

i) Symptoms: This disease is caused by *Pestalotia theae/Pestalotiopsis theae*. Mature leaves (Fig. 8.4), young shoots and bare stalks are affected by this pathogen. Infection on young shoot results in dieback of shoots. Dieback of shoots became a major problem mainly due to continuous shear harvesting. Grey blight adversely affects the health of the bushes, which in turn affects yield while dieback of young shoots directly leads to substantial crop loss. Studies indicated that the disease incidence is in its peak during July to December. The crop loss due to the disease is 17%. The economic threshold level of the disease is fixed as 18% at a sale price of Rs. 50/- per kg of made tea.

Fig. 8.4. Grey blight infection on tea leaf.

ii) Management: Spraying of carbendazim, mancozeb, or thiophanate methyl is suggested to control the disease.

8.2.6. Collar Canker, *Phomopsis theae*

i) Symptoms: A wound pathogen, *Phomopsis theae* is responsible for the disease (Fig. 8.5). It is prevalent in young tea and clones are more susceptible than seedlings. *In vitro* studies indicated that the pathogen completes its life cycle in 10-13 days. Impact of predisposing factors on disease development indicated that the nature of soil is an important factor. Disease incidence was more in gravelly soil.

Fig. 8.5. Collar canker on tea.

ii) Management: Deep planting, application of fertilizer near to plant collar and mulching with raw coir pith close to the collar predispose the disease. Soil drenching of the systemic fungicide carbendazim and application of the same as wound dressing provided satisfactory control. However, the efficacy was superior when biological control agents like *Trichoderma harzianum* and *Gliocladium virens*, were applied both to the soil around the bush and used for wound dressing.

8.2.7. Thorny Stem Blight, *Tunstallia aculeata*

i) Symptoms: The disease is caused by the fungus *Tunstallia aculeata*. The spread of the disease was higher when the pruning timing coincided with the sporulation period. There was also a direct correlation between root carbohydrate reserves and disease development.

ii) Management: An integrated management strategy combining replanting with resistant genotype, cultural control, chemical and biological control measures was found ideal.

8.2.8. Black root, *Rosellina arcuata*; Brown root, *Fomes noxius*; Red root, *Poria hypolateritia* and Root Splitting

i) Symptoms: Four types of root diseases viz., root splitting, black, red and brown root disease may occur in patches. *Rosellina arcmata* produces black mycelium on affected roots and white star shaped mycelium between bark and wood. *Fomes noxius* affected roots become soft and spongy. *Poria hypolateritia* produces blood red mycelium on affected roots.

ii) Management: Black root disease can be controlled by avoiding burial of pruning's in infested field and incorporation of *Trichoderma virens* at 200 g/plant at the time of planting. Other root diseases can be controlled by isolating the infected area, opening trenches of 1.3 m deep and 45 cm width and uprooting and burning of infected bushes. Rehabilitating the soil with Guatemala grass and use of biocontrol agents at 200 g/plant (*Trichoderma harzianum* for red root and root splitting; *T. viride* for black root; *T. harzianum, T. viride, T. hamatum, T. resei* and *T. koningii* for brown rot and *T. virens* for red, brown and root splitting) effectively check the diseases.

Soil fumigation with metham sodium and carbon disulphide gives effective control of these diseases.

8.3. COCONUT

The coconut palm (*Cocos nucifera*), originated in South East Asia, is now cultivated widely throughout the tropics of the world. The largest producers of coconut are the Philippines, Indonesia, India and Sri Lanka. In all the coconut producing countries, it is an important source of export earning despite being a regular constituent of the food. Coconut is an important small holders' plantation crop. The total area under coconut cultivation in India is 2.029 million ha with an annual production of 10.190 million tonnes and the average productivity of 5.0 tonnes/ha is the highest in the world (Bijay Kumar, 2007) (Table 8.4). It provides proteins, fats and some vitamins.

Coconut contributes 700 billion rupees to the GDP of the country. Coir and coir products exported from India earned Rs. 2260 million during the above period. The contribution of the crop to the total edible oil pool in India is around 6 per cent. It is supporting the livelihood of 10 million people across the country. The four Southern States – Kerala, Tamil Nadu, Karnataka and Andhra Pradesh together accounts for 90% of the total area and production. Coconut is predominantly grown under rain fed conditions in Kerala and parts of coastal Karnataka, Maharashtra and Tamil Nadu. In rest of the country, it is mainly grown under irrigated conditions.

Table 8.4. State-wise area, production and productivity of coconut in India (2005-06)

State	Area in '000 ha	Production in '000 MT	Productivity in MT/ha
Kerala	897.8	4352.3	4.847
Tamil Nadu	370.5	3348.5	9.037
Karnataka	467.5	832.5	1.780
Andhra Pradesh	104.0	613.7	5.900
Maharashtra	18.0	188.1	10.450
Orissa	50.8	189.0	3.720
West Bengal	24.9	222.6	8.939
Assam	19.1	141.0	7.382
Goa	25.3	86.2	3.407
Gujarat	16.4	95.1	5.798
Andaman & Nicobar	25.6	59.9	2.339
Lakshadweep	2.7	36.5	13.518
Pondicherry	2.1	19.2	9.142
Tripura	3.3	4.8	1.454
Nagaland	0.9	0.8	0.888
Total	2028.9	10190.1	5.0

8.3.1. Basal Stem Rot or Thanjavur Wilt, *Ganoderma lucidum, G. applanatum*

i) Symptoms: In Tamil Nadu and Andhra Pradesh, Thanjavur wilt, otherwise called Ganoderma wilt is a serious problem. The symptoms are presence of bleeding patches at the stem base, premature yellowing and drooping of outer whorl of leaves and gradual drying of spindle (Fig. 8.6). The other symptoms include decay of root system, flaccid spindle leaves, browning of outer leaves, and appearance of bleeding patches on the basal region of the stem. In advanced stage, bracket of fungus will be seen at the base of the tree.

Fig. 8.6. Basal stem rot in coconut.

ii) Management: Integrated control measures include phytosanitation, digging isolation trenches around diseased palms, root feeding of calixin, drenching the basin with 1% Bordeaux mixture and application of 5 kg neem cake/palm. Raise banana in diseased garden, since banana is resistant to the pathogen.

Light irrigation coupled with Bordeaux mixture drenching (40 litres of 1% Bordeaux mixture around the trunk in a radius of 1.5 m) helps in reducing the disease incidence. Raising banana as intercrop is recommended. *T. harzianum* has been found to control the multiplication of the pathogen in sick soils. When combined with phosphobacteria or plant growth promoting rhizobacteria, synergistic effects have been noticed. *T. harzianum* applied along with neem cake reduced the disease index and increased the yield of coconut.

8.3.2. Bud Rot, *Phytophthora palmivora*

i) Symptoms: Bud rot disease of coconut is characterized by withering, rotting of spindle leaf and foul smell (Fig. 8.7). The disease is caused by a fungus *Phytophthora palmivora*. Tender leaf and soft tissues of the crown rot into a slimy mass and emit foul odour.

ii) Management: To control this disease remove decayed tissues and dress with 10% Bordeaux paste. The disease can be prevented by taking up prophylactic spraying of 1% Bordeaux mixture during the pre monsoon season.

Prophylactic spraying of young palms with 1% Bordeaux mixture, improved drainage, wide spacing and weed control helps to reduce the disease severity. Removal of infected spindle showing initial yellowing symptoms and application of 1% Bordeaux paste reduce the disease incidence. *Bacillus subtilis, B. macerans, B. amyloliquifaciens* and *Myrothecium roridum* have all been found to be checking the growth of the pathogen. The hybrid Malayan Dwarf x West African Tall (Hybrid PB 12) is tolerant.

Fig. 8.7. Bud rot on coconut.

8.3.3. Stem Bleeding, *Theilaviopsis paradoxa*

i) Symptoms: The characteristic symptom of stem bleeding is the dark gummy exudation from the trunk. The causal organism is a fungus *Theilaviopsis paradoxa*. Exudation of reddish brown liquid is observed through longitudinal cracks in the trunk, generally at the base of the trunk (Fig. 8.8). Bleeding patches spread through out as the disease advances. The liquid oozing out dries up and turns black. Tissues below the lesions rot and turn yellow and then black. Premature yellowing of leaves is observed in the outer whorl. Trunk gradually tapers at apex and crown size becomes reduced.

Fig. 8.8. Stem bleeding on coconut.

Growth cracks on trunk, severe summer followed by sudden wetting, imbalanced nutrition, and excess salinity are the predisposing factors for the disease.

ii) Management: Control measures include chipping off the infected tissues, wound dressing and root feeding with 5% calixin and application of neem cake along with antagonistic fungi like *Trichoderma*. Remove the affected tissues using a chisel and apply Bordeaux paste and coal tar after 1 - 2 days (Fig. 8.9).

Fig. 8.9. Chiseling affected potion and application of Bordeaux paste and coal tar.

The integrated management practices recommended are removal of not only dead palms but also those in advanced stages of disease and/or stumps along with bole and root system and destroying; isolation of diseased palms from healthy palms by digging trenches of 1 m deep and 30 cm wide; providing adequate soil moisture through irrigation coupled with mulches; avoiding flood irrigation, overcrowding, preventing wounding of stem which predispose the palms to infection and ploughing to prevent the spread of inoculum; providing good drainage facilities; application of 50 kg FYM and 5 kg neem cake/palm/year and raising *Ganoderma* resistant crop like banana as intercrop.

Neem cake enriched with *Trichoderma virens, T. hamatum* and *T. harzianum* has been found to be very effective in reducing the population of the pathogen in the soil. These antagonistic fungi thrive well in neem cake supplemented with a small quantity of rice/wheat bran thus effecting their multiplication an easy task. *Trichoderma harzianum, T. viride, T. hamatum* and *T. virens* grow very well in rice bran and neem cake (1:1 w/w) and reduced stem bleeding when applied to the soil. Soil application of neem cake and FYM mixed with. *T. virens* showed reduction of stem bleeding up to 31.3%. Soil application of *T. virens* along with neem cake and FYM showed the least disease index and the highest yield.

West Coast Tall was found more resistant to stem bleeding.

8.3.4. Fruit Rot or Mahali, *Phytophthora palmivora*

i) Symptoms: Shedding of female flowers (buttons) and immature nuts are the symptoms of the disease. Water soaked lesions appear on the young fruits or buttons near the stalk which later develop and result in the decay of the under lying tissues (Fig. 8.10). The pathogen is more active during rainy season.

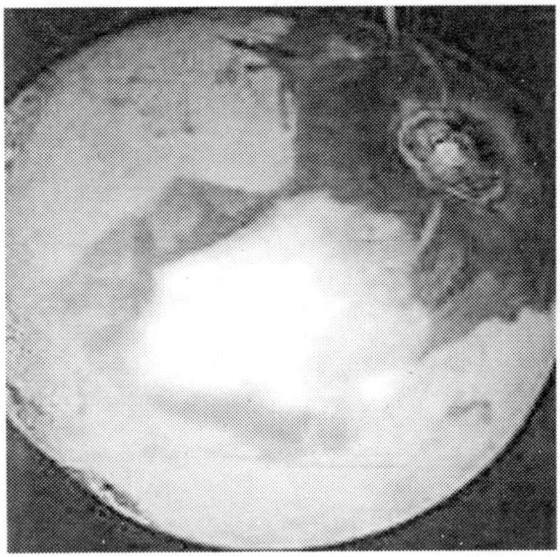

Fig. 8.10. Fruit rot of coconut.

ii) Management: Pre-monsoon spraying the crown with 1% Bordeaux mixture (or any other copper fungicide such as copper oxychloride 0.5%) will control the disease. One or two sprayings at intervals of 40 days is also advised. The shed nuts should be collected and burnt.

8.3.5. Leaf Rot, *Colletotrichum gloeosporioides, Exerohilum rostratum*

i) Symptoms: This fungal disease is seen in palms affected by root (wilt) disease. The first visible symptom is blackening and shriveling up of the distal ends of the leaflets in the spindle. Later the affected portion breaks off in bits giving the infected leaves a fan-like appearance. If not controlled, subsequent spindles also get infected.

8.4. ARECANUT

Areca nut (*Areca catechu*), otherwise known as the betel nut, has its origin in the humid regions of Asia and Malay Islands. India continues to dominate the world in area, production and productivity of areca nut and has achieved self sufficiency in areca nut production. It is an important cash crop of India. India is the largest producer of areca nut in the world covering an area of 3.811 lakh ha with annual production of 4.831 lakh tonnes (Bijay Kumar, 2007) (Table 8.5). Most of the production is domestically consumed. It is masticatory and the ripe fruits are sometimes used as an antihelminthic and astringent in Europe. It has a toning effect and stimulates the nerves. The extract after boiling the nuts contains tannins which are used for tanning leather. Areca nut industry forms the economic back bone of nearly 6 million people in India and for many of them it is the sole means of livelihood.

Table 8.5. State-wise area, production and productivity of areca nut in India (2005-06)

State	Area in '000 ha	Production in '000 MT	Productivity in MT/ha
Kerala	108.6	119.3	1.098
Tamil Nadu	4.8	10.3	2.140
Karnataka	161.6	215.7	1.334
Andhra Pradesh	0.3	0.2	0.666
Maharashtra	2.0	5.0	2.500
West Bengal	9.4	29.3	3.128
Assam	71.2	66.8	0.938
Goa	1.6	2.6	1.625
Andaman & Nicobar	4.4	4.8	1.090
Meghalaya	11.5	15.5	1.356
Pondicherry	0.1	0.1	1.000
Tripura	3.4	6.9	2.029
Mizoram	2.0	5.3	2.650
Nagaland	0.2	1.3	6.500
Total	381.1	483.1	1.3

8.4.1. Fungal Diseases

8.4.1. Koleroga/ Mahali/Fruit Rot, *Phytophthora arecae*

i) Symptoms: Koleroga caused by *Phytophthora arecae* is one of the major diseases of areca nut causing serious losses. This occurs as an epidemic in the heavy rainfall areas of Karnataka and Kerala. The disease first makes its appearance after monsoon period. The first symptom is the appearance of water-soaked lesions on the nut surface near the calyx. The patches enlarge and nuts darken and they shed in large number. The fallen nuts soon develop whitish mycelial mass all over. Nuts of all ages are attacked and if unchecked, the pathogen invades crown causing the leaves and bunches to whither. Sometimes, the infected nuts may not be shed and remain mummified in the bunches. Such type of infection is known as 'dry mahali'.

ii) Management: Removal and destruction of infected debris, shed nuts, dried bunches and dead crowns help to minimize the disease spread and development. Covering the bunches with polythene (Fig. 8.11) or leaf sheath or grass before the onset of monsoon helps to reduce the disease. Pre and post monsoon spraying with 1% Bordeaux mixture gives effective control of koleroga. Spray Bordeaux mixture 1% on all bunches three times in a year, once just before the onset of southwest monsoon and the rest at 40 days intervals. If monsoon season is prolonged give a third spray. Use rosin soda adhesive to ensure tenacity of the spray deposit on treated substrate.

Fig. 8.11. Covering the bunches with polythene film to control koleroga of areca nut.

8.4.2. Foot Rot or Anabe Disease, *Ganoderma lucidum*

i) Symptoms: This disease is caused by the fungus, *Ganoderma lucidum*. It is more prevalent in neglected gardens, causing 7% losses. Symptoms of the disease are akin to that of the draught. The initial visible symptom is yellowing of outer whorl of leaves which gradually extends to inner whorls followed by wilting and drooping. The development of inflorescence and nuts are arrested. Nuts already formed are shed. At later stages, the weakened crown topples off leaving a bare trunk. All around the base of the palm, brownish patches appear which exude a brown liquid (Fig. 8.12). The interior of the stem at the basal region is discoloured and rotten, emitting a foul smell. The infection extends to roots and gets discoloured, brittle and dry. The fungal invasion interrupts uptake of water and nutrients by the palm, leading to yellowing and wilting.

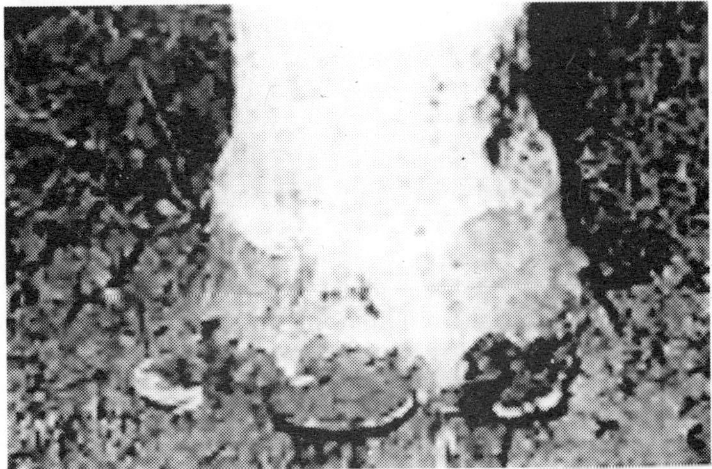

Fig. 8.12. Foot rot of areca nut.

ii) Management: Since the infected stumps and roots act as the main foci of infection, strict phyto-sanitary measures are to be adopted by removing and destroying the stumps along with roots. Isolate affected palms by digging trenches 60 cm deep and 30 cm wide around, away from the base and drench with 0.3% captan. Drench the soil with 1% Bordeaux mixture before planting. Discourage growing of collateral hosts of the fungus like *Delonix regia* and *Pongamia glabra.*

- Isolate affected palms by digging trenches 60 cm deep and 30 cm wide around, one metre away from the base and drench with captan (0.3%), calixin (0.1%) or copper oxychloride (0.3%).
- Remove and destroy all severely affected palms and stumps of dead palms.
- Drench the soil with 1% Bordeaux mixture before planting healthy seedlings.
- Follow recommended spacing of 2.7 x 2.7 m.
- Discourage growing of collateral hosts of fungus such as *Delonix regia* and *Pongamia glabra* in the vicinity of gardens.
- Apply organic manure at 15-20 kg/palm and 15-20 kg green leaf along with 2 kg neem cake per palm to reduce spread of the disease.
- Avoid flood irrigation and water flowing from infected palms to healthy palms.
- Drench the root zone of affected palm with calixin (3 ml/ l) at 15- 20 l/palm + root feeding with calixin (15 ml/l) at 125 ml/palm at quarterly intervals.
- Avoid repeated ploughing and deep digging which helps to reduce the disease incidence.

8.4.3. Bud Rot, *Phytophthora palmivora*

i) Symptoms: Bud rot is a fatal disease of areca palms caused by *Phytophthora palmivora* and characterized by rotting of terminal bud and surrounding tissues and ultimately killing the palm. Affected spindle appear yellow, later turning brown (Fig. 8.13) and finally the whole spindle rots emitting a disagreeable odour. Severe infection leads to death of palm.

ii) Management: Early detection of the disease and prompt removal of infected tissues will help in recovery of palms and also prevents the spread of the disease. In early stages of infection, scoop out affected rotten tissues by making longitudinal side splits and apply Bordeaux paste on the exposed healthy tissues or drench with 1% Bordeaux mixture. Apply 1% Bordeaux mixture to the crown of healthy palms in the vicinity of the affected ones. Avoid heavy intercropping, cultivation in low lying areas and closer planting to reduce spread and development of the disease.

Fig. 8.13. Bud rot of areca nut.

8.4.4. Inflorescence Dieback and Button Shedding, *Colletotrichum gloeosporoides*

i) Symptoms: Die-back of inflorescence due to micro-organisms is reported to be associated with the low fruit set in areca nut. About 60% of the palms in the states of Karnataka and Kerala are infected by this disease causing severe shedding of buttons. The symptoms include yellowing and drying of the rachis from the tip towards the base followed by shedding of female flowers (buttons) (Fig. 8.14).

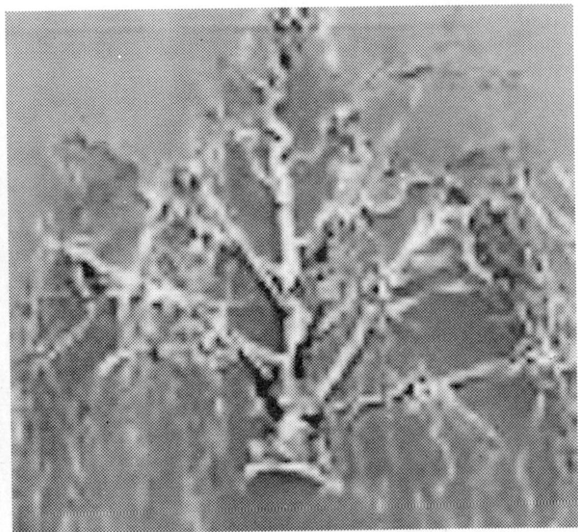

Fig. 8.14. Button shedding in areca nut.

ii) Management : Removal and burning of inflorescence helps to reduce the load of inoculum in the field. Spray zineb (4 g/1) or mancozeb (3 g/l) twice, once just after female flowers are set and again 15-28 days later. Aureofungin solution at 50 ppm concentration is also effective in controlling the disease.

8.4.5. Stem Bleeding, *Thielaviopsis paradoxa*

i) Symptoms: Stem bleeding caused by *Thielaviopsis paradoxa*, is prevalent in isolated pockets in all areca nut-growing States of South India. Younger and middle-aged palms are more susceptible. Symptoms appear mostly on lower portions of the stem as small discoloured depressions which later coalesce and cracks develop with the progress of the disease. The fibrous layer disintegrates which hollows up to varying depths and brown gummy exudates oozes out. Crowns of affected palms get reduced in size followed by reduction in yield. The disease is serious in gardens with poor drainage.

ii) Management : Improving the drainage may help in minimizing the disease incidence. Scooping out the affected portions and application of coal tar or Bordeaux paste is effective to reduce the disease incidence.

8.5. BETEL VINE

Betel vine (*Piper betel*) is a perennial, dioecious, evergreen creeper grown in India. It is a chewing stimulant of brain, lungs and heart. The leaves contain Vitamins B and C. It is an important cash crop in Andhra Pradesh, Assam, Bihar, Karnataka, Kerala, Madhya Pradesh, Maharashtra, Orissa, Tamil Nadu, Tripura, Uttar Pradesh and West Bengal with an annual turnover of about Rs.700 crores. Its leaves are exported to Pakistan, Bangladesh, Indonesia, Malaysia, Burma and Thailand.

8.5.1. Foot and Leaf Rot, *Phytophthora parasitica*

i) Symptoms: The plant loses luster, leaves droop while still green, turn yellow and drop. Leaves turn brown and later turn black and rot (Fig. 8.15). Blackish brown marks will be visible on the stem at ground level or slightly above. The stem is discoloured and later rots. Stem becomes disintegrated.

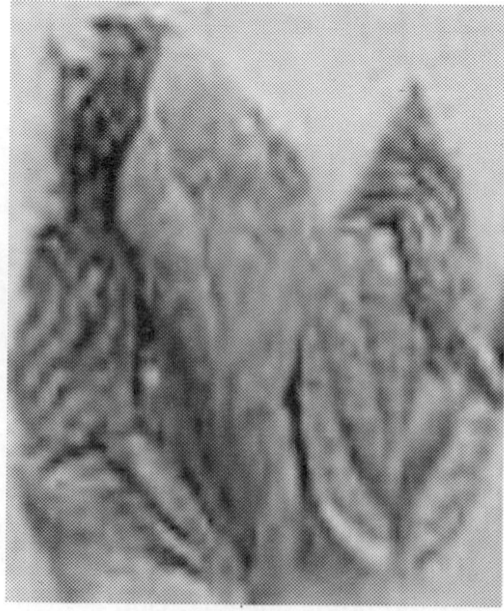

Fig. 8.15. Foot rot of betel vine.

ii) Management: Drenching the soil with 4 g copper oxychloride/litre of water gives effective control of the disease. Soil application of *T. harzianum* reduced the vine mortality (11 to 21% compared to 15 to 55% in control) and increased the disease control (46 to 75% compared to 0% in control).

8.5.2. Collar/Basal Rot, *Sclerotium rolfsii*

i) Symptoms: The plants are usually attacked at ground level. Dense white cotton like mass of threads (mycelium) are seen on stems. This causes rotting of affected portion causing wilting and ultimate death of plants.

ii) Management: Spraying 1% Bordeaux mixture or 0.2% zineb, or drenching with 0.1% wettable ceresan is effective.

Soil application of *T. harzianum* (strains MTCC Nos. 3841 to 3843) and *T. virens* (strain MTCC No. 794) increased vine height (199 to 254 cm compared to 67 to 69 cm in control), decreased plant mortality (7 to 33% compared to 73 to 87% in control), increased the disease control (64 to 83% compared to 0% in control) and increased the yield significantly. Consortium of 2 strains of *Pseudomonas fluorescens* (NBRI-N6 and NBRI-N) showed significant increase in yield of betel vine leaves.

8.5.3. Anthracnose/ Leaf Spot, *Colletotrichum piperis*

i) Symptoms: Small black circular irregular lesions appear on leaves expanding rapidly in humid conditions. Leaf spot appears as brownish black centre with yellowish halo around (Fig 8.16) and in serve causes the leaves drop owing to shrinkage of tissues.

Fig. 8.16. Leaf spots on betel vine.

ii) Management: Destroying affected leaves helps to control spread of the disease. Spraying of 1% Bordeaux mixture or 0.2% zineb gives effective control of anthracnose.

8.5.4. Powdery Mildew, *Oidium piperis*

i) Management: Destroy affected leaves to control the spread of disease.

8.6. OIL PALM

Oil palm (*Elaeis guineensis*) is the highest oil-yielding plant (4-6 tonnes oil/ha) among perennial oil-yielding crops, producing palm oil and palm kernel oil. These are used for culinary as well as industrial purposes. It occupies an area of 44,413 ha in Andhra Pradesh (80% area), Karnataka, Assam, Gujarat, Goa, Kerala, Maharashtra, Orissa, Tamil Nadu, Tripura and West Bengal. Average productivity in oil palm plantations is now 4-5 tonnes/ha which compares favourably with other countries.

Oil palm diseases include: blast (*Pythium splendens, Rhizoctonia lamellifera*), freckle (*Cercospora elaeidis*), anthracnose (*Botryodiplodia palmarum, Melanconium elaeidis, Glomerella cingulata*), seedling blight (*Curvularia eragrostidis*), yellow patch and vascular wilt (*Fusarium oxysporum*), basal trunk rot (*Ceratocystis paradoxa, Ganoderma* spp., *Armillaria mellea*); crown disease and fruit rot (*Marasmius palmivorus*). Spear (bud) rot is caused by the bacterium Erwinia spp., which has been devastating in Central Africa.

8.6.1. Basal Stem Rot, *Ganoderma boninense*

i) Symptoms: The basal stem rot (BSR) of oil palm caused by *Ganoderma boninense* is the most serious disease of oil palm in Malaysia. Under severe infestation situations, more than 50% of oil palm stand can be lost to the malady (Fig. 8.17).

Fig. 8.17. Basal stem rot on oil palm.

ii) Management : Arbuscular mycorrhizal fungi (AMF) have recently been experienced to be able to confer protection to nursery inoculated palms that were subsequently planted in *Ganoderma* endemic fields. Wide-scale use of AMF is expected to further delay if not prevent infection of palms replanted on fields previously devastated by *Ganoderma*.

8.6.2. Anthracnose, *Botryodiplodia palmarum, Melanconium elaeidis, Glomerella cingulata*

i) Symptoms: This disease occurs in the nursery. It is recognized by regular or irregular brown to black leaf blotches surrounded by yellow haloes, which develop along the margin, centre or tip of the leaves. It causes heavy seedling loss.

ii) Management : The disease can be controlled by spraying mancozeb or captan at the rate of 200 g/100 litres of water. Copper fungicides should not be used because of the extreme susceptibility of oil palm seedlings to copper burn (scorching).

8.7. CACAO

Cacao (*Theobroma cacao*) is an important commercial crop grown the world over. It is native of Amazon Valley of South America. The total area under cacao cultivation in India is 0.295 lakh ha with an annual production of 0.102 lakh tonnes with an average productivity of 0.345 tonnes/ha (Bijay Kumar, 2007) (Table 8.6). In India, it is being cultivated in Kerala, Karnataka, Andhra Pradesh and Tamil Nadu. Kerala account for about 76% of the area and 78% of the total production in the country.

Table 8.6. State-wise area, production and productivity of cacao in India (2005-06)

State	Area in '000 ha	Production in '000 MT	Productivity in MT/ha
Kerala	10.2	6.5	0.637
Tamil Nadu	1.4	0.2	0.142
Karnataka	6.0	2.8	0.445
Andhra Pradesh	11.8	0.7	0.059
Total	29.5	10.2	0.345

8.7.1. Seedling Dieback, *Phytophthora palmivora*

i) Symptoms: Leaves turn yellow and fall leaving a bare length of shoot. Xylem discoloured by brown streaks.

ii) Management: The infected and dead seedlings should be removed and destroyed. Improvement of drainage facilities in the garden followed by drenching with 1% Bordeaux mixture was also found effective.

8.7.2. Black Pod Rot, *Phytophthora palmivora*

i) Symptoms: Black pod disease is the major disease of cacao prevalent during monsoon season. The affected pods become pale brownish spoiling the quality of beans. Infection appears as chocolate brown spot which spreads very rapidly and soon occupies the entire surface of the pod (Fig. 8.18). As disease advances, a whitish growth of fungus consisting of fungal sporangia is produced over the affected pod surface. Affected pods become dark brown or black.

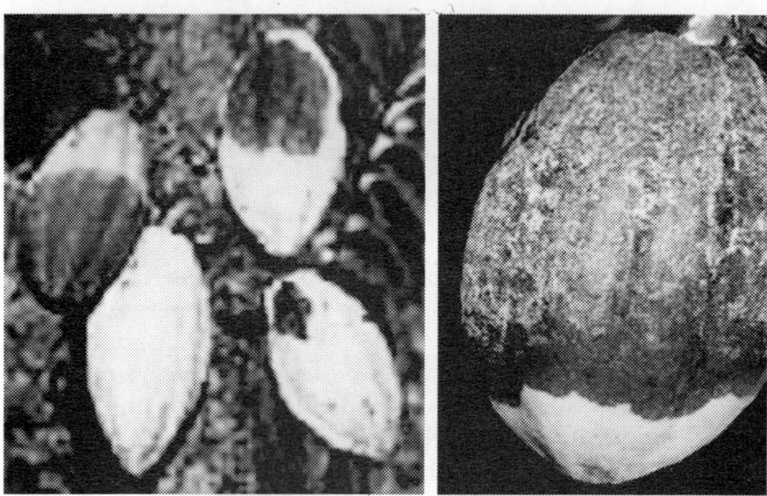

Fig. 8.18. Black pod on cacao.

ii) Management: All the diseased pods should be removed and destroyed at weekly interval especially during rainy season. Regulation of shade by proper pruning and allowing the sunlight to penetrate and thus help in minimizing the disease incidence. The disease can be prevented by phytosanitation and prophylactic spraying with 1% Bordeaux mixture before the onset of monsoon.

8.7.3. Charcoal Pod Rot, *Botryodiplodia theobromae*

i) Symptoms: Infection appears as dark brown to black coloured spots. Affected pods turn black and remain as mummified fruit. Internal tissues rot and affected beans turn black. Spores appear in masses forming a soot.

ii) Management: Spraying with 1% Bordeaux mixture was found effective.

8.7.4. Brown pod rot, *Diplodia natalensis*

i) Symptoms: The pathogen produces water soaked chocolate brown or dark lesions on pods at tip or stalk end. Development of black sooty mould takes place on affected pods.

ii) Management : Spraying of copper oxychloride or captafol at 2g/litre of water is effective for the management of disease.

8.7.5. Pink Disease, *Corticum salmonicolor*

i) Symptoms: The pathogen causes pinkish powdery coating on stem of the affected plants. It also causes wilting of shoots, shedding of leaves and ultimately drying of branches.

ii) Management : Spraying of copper oxychloride or captafol at 2g/litre of water is effective for the management of disease.

8.7.6. White Thread Blight, *Marasmius scandens*

i) Symptoms: White mycelial threads of fungus spread longitudinally and irregularly along the surface of young stem or branches. Affected leaves turn dark brown. Leaves eventually detach from the stem but remain suspended by mycelial thread.

ii) Management : Spraying of copper oxychloride or captafol at 2g/litre of water is effective for the management of disease.

8.7.7. Vascular Streak Die-back, *Oncobasidum theobromae*

i) Symptoms: The pathogen is responsible for characteristic yellowing of 1 or 2 leaves on the 2nd or 3rd flush behind growing tip. Other leaves on the shoot soon show similar symptoms. Brown streaking of woody tissue extending well beyond the region of yellow leaves is also observed.

ii) Management: Soil drenching with copper oxychloride at 2g/litre of water is recommended against wilt and dieback.

8.7.8. Stem Canker, *Phytophthora palmivora*

i) Symptoms: Stem cankers are characterized by oval to round, rusty-brown discolouration of the external bark that looks purple when scraped (Fig. 8.19). Attack of young shoots results in die-back.

Fig. 8.19. Stem canker on cacao.

ii) Management : Disease can be effectively managed by wound sealing with Bordeaux paste.

8.8. RUBBER

India is the fourth largest producer of natural rubber with a share of 8.8% in world production in 2006. The two major states with rubber plantations are Kerala and Tamil Nadu, with Kerala alone accounting for over 80% of the area under plantation. India produced an estimated 852,000 tonnes of natural rubber during 2006-2007 and exported 56,545 tonnes while it imported 89,699 tonnes. In 2006 the yield per hectare in India at 1879 kg, was amongst the highest in the world.

8.8.1. Abnormal Leaf Fall, *Phytophthora palmivora, P. meadii, P. nicotianae* var. *parasitica, P. botryosa*

The disease recur annually during southwest monsoon period. Prolonged wet weather coupled with humid atmospheric conditions favour the disease. Disease incidence is less in Kanyakumari District, but increasingly heavy northwards along the southwestern coast.

i) Symptoms: First the fruits rot, later infected leaves fall in large numbers prematurely, either green or after turning coppery red. A black lesion may develop on the petiole with a drop of latex, often coagulated, in the centre. Lesion may develop on the midrib and leaf blades also (Fig. 8.20). Heavy defoliation may lead to considerable loss of crop and die-back of terminal twigs.

Fig. 8.20. Leaf fall in rubber.

ii) Management: Prophylactic spraying of the foliage prior to the onset of southwest monsoon with either 1% Bordeaux mixture using high volume sprayers or oil-based copper oxychloride dispersed in diluents spray oil employing either low volume air blast sprayers from the ground or through aerial application. For micron spraying, based on the tree spread, foliage intensity, planting material used and age of plants, two rounds of spray using about 17 to 22 litres of fungicide oil mixture (1: 6 proportion) per hectare per round with a gap of 10 to 15 days or a single round of spray with about 30-40 litres of fungicide oil mixture (1: 5 proportion) per hectare may be necessary. For aerial spraying, 6.2 litres of 40% oil based copper oxychloride paste in 37 litres of diluents oil or 8 kg oil dispersible copper oxychloride powder in 40 litres of oil is used per hectare. Spraying should be done as close to the monsoon as possible. Rubber clones RRII 105, PB 217, GT 1 and GL 1 are showing some tolerant reaction to the disease.

8.8.2. Bird's Eye Spot, *Drechslera heveae*

This is a hot weather disease causing serious damage in the nursery during April-May. Weaker plants and plants growing under exposed situations are more susceptible.

i) Symptoms: Symptoms appear as small necrotic spots with dark/ brown margins and pale centre (Fig. 8.21). Severe infection leads to premature defoliation and dieback. Nursery seedlings are susceptible.

Fig. 8.21. Bird's eye spot on rubber leaf.

ii) Management: Repeated spraying with 1% Bordeaux mixture or 0.2% mancozeb (Mancozeb/ Indocile M 45 at 2.5 g/l) or carbendazim 0.02% (Baiting at 0.4 ml/l). Shading the nursery plants reduces the disease incidence. Seedlings should be maintained in vigorous condition through adequate balanced nutrition.

8.8.3. Black Stripe, Black Thread or Black Rot, *Phytophthora palmivora, P. meadii*

The disease is prevalent during the rainy season.

i) Symptoms: In the renewed bark region, small depressions are formed due to localized rotting and drying of bark, which gets pressed to the wood (Fig. 8.22). When scraped, deep vertical black lines are noticed running downward into the tapping bark and upwards into the renewed bark. Bark rot often reduces yield. The renewed bark becomes highly uneven.

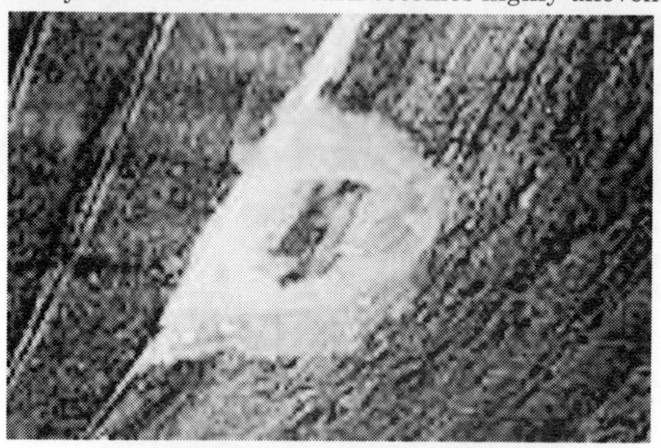

Fig. 8.22. Black stripe on rubber.

ii) Management: If tapping is regularly done during rainy season, the tapping panel should be disinfected at weekly intervals by brushing with mancozeb 0.375% a.i (Dithane/Indofil M-45 at 5 g/l). Phosphoric acid formulations at 0.08% (Akomin and Phosjet at 2 ml/l) are also effective. In the infected cases, scrape off the affected tissues and apply the fungicide. When the fungicide dries up, a wound dressing compound may be applied (Fig. 8.23).

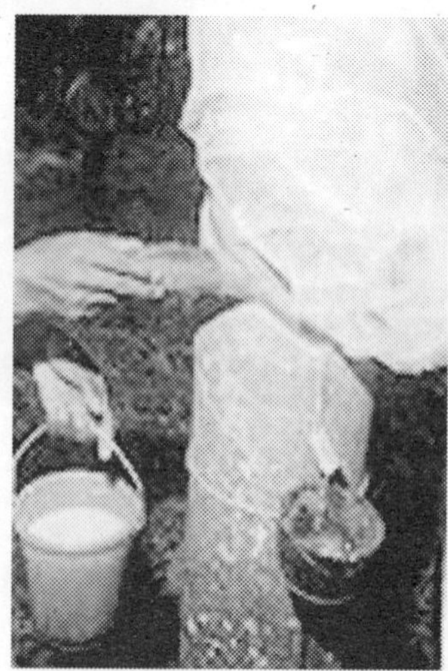

Fig. 8.23. Fungicidal application for the control of black stripe disease.

8.8.4. Dry Rot, Stump Rot, Collar Rot or Charcoal Rot, *Ustulina deusta*

Infection of the pathogen takes place during the wet weather.

i) Symptoms: Affects roots, collar, trunk and branches with copious exudation of latex from the lesions (Fig. 8.24). Affected bark and wood become soft and powdery with double black lines in the wood. Grey, crust-like flat fructifications are seen on the bark, which later coalesce and turn black. Affected trees or branches are killed or blown over. *Ustulina* root infections are also noticed.

Fig. 8.24. Dry rot disease on rubber.

ii) Management: Scrape off the fructifications, affected bark and wood showing black lines. Apply a wound dressing compound in which thiram 0.75% a.i. (Thiride at 10 g/l) incorporated. Hexaconazole 0.02% a.i. (Contaf at 4 ml/l) is also effective. Avoid accumulation of rubber at the base of the tree. For root infection, clean up affected areas by washing with Emisan solution. Scrape out the fructifications. Wash the wound again with fungicides. When it dries up, apply a wound dressing compound.

8.8.5. Leaf Spot, *Corynespora cassiicola*

Earlier the disease was confined to nurseries only. From 1996 onwards severe incidence of the disease was observed in mature plantations in the Dakshina Kannnada District of Karnataka State and in the adjoining districts of Kerala. In nurseries, disease incidence is noticed during November to May period. On mature trees, disease is observed during the refoliation period from February to May.

i) Symptoms: Appearance of large spots, with brown margins and pale centre is the common symptom (Fig. 8.25). The centre may fall off forming shot holes. On mature trees, light green leaves during refoliation are more susceptible. Several lesions coalesce to form large blighted area. Disease spreads along the veins leading to a brownish "railway track" or "fish bone" like appearance. Even a localized infection on midrib or base of a leaf causes leaf abscission. Defoliation leads to die-back of branches.

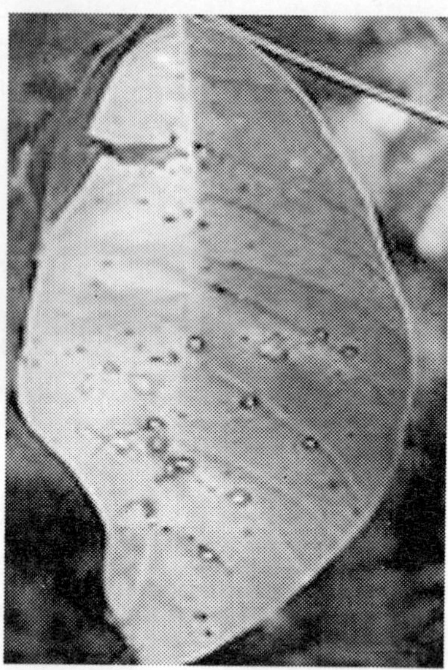

Fig. 8.25. Leaf spot on rubber.

ii) Management: Repeated spraying with 1% Bordeaux mixture or 0.2% mancozeb or 0.02% carbendazim is recommended for nursery. Shading the nursery reduces the disease incidence. Seedlings should be maintained in vigorous condition through adequate balanced nutrition. High volume spraying with mancozeb 0.2% (Dithane/Indofil M-45 at 2.66 g/l) carbendazim 0.05% (Bavistin at 1g/l) at 2-3 weeks interval during refoliation is effective in mature plantation. Micron spraying with oil dispersible copper oxychloride 56% (8kg) or oil dispersible mancozeb 70% (7 kg) dispersed in 40 litres spray oil per ha is also effective.

8.8.6. Patch Canker or Bark Canker, *Phytophthora palmivora*, *Phytophthora meadii*, *Pythium vexans*

Infection mostly occurs during the wet weather.

i) Symptoms: Infection occurs on the tapping panel region, or anywhere on the stem including the collar region and occasionally on the roots causing swelling and bark burst with amber-coloured liquid oozing out (Fig. 8.26). This may not be evident in all cases. In most cases oozing of latex is observed. The bark rots, and a coagulated rubber pad, emanating a foul smell is seen in between the wood and the rotting bark. When this is removed, a slight discolouration of the wood in this region is also noticed.

ii) Management: The affected region may be scraped to remove all the rotting bark and the coagulated rubber and the wound washed well with mancozeb 0.75% (Dithane/Indofil M-45 at 10 g/l). When the fungicide dries up, apply wound dressing compound.

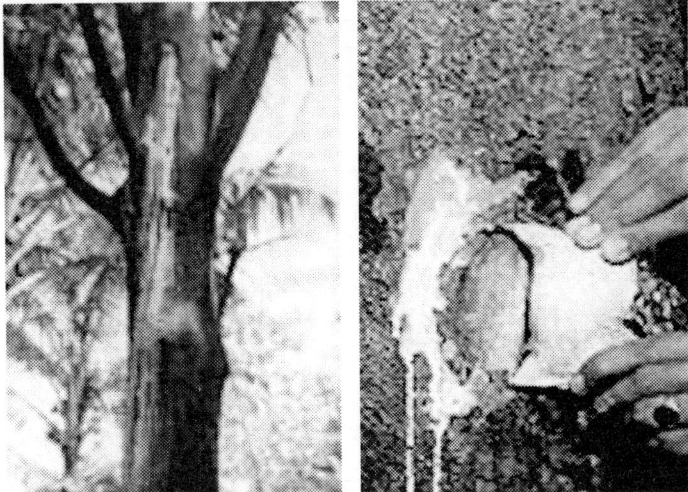

Fig. 8.26. Bark canker on rubber.

8.8.7. Pink Disease, *Corticium salmonicolor*

The disease is noticed during southwest monsoon period in the traditional rubber growing areas in South India. The disease incidence is very mild in Kanyakumari District due to the weak southwest monsoon and absence of prolonged wet conditions. Disease has been noticed in Assam, Tripura and Meghalaya in North East.

i) Symptoms: The disease is more damaging on plants in the age group of 2 to 12 years. Main seat of infection is usually the fork region. White or pink coloured cobweb mycelial growth on the bark surface with streaks of latex oozing out from the lesions can be observed (Fig. 8.27). Rotting, drying up and cracking of the affected bark follows. Sprouts develop from below the affected portion. The distal portion of branches dry and dried leaves stick to the dead branches.

ii) Management

a) Prophylactic: Prophylactic treatment for pink disease can be taken up for highly susceptible clones planted in highly disease prone areas. Two rounds of spraying with 1% Bordeaux mixture (during May and August) onto the fork and branches reduce the disease incidence. For 2-year-old plants, the top most brown portion has to be applied with Bordeaux paste in a 30 cm wide band all around. The forking region also has to be applied with the paste. In leaning plants, in addition a 30 cm wide band on the convex side of the bend has to be applied. In 3 years old plants, application has to be made on all major forking regions except the lower most one and also on all the leaders of the top most brown regions. Only one application is required. No scraping is necessary. The application can be made from ground using a long handled-brush before the onset of monsoon. During October, a single tracing and treatment need be made for infection, if any. No sticker should be added to the paste. Rubber clones like PB 86, RRIM 513, GL 1, PR 107, GT 1 and PB 260 are less susceptible.

Fig. 8.27. Pink disease on rubber.

b) Curative: Frequent tree inspection during June to October for detecting infection and application of Bordeaux paste in early stage on infected part and up to 30 cm above and below. In advanced cases, Bordeaux paste may be initially applied and then the affected portion scraped to remove mycelium and decayed bark, followed by a second Bordeaux paste application. Dried branches should be pruned after disinfection. Thiram 0.75% [Thiride (TMTD) at 10g/l] is incorporated in a wound dressing compound or tridemorph 2% (Calixin at 25 ml/l) in 1% ammoniated field latex or thiride 0.75% or tridemorph 1% or propiconazole 0.1% (Tilt at 4 ml/l) in pidyvyl, china clay and water (1: 2: 4 by volume) is also effective as rain wash is prevented.

8.8.8. Powdery Mildew, *Oidium heveae*

The disease is predominantly noticed on newly formed tender flush during the refoliation period from January to March. The disease is severe in Kanyakumari, Idukki and Wynad district of South India and North Eastern States. Cloudy days with light rains and/or misty nights with dew formation during refoliation favour serious disease outbreaks.

i) Symptoms: Tender leaves with ashy coating curl, crinkle, and edges roll inwards and fall, leaving the petioles attached to the twigs like a broom-stick. After a few days, the petioles also fall. Die-back of twigs follows. On older leaves, white patches later causing necrotic spots reduce photosynthetic efficiency. Infected flowers and tender fruits are shed, affecting seed production.

ii) Management : Dusting of 11 to 14 kg 325-mesh fine sulphur dust/ round/ hectare should be taken up during the refoliation period commencing from bud break in about 10% of the trees, giving 3 to 5 rounds at weekly to fortnightly intervals. Sulphur mixed with an inert material like talc (70: 30) is commonly used. Wettable sulphur or carbendazim at 2.5 g/l is also effective in nurseries and for young plants as a spray. Spraying 0.05% a.i. carbendazim (Bavistin at 1 g/l) is more effective than sulphur for nurseries and young rubber plantations. Alternate use of carbendazim and sulphur is recommended to avoid resistance problem. For efficiency, dusting may be carried out in the early morning hours when the leaves are moist and the atmosphere calm. Rubber clones PB 86, GT 1, GL1, PR 107, PB 5/139, RRIM 703, RRII 208 and PB 310 show some tolerance.

8.8.9. Brown Root Disease, *Phellinus noxius*

i) Symptoms: The characteristic symptoms of the disease are yellowish discolouration of foliage, black encrustation of soil, sand and fungal hyphae cemented to roots, and brown lines in the affected roots. Plants dry up in advanced cases.

ii) Management: Open up the root system. Trace and prune completely killed and dried roots. Wash partially affected and healthy roots with emisan solution. When the fungicide dries up, apply a thin coating with a wound dressing compound. Refill the soil and drench the base with 3 g copper oxychloride mixed in a litre of water.

8.8.10. Secondary Leaf Fall, *Gloeosporium alborubrum*

The disease occurs during April-October.

i) Symptoms: Infects tender leaves mostly at leaf tip region where numerous spots coalesce and dry leading to defoliation. Infected leaves crinkle and become distorted.

ii) Management: Spraying with mancozeb at 2.6 g/litre or carbendazim at 2g/litre of water gives effective control of the disease.

8.9. CASHEW NUT

India is the largest producer, importer, processor, exporter and second largest consumer of cashew. In cashew nut, area has increased from 0.176 million hectares in 1961 to 0.843 million hectares in 2005-06. The production in cashew has gone up from 0.079 million tonnes in 1961 to 0.579 million tonnes in 2005-06 (Bijay Kumar, 2007) (Table 8.7). India exported cashew kernels worth US$ 553 million during 2006-07. Export of cashew is rising at 27% per annum. These export earnings are exceeded only by coffee and rice among agri-exports.

Table 8.7. State-wise area, production and productivity of cashew nut in India (2005-06)

State	Area in '000 ha	Production in '000 MT	Productivity in MT/ha
Kerala	80.0	67.0	0.837
Tamil Nadu	121.0	56.0	0.479
Karnataka	100.0	45.0	0.450
Andhra Pradesh	170.0	92.0	0.541
Maharashtra	160.0	183.0	1.143
Orissa	120.0	78.0	0.650
West Bengal	10.0	10.0	1.000
Assam	14.0	10.0	0.642
Goa	55.0	27.0	0.490
Gujarat	4.0	4.0	1.000
Meghalaya	6.4	6.4	1.000
Pondicherry	3.0	1.0	0.333
Total	843.4	579.4	0.7

As compared to the number of pests attacking cashew and intensity of damage caused by them, the magnitude of the disease problem is rather comparatively low. Among few diseases attacking cashew, die-back and damping-off of seedlings are found to be most destructive.

Fig. 8.28. Die-back on cashew nut.

8.9.1. Dieback or Pink Disease, *Corticium salmonicolor*

i) Symptoms: This is a fungal disease caused by *Corticium salmonicolor* which usually occurs during the rainy season. The affected branches initially show white or pinkish growth on the bark; a film of silky thread or mycelium develops. Later, the fungus penetrates into the deep tissues and develops a pinkish growth which are the spores resulting in gradual death of the shoots. In advanced stages, the bark splits and peels off. The leaves turn yellow and fall off giving a barren appearance to a tree (Fig. 8.28). Affected shoots start drying up from the tip.

ii) Management: For effective control of the disease, prune all the affected branches below the spot of infection and destroy them. Protect the cut surface by applying Bordeaux paste. The disease can be prevented by giving prophylactic spray of 1% Bordeaux mixture twice a year, first in May-June before the onset of monsoon and second in October.

8.9.2. Damping-Off

i) Symptoms: It is a fungal disease caused by *Fusarium* sp., *Pythium* sp., *Phytopthora palmivora* and *Cylindrocladium scoparium* and affects the cashew seedlings in nursery especially when the soil medium gets too wet. The fungi attack either the collar region or the root zone of the tender seedlings that turn pale with water-soaked discoloured areas around the stem resulting in rotting and collapse of seedlings causing severe loss at the nursery stage (Fig. 8.29).

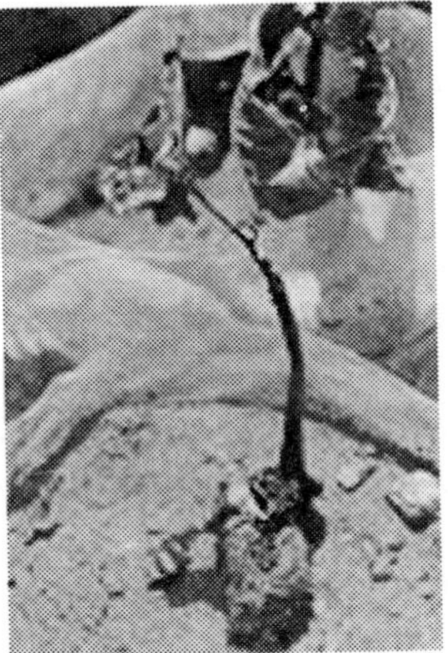

Fig. 8.29. Damping-off of cashew seedling.

ii) Management: Seeds for planting should be treated with Arasan before sowing. Soil media for potting should be treated with soil fumigants.

The disease incidence is severe in nursery where drainage conditions are poor. Therefore, adequate drainage should be provided during the rainy season. The disease can be controlled by drenching the bags with 1% Bordeaux mixture or copper oxychloride at 3 g/l of water.

8.9.3. Anthracnose, *Collectorichum gloeosporoides*

i) Symptoms: The disease attacks young plant tissues and can cause severe crop loss. This disease is caused by the fungus *Collectorichum gloeosp-*

oroides that usually infect tender leaves, shoots, inflorescences, young fruits (apples) and young nuts (Fig. 8.30). This disease is most prevalent when there is excessive rainfall coinciding with the appearance of new growth and flowering. Infected parts in their early stage show shiny, water-soaked lesions which later turn reddish-brown. At the lesion site, resinous exudation can be seen. As the disease progresses, the lesions enlarge in size, all affected tender leaves wrinkle, and the young apples and nuts become shriveled. Inflorescences become black. The disease is promoted by warm, damp conditions.

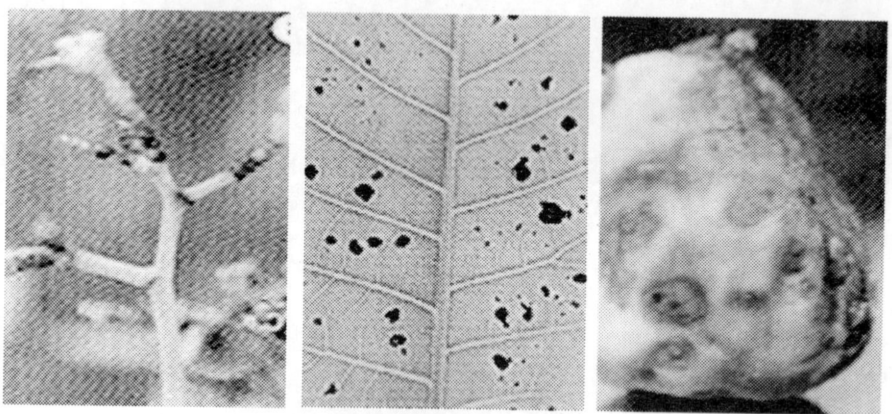

Fig. 8.30. Anthracnose on cashew nut.

ii) Management

- Prune dead branches and twigs and remove from the field. They constitute the principal source of infection.
- Though pesticide application on cashew is not practiced, copper based fungicides are known to be effective against anthracnose.
- Spray 1% Bordeaux mixture at new flush and blossom stages.

8.9.4. Powdery Mildew, *Oidium anacardii*

i) Symptoms: This disease is particularly serious in coastal areas. Infected panicles and leaves are coated with white, powdery fungal growth. In severe attacks the entire panicle may be infected and the fruit and nuts fail to set.

ii) Management: Pesticide application is not practiced in cashew fields. However, sprays of potassium/sodium bicarbonate and sulphur provide good control of powdery mildew.

8.9.5. Gummosis, *Diplodia natalensis, Ceratocystis* sp.

i) Symptoms: The pathogen causes exudation of a brown liquid from longitudinal cracks on the main stem and branches. Later the exudates turns black (Fig. 8.31).

Fig. 8.31. Gummosis on cashew nut.

ii) Management: The disease can be managed by scraping the wound in the bark and applying Bordeaux paste.

8.9.6. Red Rust, *Cephaleuros virescens*

i) Symptoms: The pathogen causes circular and slightly raised red rust spots on leaves (Fig. 8.32). Spots on upper surface of leaves consist of numerous filaments, which project through cuticle. Release of spores results in fading of spots to cream to white texture.

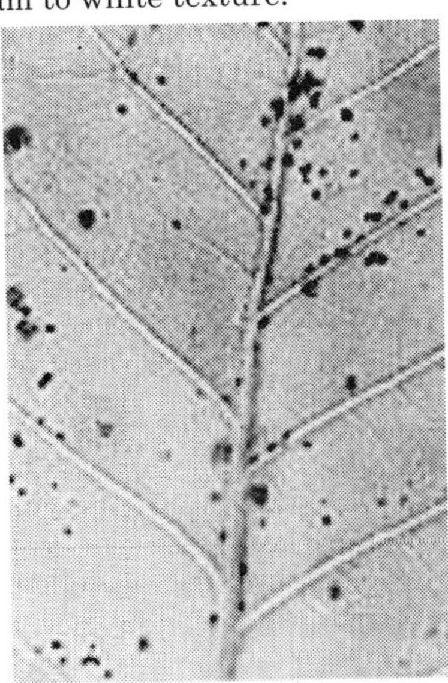

Fig. 8.32. Red rust on leaf of cashew nut.

ii) Management: Spraying of copper oxychlodide at 3g/litre of water gives effective management of the disease.

8.9.7. Blossom and Twig Blight

i) Management: Spray 1% Bordeaux mixture to seedlings to manage the disease.

Table 8.8. Disease management in plantation crops

Crop	Disease	Symptoms	Management
Coffee	Leaf rust, *Haemeleia vastastrix*	Small pale-yellow spots on the lower surface of infected leaves, orange-yellow spore mass appears, defoliation and die-back. Results in serious crop loss and causes fluctuations in production.	3 applications of 0.5% Bordeaux mixture for susceptible varieties.
	Black rot, *Koleroga noxia*	Blackening and rotting of affected leaves, young twigs and berries. Affected leaves get detached and hang down by means of slimy fungal strands. Defoliation and berry drop occur.	Remove and burn affected parts. Apply 1% Bordeaux mixture close to the south westerly monsoon if needed. Centre the coffee bushes, regulate the overhead canopy.
	Anthracnose, *Colletotrichum gloeosporioides*	A weak parasite but becomes vigorous under favourable environmental conditions. Causes 3 diseases: 1. Radcliffe disease: Yellowing of one or more functional leaves, necrosis of nodes, leaf-fall, die-back of twigs and death of many branches. 2. Stalk rot of berries and leaves. Necrosis of nodes and internodes from the junction of brown wood and green wood towards apex followed by berry drop and defoliation due to decaying of stalk portion. 3. Brown blight: roundish spots are first brown, then turn grey making the leaf look blighted.	Provide good drainage to remove excess water from the root zone. Spray with 0.5% Bordeaux mixture to prevent defoliation. Maintain adequate shade to avoid sun scald.
	Root rots, *Fomes* spp., *Fusarium* spp.	1. Stump or brown root-rots: affected bushes show a gradual yellowing of leaves and defoliation and death. Affected roots are brittle and show dark brown wavy lines	Burn the plants at site, isolate affected patches by digging deep trenches, ring the bark and poison the shade trees while thinning. Remove the affected plants

		of the fungus. 2. Fusarial root rot or wilt: a sudden wilting and yellowing of leaves leading to defoliation and death of aerial parts. Roots turn brown to pinkish.	and treat the soil at 1kg lime per plant to raise the soil pH. Treat the soil around the affected area with PCNB (brassicol) at 0.4%. Maintain the vigour of plants.
	Berry blotch	Necrotic spots on the exposed surface of green berries enlarge and cover the major portion. Fruit skin shrivels and sticks fast.	Spray 1% Bordeaux mixture during June and late August, maintain medium shade overhead.
	Nursery disease (Collar rot or damping-off)	Pre-emergence damping-off in germinating beds. Collar rot in primary or secondary beds. Seedlings show wilting, rotting extends from the collar all around and seedlings collapse.	Apply PCNB (brassicol) 0.4% or terrachlor or rhizoctol or Flit 406 or ferbam or dithane M 22 at 0.4%. Maintain good shade and sanitation. Remove and burn affected seedlings.
	Brown eye spot.	Usually severe in the nursery but found on young coffee plants also, small circular necrotic spots appear with dark brown margin and pale white centre. Leaves turn yellow and are shed.	Spray nursery seedlings with 0.4% flit 406 and young coffee plants with 1% Bordeaux mixture during April-May and Sept-Oct, avoid exposing the plants to direct sun.
Tea	Blister blight	Affect pluckable shoots. Optimum conditions include cloudy weather, 80% relative humidity and temperature below 25°C.	Copper oxychloride, Bordeaux mixture
	Black root, Red root	Occurs in new area brought under cultivation after clearing jungles. The disease is severe during damp weather.	Soil fumigation with Metham sodium and Carbon disulphide
	Brown root	Common in low elevation. Occurs in new area brought under cultivation after clearing jungles.	Soil fumigation with Metham sodium and Carbon disulphide
Coconut	Stem bleeding, *Ceratocystis paradoxa*	Exudation of the brownish black liquid through cracks on the trunk. This fluid turns black as it dries. Tissues below the bleeding patch become decayed. In young palms tissues rot internally and on piercing the stem a yellow fluid rushes out.	Chisel out the rotten portion. Treat the area with Bordeaux paste or coal tar.

	Bud rot, *Phytophthora Palmivora*	The initial symptom is the withering of spindle leaf. The leaf later dries up and droops down. If the leaf is pulled, it comes out easily as the base is completely rotten. The rotting proceeds downwards killing the bud which results in the death of the palm.	Clean the rotten and drench the bud and its surrounding portions with 1% Bordeaux mixture. Burn the rotten parts. Spray 1% Bordeaux mixture on the surrounding palms.
	Basal Stem Rot/Ganoderma wilt, *Ganoderma lucidum*	Initially infects the roots. The first visual symptom is the appearance of oozing of reddish brown fluid from growth cracks at the base of the trunk. The lesions spread linearly and sideways. Meanwhile the lower whorls of leaves start drooping and dry prematurely. As the lesions spread towards the crown, leaves dry up and ultimately the whole crown dries up and the tree dies.	Cut and remove all dead roots and a few surrounding healthy roots. Drench the cut ends of roots with a mixture of 1.5g of Aureofungin. +1 g Copper sulphate in 5 lit of water and cover the root system. Repeat the treatment after 15 days. Also treat healthy palms in the vicinity as a preventive measure.
Areca nut	Koleroga or Mahali, *Phytophthora arecae*	Heavy shedding of nuts during rainy season – water soaked lesions are formed near the perianth end and gradually covering the entire surface of the nut – the nut becomes dark green – finally leads to nut shedding.	Spray 1% Bordeaux mixture to the bunches at least two times at an interval of 45 days. The first spray should be given immediately after the first few monsoon showers. If the monsoon prolongs, a third spray is essential. Collect all the infected nuts and other plant parts and destroy it. Covering the bunches with polybags gives a complete control.
	Bud rot, *Phytophthora arecae*	Yellowing of spindle leaf – rotting of growing bud and surrounding tissues – palm emits a disagreeable odour.	Remove the infected tissue completely and treat the wound with Bordeaux paste. Spray 1% Bordeaux mixture to the crown of healthy palms which are in the vicinity of the affected palm. Remove koleroga affected bunches and destroy them.
	Inflorescence dieback and button	Yellowing and drying of rachis from the tip towards	To control the fungus, spray mancozeb at 3g/litre of water

shedding, *Collectotrichum gloeosporioides*	the base followed by shedding of female flowers or buttons – caused by several factors. *Collectotrichum gloeosporioides*, a fungus is also associated with this disease.	at the time of opening of female flowers in most of the inflorescences. If necessary, second spray should be given 25 days after the first spray. Remove the fully affected inflorescences and destroy them by burning to prevent the spread and severity of the disease.
Anabe roga or foot rot, *Ganoderma lucidum*	Yellowing of outer whorl of leaves which gradually extends to the inner whorls.	Proper management of the garden is the best way to check the occurrence of the disease. Improve the drainage. Drench the root zone of the affected palm with 0.3% calixin at quarterly intervals at 15-20 litres/palm + root feeding of calixin (125ml/palm) during January, March, July and October. Apply 2 kg neem cake/palm/year–phytosanitory measures like cutting and burning of the dead palms along with bole and roots should be followed strictly.

Chapter 9

SPICE CROPS

From time immemorial, India is regarded by the rest of the world as the "magic land of spices". Spices are natural plant products used to improve flavour, aroma, taste and colour of food products, beverages, liquors, pharmaceutical, cosmetic and perfumery products. No other country in the world has such a diverse variety of spice crops as in India. Indian spices are also noted for their excellent aroma, flavour and pungency not so easily matched by any other country. Even in minute quantities, spices are a real delight to all our sensory organs, making the food dishes more palatable. Usage of spices in food processing industries is increasing quite fast in all the countries. Hence, the world demand for spices in recent years is continuously on the rise.

India is the largest producer, consumer and exporter of spices in the world. India is the largest producer in the world of ginger, turmeric and black pepper. The annual production of spices is more than 5 million tonnes valued at US$ 1,500 million. India's share of the world spices trade is estimated at 45 to 50% by volume and 25% by value. The present annual production of spices in the country is 5.107 million tonnes from over 2.857 million hectares (Bijay Kumar, 2007) (Table 9.1). The spices contributed 2.76% of total horticulture production in 2005-2006. About 8.5% of India's export earnings from agricultural and allied products come from spices. Indian spices flavour foods in over 134 countries. The important spice crops include black pepper, cardamom, ginger, garlic, turmeric and chilli.

Table 9.1. State-wise area, production and productivity of major spices in India (2005-06)

State	Area in '000 ha	Production in '000 MT	Productivity in MT/ha
Andhra Pradesh	326.9	920.5	2.8
Tamil Nadu	130.0	784.0	6.0
Karnataka	258.8	619.8	2.4
Gujarat	426.1	505.6	1.2
Madhya Pradesh	292.3	345.7	1.2

Rajasthan	347.9	297.1	0.9
Uttar Pradesh	91.0	283.8	3.1
West Bengal	108.0	226.7	2.1
Orissa	145.6	195.3	1.3
Kerala	428.1	191.8	0.4
Chhattisgarh	28.9	168.2	5.8
Maharashtra	112.5	106.0	0.9
Others	162.1	463.2	2.9
Total	2857.7	5107.7	1.8

Since spices form a part of traditional/ethnic medicines, the demand for organically produced spices is growing. Export of organic spices from India has started in right earnest. The country at present exports around 50 tonnes of different varieties of organic spices. Export will get a significant boost in the coming years as more and more farmers switch to organic methods.

Export of Spice Produce

India is the largest producer, consumer and exporter of spices and spice products in the world and produces more than 50 spices. Spice export figures for the year 1996-97 and 2006-07 shows the phenomenal growth of the Indian spice industry. The Indian spice export for the year 1996-97 was 2, 25,000 metric tonnes which was worth $350 million; in the year 2006-07 the spice exports touched the figure of 3,75,000 MT which was worth US$ 800 million. Thus, these figures show the exponential growth of the Indian spice industry in a very short time. Spice exports from India have touched US $ one billion mark during 2007-08. The major destination of Indian spice exports is USA followed by European Union, Sri Lanka, Japan and Middle East Countries.

9.1. BLACK PEPPER

Black pepper (*Piper nigrum*), known as the "King of spices", has its origin in the South-Western hills of India. It is a branching and climbing perennial shrub belonging to the family Piperaceae and is grown in the hot and humid parts of the world. Pepper is loved for its spicy factor, piperine. It is used in various culinary seasonings, as a preservative for meat and fish, in medicine and perfumery (pepper oil). In India, it is a common practice to use live trees as standards for climbing pepper. Pepper is the most important spice produced in the country and accounts for 50-60% of the total export earnings. It is being cultivated in 0.223 million hectares producing 0.070 million tonnes of black pepper with an average yield of 1.3 tonnes per hectare.

9.1.1. Quick Wilt/Foot Rot, *Phytophthora capsici, P. palmivora*

i) Symptoms: The pathogen causes black spots on leaves with fine fibre like projection from advancing margins which rapidly enlarge and cause defoliation (Fig. 9.1). Tender leaves and succulent shoot tips of new runner shoots trailing on the soil turn black. When the main stem is affected, the entire vine wilts followed by shedding of spikes with or without black spots.

Fig. 9.1. Foot rot infected plants of black pepper.

ii) Management: Adoption of phytosanitory measures like removal and destruction of dead vines along with root systems from garden as well as use of disease-free planting materials, keeping tillage operations to the minimum to avoid soil disturbance and root damage, providing adequate drainage to reduce water stagnation, providing green legume cover / grass cover to avoid rain splash spread to adjoining vines, application of *Trichoderma* spp. at 75 g/vine along with neem cake at 2 kg/vine during May-June (pre-monsoon) with suitable carrier medium such as coffee husk and well rotten cow dung followed by second round of application during August-September (post-monsoon) gives effective control of the disease. Prophylactic application of Bordeaux paste to collar region during May-June (pre-monsoon) and spraying with 1% Bordeaux mixture or 0.2% copper oxychloride or 0.01% metalaxyl or 0.3% potassium phosphonate helped in reduction of the disease. Spraying of modified panchagavya at 10^1 dilution at 2-3 litres/vine twice as pre-monsoon (May-June) and post-monsoon (August-September) is recommended.

The freshly emerging runner shoots should not be allowed to trail on the ground. They must be either tied back to the standard or pruned off. The branches of support trees should be pruned at the onset of monsoon to avoid build up of humidity and for better penetration of sunlight. Reduced humidity and presence of sunlight reduces the intensity of leaf infection. At the onset of monsoon (May-June), application of *T. harzianum* (10^{10} cfu/g) around the base of the vine is recommended. A second application of *T. harzianum* during August-September is also recommended.

In disease affected gardens, plants treated with *T. harzianum* (IISR 1369 strain with 10^{10} cfu/g) at 50g/vine along with 10 kg of FYM or 1 kg neem cake during May – June period immediately after the receipt of first early monsoon shower and second dose of inoculum applied during August-September gave effective control of foot rot and slow decline (*Meloidogyne incognita, Radopholus similis*).

Five strains of *Pseudomonas fluorescens* (IISR-11, 8, 51, 6 and 13) gave protection against foot rot (above 55% over control) to the plants, and were also found to be efficient in growth promotion. Maximum disease suppression was obtained in the plants treated with strain IISR-51 (86%).

Integration of *T. harzianum* (IISR 1369) with *Pseudomonas fluorescens* (IISR-41) gave very effective control of foot rot (10% disease incidence as compared to 90% in control) and increased plant height under field conditions. Integration of *T. harzianum* (IISR 1369) with *Pseudomonas fluorescens* (IISR-11 or 6) suppressed the root rot to the extent of 63% over control and improved the vigour of pepper vines.

T. harzianum culture mixed with pre-wetted neem cake or FYM at 1 kg/100 kg and incubated for 2 days. Pre-monsoon application of this mixture at 5 kg/vine below 10 years and 10 kg/vine above 10 years gave good control of wilt disease.

An integrated management strategy involving phytosanitation, cultural practices and application of Bordeaux mixture (1%) and copper oxychloride or potassium phosphonate (Akomin) is effective against foot rot disease. Field trials showed effectiveness of *Trichoderma harzianum* and *T. virens* to manage *Phytophthora* foot rot.

Raising plants in solarized nursery mixture, fortified with *Trichoderma* spp. and endomycorrhizae, has been found effective in suppressing root rot. Endomycorrhizae and *T. harzianum* are effective against *Phytophthora* spp. and *Meloidogyne* spp.

Foliar spray and drenching of root zone with copper oxychloride + metalaxyl (3 g/litre of water) prior to and after monsoon is effective.

Among the several black pepper lines screened for their reaction to *Phytophthora capsici*, the casual agent of foot rot disease, P24, an open pollinated progeny of Perambramundij has been identified as tolerant.

9.1.2. Leaf Rot, *Rhizoctonia solani*

i) Symptoms: The disease is often serious in nurseries during April-May when warm humid conditions prevail. The fungus infects both leaves and stems. Grayish sunken spots and mycelial threads appear on the leaves. The infected leaves are attached to one another with the mycelial threads. On stem, the infection occurs as dark brown lesions, which spread both upwards and downwards. The new flushes subtending the points of infection gradually droop and dry up.

ii) Management: Removal and destruction of the infected berries to avoid repeated overlapping of infections is recommended. A prophylactic spray with 1% Bordeaux mixture prevents the disease. Spraying with 1% Bordeaux mixture during June-July, late July and late August was found effective in management of the disease. Foliar spray with liquid formulation of *Pseudomonas fluorescens* and drenching at both pre- and post-monsoon treatments provided good protection. The treated plants had 35% disease index as compared to 61% in control and the yield was 2.34 kg/vine as compared to 0.78 kg/vine in control. However, this was on par with Bordeaux mixture spray with 29% disease index and 3.65 kg yield/vine.

9.1.3. Basal Wilt, *Sclerotium rolfsii*

i) Symptoms: The disease is mainly noticed in nurseries during June to September. Grayish lesions appear on stems and leaves. On the leaves, white mycelium is seen at the advancing edges of lesions. The mycelial threads later girdle the stem resulting in drooping of leaves beyond the point of infection and in advanced stages, the rooted cuttings dry up. Small whitish to cream coloured grain like sclerotial bodies appear on the mature lesions.

ii) Management: The affected cuttings along with defoliated leaves should be removed and destroyed. Later all the cuttings should be sprayed with 1% Bordeaux mixture or 0.2% carbendazim.

9.1.4. Pollu Disease, *Colletotrichum gloeosporioides*

i) Symptoms: The pathogen infects the leaves causing angular to irregular yellowish brown to dark brown irregular necrotic leaf spots with a chlorotic halo. Affected berries show brown sunken patches during early stages and further development is affected. In later stages, the discolouration gradually increases and berries show the characteristic cross splitting. Berries turn black and dry.

ii) Management: Spraying 1% Bordeaux mixture or 0.2% copper oxychloride gives effective control of the disease.

9.2. CARDAMOM

Cardamom (*Elettaria cardamomum*), also known as the "Queen of Spices" enjoys a unique position in the world market has its origin in the evergreen rain forests of Western Ghats of South India. India enjoyed near monopoly in the field till recently. Guatemala has emerged as a competitor to India. India and Guatemala are the main producers and exporters of cardamom. Other cardamom growing countries are Tanzania, Sri Lanka, El Salvador, Vietnam, Laos, Kampuchea and Papua New Guinea. The area under cultivation in India during 2004-05 was 73,237 ha, with an annual production of 11,415 metric tonnes. The export earning during 2006-07 was Rs. 274.7 million by exporting 650 million tonnes of small cardamom. Kerala, Karnataka, and Tamil Nadu produce cardamom in India. While Kerala accounts for 60 per cent of the production, Karnataka contributes 30 per cent, and Tamil Nadu 10 per cent. In Karnataka, cardamom is cultivated mostly in Kodagu, Hassan, and Chikmagalur districts. The plants last for 15 years. They grow in moist, elevated, and shady forest loams. They require a fairly distributed annual rainfall ranging from 2,500 mm to 5,000 mm and are found within an altitude ranging between 600 m and 1,200 m above the mean sea level. It is used for flavouring various food preparations, confectionery, beverages, liquors and medicines.

9.2.1. Damping-Off, *Pythium vexans, Rhizoctonia solani*

i) Symptoms: This disease is caused by *Pythium vexans* and *Rhizoctonia solani*. Infection is observed at the collar region.

ii) Management: Provide good drainage, and spray and drench the nursery with 1% Bordeaux mixture or 0.2% copper oxychloride.

9.2.2. Nursery Leaf Spot, *Phyllosticta elettariae*

i) Symptoms: This disease is caused by *Phyllosticta elettariae*. Pale specks appear on the leaf lamina, which dry up and become paper white.

ii) Management: Spraying the plants with mancozeb 0.25% at fortnightly intervals is effective in controlling the disease.

9.2.3. Capsule Rot/Azhukal, *Phytophthora* spp.

i) Symptoms: This is a fungal disease caused by *Phytophthora* spp. occurring during the rainy season. It affects the leaves, tender shoots, panicles and capsules. On the infected leaves, water soaked lesions appear first and rotting and shedding of leaves along the veins occur there after. The infected capsules become dull greenish brown and decay. This emits a foul smell and subsequently shed. Infection spreads to the panicles also.

ii) Management: Trashing and destruction of the infected parts should be done as a phytosanitary measure just prior to the onset of southwest monsoon. Remove the trash (dried leaves and leaf sheaths) from the basal region of the plant to the extent possible.

Spray the shoots with 1% Bordeaux mixture with adhesive (rosin soda or any other sticker) by the commencement of the monsoon and continue the spraying operation two or three times up to November-December according to the intensity of the disease and rainfall. Give a copious spray to the panicle with 1% Bordeaux mixture at 3 litres/plant during July-August when the disease intensity is maximum.

Trichoderma spp. can be used along with cow dung for controlling this disease.

9.2.4. Rhizome Rot or Clump Rot, *Pythium vexans, Fusarium oxysporum, Rhizoctonia solani*

i) Symptoms: The disease occurs during South West monsoon. The symptoms include pale yellow colour on the foliage and premature death of older leaves. Collar portion of the aerial shoots becomes brittle and tiller breaks easily from the rhizome at the bulbous base (Fig. 9.2). Water soaked lesions appear first on leaves. The pathogen causes rotting and shredding of leaves along the veins. Rotting develops in the collar region emitting a foul smell. Capsules turn dull greenish brown emitting a foul smell.

Fig. 9.2. Clump rot of cardamom.

ii) Management: Soil drenching with 0.2% copper oxychloride (2-3 litres per plant) and repeating this two times at monthly intervals gives effective control of the disease. Drenching the soil with 0.1% Ridomil (Metalaxyl + mancozeb) or applying *Trichoderma viride* (native strain isolated from cardamom plantation) as a prophylactic measure is effective.

As a bio-control measure, inoculate seedlings with native arbuscular mycorrhiza, *Trichoderma* and *Pseudomonas fluorescens* at the time of planting in the nursery and main field, and apply during pre-monsoon period in established plantations.

Solarization of nursery beds before sowing seeds and incorporation of *T. harzianum* has resulted in prevention of damping off and production of pathogen-free healthy seedlings. Soil solarization can be done for sterilizing the nursery mixture. To the sterilized mixture, biocontrol agents such as arbuscular mycorrhizal fungus at 100 g/kg and *T. harzianum* (10^{10} cfu/g) may be added at the time of filling of nursery mixture in polybags. Since the biocontrol agents protect the root system only, the aerial portion may be protected with 1% Bordeaux mixture spray.

Removal and burning of infected panicles and rhizomes from the plantation reduced the disease incidence. Spraying the panicles during pre-monsoon and post-monsoon is recommended. Incorporation of *Trichoderma* spp. multiplied in suitable organic medium (1 kg/clump) prior to the onset of monsoon season is a prophylactic operation. Spraying of 1% Bordeaux mixture at 3 litres/plant with an adhesive by the commencement of monsoon and continuing the spraying operation up to November-December is effective. Integration of *T. harzianum* (IISR 1369) with *Pseudomonas fluorescens* (IISR-11 or 6) suppressed the clump rot to the extent of 36% over control. Soil application of *T. harzianum* or *T. viride* to nursery beds gave disease control anging from 53 to 84%. Soil application of *T. harzianum* or *T. viride* to the main field resulted in high disease control ranging from 62.3 to 64.8%.

Cardamom cvs. ICRI-2 (Mysore) and ICRI-3 (Malbar) are tolerant to apsule rot and rhizome rot disease.

9.2.5. Leaf Blotch, *Phaeodactylium venkatesanum*

i) Symptoms: The fungus *Phaeodactylium venkatesanum* causes this disease. The disease is characterized by the appearance of large blotches of irregular lesions with alternating shades of light and dark brown necrotic tissues. This is mainly observed on mature leaves. On the lower surface of the lesions ash coloured white superficial growth of the fungus appears during moist weather conditions.

ii) Management: The fungicides, Bordeaux mixture (1%), mancozeb (0.3%) and carbendazim (0.1%) are effective in controlling the disease.

9.2.6. Leaf Blight or Chenthal, *Colletotrichum gloeosporoides*

i) Symptoms: The disease occurs during the post-monsoon period. It causes water soaked rectangular lesions on the foliage, which elongate to form parallelly arranged streaks measuring 5 cm. Severely infected plants show a burnt appearance. The lesions are clearly visible even on dried leaves. The incidence of the disease appears to be more severe in areas, which do not have proper shade.

ii) Management: Providing adequate shade is recommended. Spraying carbendazim or mancozeb or copper oxychloride at 3 g/litre of water at monthly intervals gives effective control of the disease.

Waiting Period for Fungicides

Mancozeb 30 days

Integrated Disease Management

- Mancozeb (0.25%) + Ridomil (0.1%) were found to be the most effective fungicides for controlling damping off and leaf spot diseases in cardamom nursery.
- Clump rot disease could be effectively checked by drenching the affected and neighbouring clumps with 1% Bordeaux mixture once before the onset of monsoon (June) and again after the end of monsoon (August).
- Katte disease spread was found to be effectively checked by removing affected clumps, old and decayed suckers and plant parts by regular surveys and by avoiding planting of clones from affected gardens.
- Use disease tolerant varieties for planting [Cardamom cvs. ICRI-2 (Mysore) against capsule rot, and ICRI-3 (Malabar) against rhizome rot].

- Destroy wild hosts of katte disease like *Amomum, Alpinia, Curcuma Colocasia, Caladium,* etc.
- Destroy capsule rot affected portions and plant debris.
- Apply *Trichoderma viride/T. harzianum* in plant basins to control capsule rot and rhizome rot diseases.

9.3. GINGER

Ginger (*Zingiber officinale*), an herbaceous perennial, belonging to the family Zingiberaceae, is valued for its rhizome. The country of its origin is not definitely known but is presumed to be either from India or China. India is the largest producer (3,05,900 metric tonnes) and exporter of dry ginger in the world (export earning of Rs.191 million). The other ginger producing countries are Jamaica, Sierra Leone, Nigeria, Southern China, Japan Taiwan and Australia. In India, It is being cultivated in 85,100 hectares producing 0.306 million tonnes of ginger with an average yield of 3.6 tonnes per hectare.

9.3.1. Rhizome Rot, *Pythium aphanidermatum*

i) Symptoms: Soft rot is the most serious disease of ginger in India and in some other countries. It is caused by *Pythium* spp., of which *P. aphanidermatum* is the principal species in India, although *P. butleri, P. gracile, P. myriotylum, P. nigriotilum*, and *P. vexans* have also been recorded. The bases of the aerial shoots become soft, watery and then rot. The affected plants become pale; the tips of the leaves turn yellow, followed by complete yellowing and drying up of the leaves (Fig. 9.3). The shoots fall and cease to produce rhizomes. The infection extends to the rhizomes; the inner tissues being reduced to a soft and black, putrefying mass. Losses can be high. The disease is favoured by high moisture content of the soil with insufficient drainage.

ii) Management: Selection of well drained soils for planting is important for managing the disease, since stagnation of water predisposes the plant to infection. Planting of ginger rhizomes in raised beds instead of flat beds is recommended. It reduces the disease incidence and increases the yield which can fetch Rs. 2,000/- more without incurring any extra expenditure. Disease-free rhizomes should be used for planting. Rhizomes may be treated in a mixture of 250 g of mancozeb (Dithane M-45) and 100 g of carbendazim (Bavistin) in 100 litres of water for 60 min followed by drying in shade for 48 hr before planting. Seed treatment with 0.25% thiram and later spraying with the same chemical also gives effective control of rhizome rot disease. Once the disease is located in the field, removal of infected clumps and drenching with mancozeb or copper oxychloride each at 0.3% during August in mid-hills and during September in low-hills checks the spread of the disease. Application of *Trichoderma harzianum* along with neem cake at 1 kg/bed helps in preventing the disease.

When harvesting ginger for seed purposes, do not commence until tops have died down naturally, as the stalk scar is an ideal point for infection to occur. Only harvest when rhizomes are dry and store in new bags in a dry and well-ventilated shed.

Fig. 9.3. Rhizome rot of ginger.

Integrated disease management involving phytosanitation, soil solarization and seed treatment was found effective in reducing rhizome rot and increasing ginger yield.

9.3.2. *Alternaria* Leaf Spot, *Alternaria* sp.

i) Symptoms: The pathogen causes oval brown spots with white or grey centre on upper side of leaf. Rhizomes do not develop.

ii) Management: Disease-free rhizomes should be used for planting. The rhizomes can be treated in a mixture of 250 g of mancozeb (Dithane M-45) and 100 g of carbendazim (Bavistin) in 100 litres of water for 60 min and drying in shade for 48 hr before sowing. Soil drenching of the affected plants with 0.3% copper oxychloride during August in mid-hills and during September in low-hills is also effective. Spraying 0.2% mancozeb or 0.3% copper oxychloride or 1% Bordeaux mixture after the first appearance of leaf spot is effective for the management of the disease.

9.3.3. *Phyllosticta* Leaf Spot, *Phyllosticta zingiberi*

i) Symptoms: A leaf-spot disease has been reported from Himachal Pradesh, caused by *Phyllosticta zingiberi*. The same disease was reported in Andhra Pradesh also. But it was not so serious. Oval water-soaked spots occur on leaves, later becoming papery, and may coalesce, causing extensive discolouration and desiccation.

ii) Management: A 1 per cent Bordeaux mixture spray provides protection against leaf spot disease.

9.3.4. *Colletotrichum* Leaf spot, *Colletotrichum zingiberis*

i) Symptoms: A leaf spot caused by *Colletotrichum zingiberis* is of common occurrence in different parts of India. The fungus attacks the leaves only, forming round to oval light-yellow spots which may coalesce to form large discoloured areas (Fig. 9.4), often drying up in the centre to form a hole. The plants become stunted in growth and the development of the rhizomes is adversely affected.

Fig. 9.4. *Colletotrichum* leaf spot on ginger leaf.

ii) Management: Removing and burning the affected parts and prophylactic spraying of 5-5-50 Bordeaux mixture can control the disease.

9.3.5. Yellows/Rhizome Rot, *Fusarium oxysporum* f. sp. *zingiberi*

i) Symptoms: Rhizome rot is a very common and serious fungal disease which is specific to ginger. Infected plants are stunted and yellow, lower leaves dry out and turn brown. Eventually all above-ground shoots dry out completely. Plant collapse is very slow (up to several weeks) compared with the rapid collapse associated with bacterial wilt infection. Diseased rhizomes show a brown internal discolouration, are normally shriveled in appearance and eventually decay leaving the outer shell intact with fibrous internal tissue remaining. Increased nematode infestations are usually associated with *Fusarium* rhizome rot, accentuating yield losses. *Fusarium* is also responsible for serious loss of planting pieces and poor germination. The disease spreads rapidly in the field during wet weather.

ii) Management : Disease-free rhizomes should be used for planting. When cutting up seed, discard any pieces showing shriveling or brown discolouration and regularly dip the cutting knife in methylated sprit or a commercial disinfectant solution. Rhizomes may be treated in a mixture of 250 g of mancozeb (Dithane M-45) and 100 g of carbendazim (Bavistin) in 100 litres of water for 60 min followed by drying in shade for 48 hr before planting. Affected plants may be drenched with 0.3% copper oxychloride during August in mid-hills and during September in low-hills. Crop rotation of at least 2-5 years between ginger crops will help to reduce *Fusarium* in the soil. Dipping the seed rhizomes in ethoxyethyl mercury chloride can control it. Avoid areas with heavy nematode infestations.

9.3.6. Leaf Blotch, *Taphrina maculans*

i) Symptoms: Small oval or irregular brown spots develop on either side of leaves. Leaves become yellow and plants present scorched appearance in severe cases (Fig. 9.5).

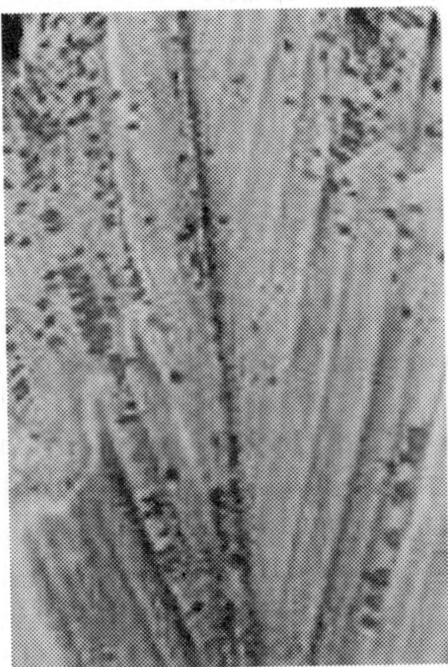

Fig. 9.5. *Taphrina* leaf spot on ginger leaf.

ii) Management: Spraying mancozeb at 2 g/litre of water gives effective management of the disease.

9.3.7. Banded Leaf and Sheath Blight

i) Symptoms: Large discoloured areas develop alternating with irregular bands. Blotching of leaf and leaf sheath occurs in case of severe infection (Fig. 9.6).

Fig. 9.6. Leaf and sheath blight on ginger.

ii) Management: Dipping seed rhizome in a solution containing Streptocycline at 2 g in 10 litres of water for 20 min is effective.

9.3.8. Thread blight, *Pellicularia filamenlosa*

i) Symptoms: Thread blight is caused by the commonly called shoe string (boot-lace) fungus. A common parasite of forest trees, it normally only occurs in recently cleared forest country and causes yellowing of the stem very similar to *Fusarium* symptoms. When infected plants are examined, black string like threads can be seen adhering to the rhizomes.

ii) Management: All roots and stumps should be removed before planting newly cleared land to prevent infection.

9.3.9. White Fungus

i) Symptoms: This disease has caused serious losses on some farms. Infection first occurs beneath the scales on the rhizomes and may progress to produce a deep brown rot over the entire surface. Affected rhizomes become enveloped in a white fungal mycelial growth which may extend up to the basal stem. Mild infections usually produce no above ground symptoms. Severe infections cause the stems to turn yellow and slowly dry out.

ii) Management: Prepare the ground well in advance of the ginger crop as this fungus will persist on crop residues. The most serious outbreaks of this disease have been encountered in areas previously cropped with sugarcane where a large amount of undecomposed cane trash remained in the soil.

Integrated Disease Management

Selection of healthy planting material from disease-free gardens (soft rot, *Pythium aphanidermatum, P. myriotylum*; bacterial wilt, *Ralstonia solanacearum*; and nematodes, *Meloidogyne* spp., *Pratylenchus* spp.) is essential. Hot water treatment of the seed rhizomes at 51°C for 10 min eliminated nematodes and surface bound pathogens also. Strict phytosanitation, provision of better drainage and cultivating ginger in raised beds are some of the important cultural operations to be adopted to reduce crop losses due to the biotic factors.

As a seed disinfection procedure, a novel technique called rhizome solarization has been devised at the Indian Institute of Spices Research, Calicut. The disinfected rhizomes when treated with *T. harzianum* and rhizobacterial strain consortia as seed treatment and soil application resulted in higher yields and growth promotion and soft rot suppression and minimal bacterial wilt. In addition, in combination with *Glomus* spp., the disease was absent probably through growth mediation. *T. harzianum* in combination with *Pseudomonas fluorescens* showed a synergistic effect in reducing the soft rot infection.

Application of *T. harzianum, T. viride, T. hamatum* and *T. virens* as seed treatment and soil application along with neem cake at 100 kg/ha gave protection and increased yield. Application of *T. viride* along with saw dust/ neem cake was found to be highly effective in suppression of rhizome rot. There was synergistic effect in protection where soil solarization followed by biocontrol was adopted. *Trichoderma* spp. and *T. viriens* were found to be highly effective in suppressing *Fusarium* yellows and *Pratylenchus coffeae.*

- Disease-free, healthy seed rhizomes should be used for planting.
- Rhizomes may be treated in a mixture of 250 g of mancozeb (Dithane M-45) and 100 g of carbendazim (Bavistin) in 100 litres of water for 60 min followed by drying in shade for 48 hr before planting.
- Adequate drainage should be provided in the field.
- The disease-affected clumps should be removed and the soil around the affected plants drenched with 0.3% copper oxychloride during August in mid-hills and during September in low-hills.
- Spraying with 0.3% copper oxychloride at 10 days interval with the appearance of leaf spot disease.

9.4. TURMERIC

Turmeric (*Curcuma longa*) is best known as an important spice crop. It is believed to have its origin in South-East Asia. Turmeric is valued for its rhizomes, often processed and used commercially. It is mainly cultivated in India and to a small extent in China, Indonesia, Peru and Jamaica. It is an important source of yellow dye and is also used as a colouring matter in food industries, confectionery and drugs. In India, it is cultivated in 1,49,400 ha with annual production of 5,29,000 tonnes during 2004-05. India's export earning from turmeric was Rs. 92 million annually.

9.4.1. Rhizome Rot, *Pythium aphanidermatum*

i) Symptoms: Rhizome rot caused by *Pythium aphanidermatum* shows progressive drying-up of the leaves of infected plants. The base of the aerial shoots shows water-soaked soft lesions. As the disease progresses infection gradually passes to the rhizomes, which begin to rot and become soft. The bright orange colour of the rhizomes changes into brown. The disease may be confined to a few isolated plants or may occur in patches (Fig. 9.7). In severe attacks the yield is considerably reduced.

ii) Management: Use disease-free rhizomes for seed purpose. One of the effective control methods is eradication and burning of infected plants. In serious cases it may be advisable to disinfect the soil with an authorized fungicide (e.g. metalaxyl and mancozeb). Rotting of rhizomes can be minimized by drenching the soil with 0.3% chestnut compound or 0.3% mancozeb at 15 days interval soon after the disease is noticed. Treat the rhizomes by dipping in a mixture of 250 g mancozeb (Dithane M-45) and 100 g carbendazim (Bavistin) in 100 litres of water for 1 hr before storage as well as before sowing. Treatment of rhizomes with hot water at 51°C for 10 min and soil application of *Trichoderma harzianum* mixed with neem cake resulted in minimum incidence of rhizome rot and maximum yield.

Fig. 9.7. Rhizome rot of turmeric.

Some cultivars (e.g. 'Suvarna' in India) have shown field tolerance to rhizome rot disease.

9.4.2. Wilt, *Fusarium solani*

i) Symptoms: The pathogen is responsible for progressive drying up of leaves, which first proceeds along the margin and later the entire leaf dries up. The root system is adversely affected and is reduced to a few decaying brown roots. Infection gradually spreads to the rhizome which begins to rot and becomes soft. The bright orange colour of the rhizome is changed into different shades of brown.

ii) Management: Drenching the soil at the root region with copper oxychloride 2.5 g/l of water or metalaxyl 1 g/l of water is effective against the disease.

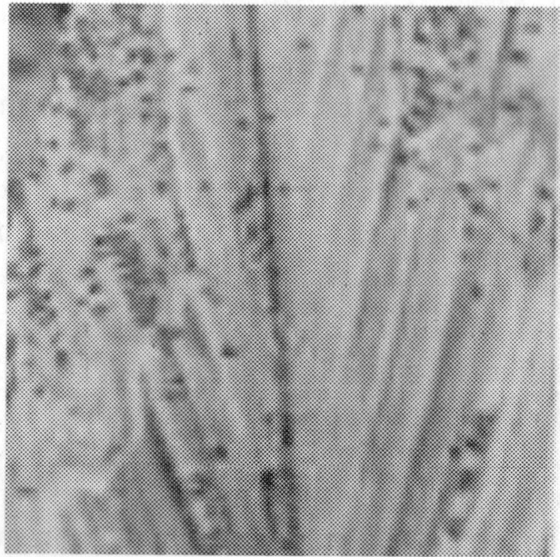

Fig. 9.8. *Taphrina* leaf spot of turmeric.

9.4.3. Leaf Spot or Blotch, *Taphrina maculans*

i) Symptoms: Leaf spot or leaf blotch, caused by *Taphrina maculans*, is characterized by the appearance of large number of spots on both surfaces of the leaves, 1–2 mm in diameter, coalescing freely. The spots first appear as pale yellow discolorations which gradually turn to reddish brown. Infected leaves are distorted, have a reddish brown appearance and soon become yellow (Fig. 9.8).

ii) Management: Spraying the crop with 0.25% mancozeb or 0.3% zineb or 1% Bordeaux mixture at 10 days interval gives effective control of the disease.

9.4.4. Leaf Spot, *Colletotrichum capsici*

i) Symptoms: Another leaf spot disease is caused by *Colletotrichum capsici*, provoking spots of variable size, enlarging to 4–5 cm × 3 cm and frequently coalescing over most of the leaf, which then dries up. Mature spots have a grayish white centre with dark brown margin surrounded by a yellow halo (Fig. 9.9). The central portion of the spot becomes thin and papery. Small spots may coalesce to form irregular necrotic patches. In very severe infestation, most leaves dry up, presenting a scorched appearance, resulting in yield losses of more than 50%. Excess shade and intercropping favours the disease development.

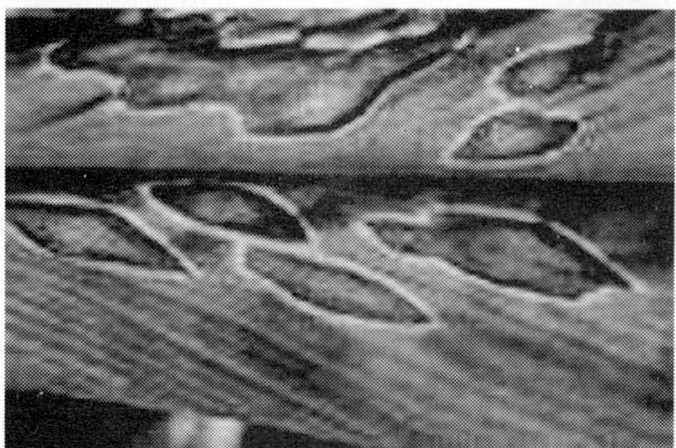

Fig. 9.9. *Colletotrichum* leaf spot on turmeric.

ii) Management: The disease can be checked by spraying Bordeaux mixture once before the symptoms appear. Captan and zineb, applied at monthly intervals, control the disease adequately. Planting materials should be selected from disease-free areas and treated with an authorized fungicide before planting. Spraying mancozeb or copper oxychloride at 3 g/l of water gives effective control of the disease.

Integrated Disease Management

Selection of healthy planting material from disease free gardens (soft rot, *Pythium aphanidermatum, P. myriotylum*; bacterial wilt, *Ralstonia solanacearum* and nematodes, *Meloidogyne* spp., *Pratylenchus* spp.) is essential. Hot water treatment of the seed rhizomes at 51°C for 10 min eliminated nematodes and surface bound pathogens also. Strict phytosanitation, provision of better drainage and cultivating turmeric in raised beds are some of the important cultural operations to be adopted to reduce crop losses due to the biotic factors.

As a seed disinfection procedure, a novel technique called rhizome solarization has been devised at the Indian Institute of Spices Research, Calicut. The disinfected rhizomes when treated with *T. harzianum* and rhizobacterial strain consortia as seed treatment and soil application resulted in higher yields and growth promotion and soft rot suppression and minimal bacterial wilt. In addition, in combination with *Glomus* spp., the disease was absent probably through growth mediation. *T. harzianum* in combination with *Pseudomonas fluorescens* showed a synergistic effect in reducing the soft rot infection.

Application of *T. harzianum, T. viride, T. hamatum* and *T. virens* as seed treatment and soil application along with neem cake at 100 kg/ha gave protection and increased yield. Application of *T. viride* along with saw dust/neem cake was found to be highly effective in suppression of rhizome rot. There was synergistic effect in protection where soil solarization followed by biocontrol was adopted. *Trichoderma* spp. and *T. viriens* were found to be highly effective in suppressing *Fusarium* yellows and *Pratylenchus coffeae*.

9.5. CINNAMON

Cinnamon (*Cinnamomum cassia*) is a thin inner bark of a cinnamon tree. The bark contains about 1% essential oil. The active principles are eugenol, cineole and cinnamaldehyde. It is used in stick form in fruit preserves. It is used ground for cakes, cookies and puddings. It is also used in spicing sauces, pickles, pulav and biryani.

9.5.1. Leaf Spot/Blight and Dieback, *Colletotrichum gloeosporoides*

The leaf spots/blight caused by a fungus *Colletotrichum gloeosporioides* (Anandaraj and Devasahayam, 2004; Kumara, 1999) is found in almost all cinnamon growing areas in Sri Lanka mainly affecting the foliage. It appears in all stages of the plant growth from seedling stage to maturity and the entire foliage can be affected in severe infections. Kumara (1999) reported that 18% foliar damage due to this disease in Matara, Sri Lanka with no significant correlation with cinnamon yield and the disease severity. Considerably higher disease incidence was observed in lands with high shades, high planting densities, poor weeding, and improper pruning Kumara (1999a, b).

i) Symptoms: The symptoms in young seedlings include, small brown specks on leaf laminae which, later coalesce to form irregular patches. Bhat *et al.* (1988) reported that seedlings of *Cinnamomum zeylanicum* in India affected by a leaf spot disease leading to partial drying of the plants and those seedlings yielded *C. gloeosporioides (Glomerella cingulata)*. Symptoms were reproduced 7 days after inoculation of healthy seedlings.

Small, brownish, leaf spots on older leaves can be seen, especially in shaded areas. These small specks coalesce to form large necrotic blotches giving scorching appearance. Spots later become papery with dark brown margins. In other cases the central portion is shed, forming shot hole appearance. Lesions may extend from tip of the leaf or from the margins. In severe infections, lesions may be larger than the half of the leaf. In some seedlings, the infection spreads to the stem causing a dieback (Karunakaran and Nair, 1980).

In India, reddish elongated spots were observed arising from the margin and resulting in defoliation (Prakasam, 1991). This also caused shot hole symptoms in later stages. However, the fungus causing this disease has been identified as *Phytophthora capsici* (Prakasam, 1991). Pu and Chang (1999) also reported brown to black spots on *C. verum* leaves in Thaiwan. These spots later coalesced and the infected leaves were shed. The pathogen was identified as *C. gloeosporioides* (Syn. *Glomerella cingulata).*

ii) Management: Pruning the affected branches and spraying 1% Bordeaux mixture are recommended for the control of the disease.

Copper based fungicides are being used by growers in Sri Lanka to control this disease. For other crops for this type of diseases caused by *C. gloeosporioides,* a range of fungicides such as chlorothalonil, carbendazim, mancozeb or thiphanate-methyl have been recommended at 7-l0 days intervals. Reducing shade especially during nursery stages may reduce the disease severity.

9.5.2. Grey Leaf Spots/Blight, *Pestalotia* spp.

Pestalotia cinnamomi causing grey blight is also considered as one of the commonest disease of cinnamon (Anon, 1996). In India, the disease was reported to be caused by *P. palmarum* on *C. verum* causing foliar damage up to 90% (Karunakaran *et al.*, 1993). *Pestalotia furierea* causing similar leaf spot symptoms were reported from Dominican Republic and also from Pakistan.

i) Symptoms: Small yellow brown spots appear on the cinnamon leaves, at first. Later, these spots turn grey with a border and spread to the lamina. Dark acervuli are produced in older lesions appearing as black dots in the centre. This disease can cause severe damage and defoliation

ii) Management: Pruning the die-back affected branches and spraying with 1% Bordeaux mixture are recommended for the control of the disease.

9.5.3. Stripe Canker, *Phytophthora cinnamomi*

Phytophthora cinnamomi causing stripe canker attacks shoots and young stems of cinnamon. Rands (1922) first reported *P. cinnamomi* causing severe losses to forest trees and avocado, as causal agent of stripe canker of *C. verum*. The fungus also affect *C. campora, C. culitlawan* and *C. sintok*. The fungus *Phytophthora cinnamomi* from pineapple was also reported pathogenic on cinnamon, but with reduced virulence (Anandaraj and Devasahayam, 2004).

i) Symptoms: Stripe canker is found on the trunks and branches, particularly of young trees of *C. verum* or more generally on *C. burmannii* in Indonesia (Mehrlich, 1934). Vertical stripes are seen on the stems with amber colour exudates at the advancing margins which harden later. Vertical stripes of dead bark are most numerous near ground level. The disease is prevalent on ill-drained soils and causes up to 42% damage. Fungus produces chlamydospores and characterized by non-papillate sporangia which can be obtained by incubating a non-sterile percolate of field soil.

ii) Management: Improving soil drainage conditions helps to keep the disease incidence at a lower level. Phytosanitation and wound dressing with tar have also been recommended as control measures for this disease (Anandaraj and Devasahayam, 2004).

9.5.4. Pink Disease, *Corticium salmonicolor*

Pink disease caused by *Corticium salmonicolor* (syn. *C. javanicum)* has been reported on cinnamon and considered as an important disease in Sri Lanka, India and Indonesia (Weiss, 2002).

i) Symptoms: Formation of pale pinkish white encrusted areas on stems or branches is the visible symptoms at the beginning. Later infection spreads destroying bark and finally leading to the death of the smaller shoots. Pathogen also reported to attack mango, jackfruit, custard apple and other fruit trees often grown in the vicinity of cinnamon plantations.

ii) Management: Burning of affected pruning's and other plant parts need to be practiced to reduce the disease incidence.

9.5.5. Seedling blight, *Diplodia* sp.

i) Symptoms: The disease occurs on the seedlings in the nursery. The pathogen causes light brown patches on the stem, which girdles the stem resulting in mortality of seedlings.

ii) Management: Pruning the die-back affected branches and spraying with 1% Bordeaux mixture are recommended for the control of the disease.

9.6. VANILLA, *Vanilla planifolia, V. andamanica*

9.6.1. Wilt, *Fusarium oxysporum* f. sp. *vanillae*

i) Symptoms: *Fusarium* wilt is the most serious disease of vanilla. The disease is more prevalent in younger plantation especially during the monsoon season. The infection starts at leaf axil and spreads to the inter nodal region resulting in rotting and drying of the stem above the point of infection. The fungus also causes leaf rot on the plant.

ii) Management: Phytosanitation such as removal and destruction of dead and diseased tissues and provision of adequate drainage in the plantation is important. Spraying and drenching with 0.1% carbendazim is effective against the disease. Addition of organic amendments, which encourages growth of antagonistic organisms, also reduces the intensity of the disease in the plantation.

Soil application of compost mounds enriched with *T. harzianum* twice resulted in substantial reduction in wilt. Soil application of rhizobacterial strain consortia (IISR-147 and IISR-148) with *T. harzianum* was effective in disease suppression of both *Phyophthora meadii* and *F. o.* f. sp. *vanillae.*

9.6.2. *Phytophthora* Rot, *Phytophthora meadii*

i) Symptoms: The pathogen causes rotting of beans, leaves and stems. In severe cases, all the beans in a bunch are completely rotten. The disease is more severe during the monsoon especially in shaded plantations and poorly drained soils.

ii) Management: Phytosanitation, spraying of 1% Bordeaux mixture and soil drenching with 0.2% copper oxychloride is effective against rot caused by *Phytophthora meadii.* Soil application of compost mounds enriched with *T. harzianum* twice resulted in substantial reduction in *Phytophthora* rot. Soil application of rhizobacterial strain consortia (IISR-147 and IISR-148) with *T. harzianum* was effective in disease suppression of both *P. meadii* and *F. o.* f. sp. *vanillae.*

9.6.3. Root Rot, *Sclerotium rolfsii*

i) Symptoms: Root rot is generally seen during the onset of monsoon rains. Initially, rotting of root tips occurs and later the rotting extends to the whole root system followed by yellowing and wilting of vines.

The disease also affects leaves and beans. The infected beans develop reddish brown sunken lesions initially and later the entire bean rots. In severe cases, all the beans in the bunch are covered with a thick white feathery fungal mat.

ii) Management: Soil drenching with 0.1% carbendazim is effective for the management of the disease.

9.6.4. Shoot Tip Rot, *Colletotrichum vanillae*

i) Symptoms: The disease is characterized by rotting of shoot tips and is more common during monsoon season.

ii) Management: Soil drenching with 0.1% carbendazim is effective for the management of the disease.

9.6.5. Dry Rot, *Fusarium* sp, *Rhizoctonia* sp.

i) Symptoms: Dry rot affects all parts of the plant. Shriveling of stems, roots and leaves followed by yellowing are the common symptoms of the disease. The disease is more severe during summer. Intermittent drying of stems and leaves occurs at advanced stages of the disease.

ii) Management : For control of dry rot disease, remove diseased plants and weeds along with surrounding soil where the disease is observed, mulch the base of stem with dry leaves before and after monsoon, avoid injury to roots during cultivation and drench soil around the base of the stem with 0.5% Bordeaux mixture.

9.7. NUTMEG, *Myristica fragrans*

Nutmeg is the dry hard wrinkled seed or pit of the nutmeg fruit. Mace is the orange red fleshy covering of the nutmeg kernel. They contain 7-14% essential oil. This essential oil contains highly toxic compound known as myristicin, used in small amounts to flavour puddings and fruit pie.

Myristicin present in nutmeg seeds could lead to delirium and deep stupor. Smaller amounts of nutmeg may cause vomiting and colic. Nutmeg has antimicrobial property. The volatile components of these spices are believed to be responsible for this property.

9.7.1. Dieback, *Diplodia natalensis*

i) Symptoms: The disease is characterized by drying up of mature and immature branches from the tip downwards.

ii) Management: The infected branches should be cut and removed and the cut ends pasted with 1% Bordeaux mixture or 0.2% copper oxychloride.

9.7.2. Thread Blight, *Marsimus pulchermia, M. equicrinus*

i) Symptoms: Two types of blights are noticed in nutmeg. The first is a white thread blight caused by *Marsimus pulchermia* wherein white hyphae aggregate to form fungal thread that traverses along the stem underneath the leaves in a fan shaped or irregular manner causing blight in the affected portions. The dried up leaves with mycelium form a major source of inoculum for the spread of the disease.

The second type of blight is called horse hair blight caused by *M. equicrinus.* Fine black silky thread of the fungus form an irregular, loose network on the stems and leaves. These strands cause blight of stems and leaves. However, these threads hold up the detached, dried leaves on the tree, giving the appearance of a bird's nest, when viewed from a distance.

Both the diseases are severe under heavy shade.

ii) Management: These diseases can be managed by adopting phytosanitation and shade regulation. Prune the affected twigs and dried hanging leaves. In severely affected gardens, spraying 1% Bordeaux mixture or 0.2% copper oxychloride may be undertaken in addition to cultural practices.

9.7.3. Fruit Rot, *Phytophthora* sp., *Diplodia natalensis*

i) Symptoms: Immature fruit split, fruit rot and fruit drop are serious in a majority of nutmeg gardens in Kerala. Immature fruit splitting and shedding are noticed in some trees without any apparent infection. In case of fruit rot, the infection starts from the pedicel as dark lesions and gradually spread to the fruit, causing brown discolouraton of the rind resulting in rotting. In advanced stages, the mace also rots emitting a foul smell. The fallen fruits become enveloped with the growth of the organism.

ii) Management: Spraying 1% Bordeaux mixture or 0.2% copper oxychloride when the fruits are half mature helps to reduce the disease incidence.

9.7.4. Shot Hole, *Colletotrichum gloeosporoides*

i) Symptoms: Sunken spots surrounded by a yellow halo are the initial symptoms. Subsequently the central portion of the necrotic region drops off resulting in shot hole symptoms. Dieback symptoms are also observed in some of the mature branches. On young seedlings drying of the leaves and subsequent defoliation are seen. In advanced stages, the necrotic spots become brittle and fall off resulting in shot holes.

ii) Management: The disease can be controlled by spraying 1% Bordeaux mixture or 0.2% copper oxychloride two or three times during rainy season.

9.8. CORIANDER

Coriander (*Coriandrum sativum*) is used as a common flavouring substance. The stems, leaves and fruits have a pleasant aroma. The whole young plant is used in preparing chutney. Its leaves are used for flavouring curries, sauces and soups. Dry seeds are extensively used in preparation of curry powder, pickling spices, sauces and seasonings. The seeds contain 0.5-1% essential oil which has an active principle, an isomer of geraniol. This plant contains Vitamin A. In medicines, its seeds are used as a carminative, refrigerant, diuretic, flatulence, vomiting and intestinal disorders. The decoction is used in indigestion, cold and other diseases. The oil from seeds is used in sweet meats, cocoa eatables, liquor and for agreeable smell and taste. Coriander seeds contain thalides which increase the levels of anticancer protective enzymes. The oil is also useful as an insecticide.

It is being cultivated in 0.282 million hectares producing 0.172 million tonnes of coriander seeds with an average yield of 0.6 tonnes per hectare. In India, coriander is cultivated in Andhra Pradesh, Madhya Pradesh, Karnataka, Tamil Nadu and Uttar Pradesh.

9.8.1. Wilt, *Fusarium* sp.

i) Symptoms: Of late, coriander wilt is becoming a serious disease which causes significant yield loss.

ii) Management: Follow crop rotation. Deep ploughing during summer should be taken up. Seed treatment and soil application with *T. harzianum* were effective for the management of the disease.

Seed treatment with *P. fluorescens* at 10 g/kg gave the lowest Fusarial wilt incidence of 14.4% (36.6% in control) with an yield of 603 kg/ha (423 kg/ha in control). *T. viride* as seed treatment (4 g/kg) and soil application (5 kg/ha) reduced the Fusarial wilt incidence by 58% with an yield of 575 kg/ha compared to 433 kg/ha in control.

9.8.2. Powdery Mildew, *Erysiphe* sp.

i) Symptoms: Powdery mildew causes about 50% loss in yield as it attacks flowers and seeds. Heavily infected flowers fail to form seeds and dry up.

ii) Management: The disease can be effectively controlled by soil application of wettable sulphur twice (once at the time of flower initiation and repeated after 15 days). The disease can also be controlled by three sprayings of 0.05% dinocap (ICBR Rs. 30.10) or 3 dustings of sulphur at 25 kg/ha (ICBR Rs. 28.32). The first application is carried after 45 days of sowing followed by 2 subsequent applications at 15 days interval.

9.8.3. Leaf Spot, *Alternaria alternata*

i) Symptoms: Once this disease becomes evident, seed losses are inevitable. This disease is seed-borne and the best control is to use clean seed lines. This disease is usually transferred from crop to crop.

ii) Management: Preventative control of this fungal disease is important. However, it has become common practice to use 2–3 kg/ha of mancozeb product at 10% flowering to control *Alternaria alternata*. Fortnightly to three weekly treatments are applied during the flowering period to protect new growth, especially where irrigation is used.

9.9. FENUGREEK

Fenugreek seed is a hard lentil seed. Its colour is dark fawn and has astringent aroma. Fenugreek seeds contain 5% bitter fixed oil. The oil has strong celery-like odour. Fenugreek seeds aid in maintaining the blood glucose levels in non-insulin dependent diabetes. The fibre present in the seed may be responsible for this. It is also used with the butter milk in the treatment of dysentery.

9.9.1. Powdery mildew

i) Symptoms: This disease appears usually in later stage of crop and becomes serious when pod formation takes place. In this disease, white powdery patches appear on the lower and upper surface of leaves and other parts of plant.

ii) Management: Crop should be dusted with 300 mesh Sulphur dust at 25 kg/ha to control this disease as soon as the symptoms are noticed. Spraying of wettable sulphur or dinocap (Karathane or Thiovit) can also be used to control the disease at 20-25 g per 10 litre of water at the initial stage of this disease. If needed two more sprays should be given at an interval of 15 days after first spray.

For control of powdery mildew in seed production of fenugreek (Methi), two sprays of propiconazole 0.1% or penconazole 0.05% or carbandazim 0.1% at 15 days interval from the appearance of disease are recommended.

9.9.2. Downy Mildew

i) Symptoms: This disease occurs during February and March. Yellow patches on the upper surface of leaves appear in the infected plants and white cottony mycelium on the lower surface of leaves.

ii) Management: This disease can be controlled by spraying of 0.2% solution of difolatan or any other copper fungicide.

9.9.3. Cercospora Leaf Spot

i) Management: A fenugreek variety '*Rajendra Kanti*' with golden yellow seeds is resistant to *Cercospora* leaf spot with yield potentials of 1.5-1.8-MT/ha and 40-42% higher yield over local variety. It is widely accepted by the farmers and has a greater coverage in Bihar and other states. It also builds upsoil health because of profuse nodulation.

9.10. CUMIN, *Cuminum cyminum*

Cumin seed is pungent, sharp and astringent. It contains 2-4% essential oil. The active principle is an aldehyde cumino. It is used as a stimulant and carminative agent. Cumin seeds contain bio-active substances called pthal-ides which increase the levels of anticancer protective enzymes in the body.

9.10.1. *Fusarium* wilt

i) Symptoms: Infected plants show peculiar symptoms of drooping of tips and leaves, leading to mortality of the entire plant. Attack of wilt is severe in younger plants.

ii) Management : Crop rotation and use of neem cake are helpful in checking spread of the fungus vis-à-vis disease. Seeds collected from disease free plots should only be used for sowing.

T. harzianum with neem cake as soil application reduced the incidence of wilt. Seed and soil treatment with *T. harzianum* was also found significantly effective for wilt disease.

9.10.2. *Alternaria* Blight

i) Symptoms: The blight affected plants show very minute brownish necrotic spots, which later turn to blackish. Mostly diseased plants fail to produce seeds. If seeds are produced they remain shriveled, light in weight and dark in colour.

ii) Management : For the control of this disease, seed treatment and spraying of 0.2% solution of mancozeb 4 times at 10 days interval commen-cing from 40 days after sowing is recommended. Add 1 ml soap solution/litre of water for better efficiency of fungicide. The crop should be kept free from weeds. Crops requiring more irrigation and mustard crop should not be grown in vicinity of this crop.

9.10.3. Powdery mildew, *Erysiphe* sp.

i) Symptoms: Affected plants in early stages show minute whitish spots on leaves, petiole, stem pedicel and seeds. In severe condition, it looks as the plants have been dusted with white powder. At later stages of attack seeds become white and shriveled and light in weight.

ii) Management: Crop should be dusted with 300 mesh sulphur dust at 25 kg/ha to control this disease as soon as the symptoms are noticed and repeated twice at 15 days interval.

Spraying of wettable sulphur or dinocap (Karathane or Thiovit) can also be used at 20-25 g per 10 litres of water at the initial stage of this disease. If needed second spray should be given 15-20 day after first spray.

9.11. CLOVE, *Eugenia caryophyllata*

It is the dried flower bud of the clove tree. It contains about 15% essential oil. The chief active principle is eugenol. Cloves are used whole in meats, pickling and fish. It is used ground in cakes, cookies and puddings. Used as flavouring agent in pulav, rice puttu and fruit cake. Cloves contain eugenol which has anti mutagenic effect. It is used for tooth pain.

9.11.1. Seedling Wilt, *Cylindrocladium* sp., *Fusarium* sp., *Rhizoctonia* sp.

i) Symptoms: Leaves of affected seedlings lose their natural luster, droop and ultimately die. Root system and collar region of seedling show varying degree of discolouration and decay.

ii) Management: Since the infected plants promote further spread of the disease, they are to be removed and the remaining seedlings should be treated with 0.1% carbendazim both as spray and soil drench. Alternatively, foliar spraying with 1% Bordeaux mixture and soil drench with 0.2% copper oxychloride is also effective for the management of the disease.

9.11.2. Leaf Rot, *Cylindrocorpon quinqueseptatum*

i) Symptoms: Leaf rot is noticed in mature trees and seedlings. The infection starts as dark diffuse patches at the leaf tip or margin and later the whole leaf rots, resulting in severe defoliation.

ii) Management: The foliage of affected trees should be sprayed with 0.1% carbandazim. Prophylactic sprays with 1% Bordeaux mixture also prevent the disease.

9.11.3. Leaf Spot and Bud Shedding, *Colletotrichum gloeosporoides*

i) Symptoms: The disease is characterized by dark brown spots with a yellow halo on leaves. Such spots also appear on the buds resulting in their shedding.

ii) Management: Prophylactic spraying with 1% Bordeaux mixture or 0.2% copper oxychloride prevents the above diseases.

9.12. FENNEL, *Foeniculum vulgare*

9.12.1. Leaf Blight, *Cercosporidium punctum*

i) Symptoms: Fennel Leaf blight caused by *Cercosporidium punctum,* is the major disease that affects fennel and can cause serious yield losses if not controlled. The pathogen causes heavy leaf loss and damage to the developing flowers. High humidity during flowering will promote *Cercosporidium* development, such that heavy leaf loss and damage to developing flowers and seed will be sustained.

ii) Management: This can be managed with early preventative fungicide applications to reduce the level of inoculum. Mancozeb is currently the product used for the control of this disease.

9.12.2. Stem Rot, *Sclerotinia minor, S. sclerotiorum*

i) Symptoms: Stem rot, commonly referred to as "drop", is an extremely important disease of fennel, and is very prevalent in the coastal production areas. The sclerotia of *S. minor* populations may remain viable in the soil for prolonged periods (up to 8 to 10 years). The crop losses from this disease range from 5 to 20%. Individual stalks begin to collapse and wilt. The interior of dead stems frequently reveals small, irregularly-shaped black structures called sclerotia.

ii) Management : Since the fungus lives in the soil, a two-year rotation with non-susceptible plants, such as corn, to prevent the buildup of pathogenic organisms can reduce the incidence of disease. Foliar application of fosetyl-AL at 6 kg a.i./ha, mancozeb at 4 kg a.i./ha or dicloran at 5 kg a.i./ha gives effective control of the disease.

9.12.3. Damping-Off, *Rhizoctonia solani, Pythium* spp.

i) Symptoms: Seedlings either rot and seeds do not germinate or germinate and collapse. The disease frequently attacks the base of the seedling and leaves a shriveled, dried-up stem at the base of the seedling.

ii) Management : Control can be achieved by using sterile potting media and fungicide-treated seed. If seedlings become symptomatic, rogue and discard all of the seedlings since asymptomatic seedlings may also be infected.

9.12.4. Root and Crown Rot, *Rhizoctonia solani, Pythium* spp., *Fusarium* spp.

i) Symptoms: Symptoms appear as wilting and a slow or rapid collapse of the plant. The roots can appear brown and water-soaked instead of white. A water-soaked lesion can often appear at the base of the stem.

ii) Management: Control can be achieved by using sterile potting media, clean pots, fungicide-treated seed and using a two-year rotation with non-susceptible plants, such as corn, to prevent the buildup of pathogenic organisms. Affected plants should be rouged out and discarded.

9.12.5. Botrytis Blight, *Botrytis cinerea*

i) Symptoms: *Botrytis* blight occurs on many different kinds of herbs. The disease appears when plants are grown under conditions of high humidity. The disease appears as tan to brown spots on leaves that can be associated with a grayish mold colonizing the damaged area. Senescing flowers are particularly susceptible and frequently have papery spots on them.

ii) Management: *Botrytis* blight can be suppressed by avoiding overhead irrigation and watering in the morning to allow time for leaves and flowers to dry. Proper plant spacing can improve air drainage and rapid leaf drying. Control can also be achieved with the use of fungicide (iprodione) sprays applied as soon as symptoms are visible.

Table 9.2. Biological control of diseases in spice crops

Spice crop	Disease/Pest	Causal organism	Effective biocontrol agents
Black pepper	Foot rot	*Phytophthora capsici*	Soil application of *Trichoderma harzianum, T. virens, Glomus fasciculatum*, spray *Pseudomonas fluorescens, Bacillus* sp.
	Quick wilt,	*Phytophthora* sp.	Soil treatment with *T. harzianum* (IISR P-26) + *P. fluorescens* (TNAU PF-1).
	Anthracnose	*Colletotrichum gloeosporoides*	Spray *P. fluorescens.*
Cardamom	Damping off	*Pythium vexans*	*T. harzianum, T. viride* in solarized nursery beds.
	Clump rot/ Rhizome rot	*Pythium vexans, Rhizoctonia solani*	Soil appln. of *T. harzianum.*
	Capsule rot	*Phytophthora meadii, P. nicotianae* var. *nicotianae*	Soil appln. of *T. harzianum, T. viride, T. virens, T. hamatum.*
Ginger	Rhizome rot	*Pythium aphanidermatum, P. myriotylum*	Soil solarization + soil appln. of *T. harzianum,* seed treatment with *P. fluorescens.*
	Yellows	*Fusarium* spp.	Soil solarization + appln. of *T. harzianum.*
Turmeric	Rhizome rot	*Fusarium* sp., *Pythium graminicolum, P. aphanidermatum*	Soil treatment with *T. harzianum, T. viride*
Fenugreek	Root rot	*R. solani*	Seed treatment with *T. viride, P. fluorescens.*
Coriander	Root rot/wilt	*Fusarium oxysporum* f. sp. *corianderii*	Seed treatment with *T. viride, T. harzianum, Streptomyces* sp.
Cumin	Wilt	*F. oxysporum* f.sp. *cumini*	Soil treatment with *Trichoderma* spp.
Vanilla	Root rot	*Phytophthora meadii, F. oxysporum* f.sp. *vanille*	Soil treatment with *T. harzianum, P. fluorescens*

REFERENCES

Agarwal, G.P. and Hasija, S.K. 1967. *Alternaria* rot of citrus fruits. *Indian Phytopath.* **20**: 259-260.

Alam, M. and Husain, A. 1976. Leaf blight of Java citronella (*Cymbopogon winterianus* Jowiff) caused by *Curvularia andropogonis* (Zimm.) Boedigin. *New Botanist* **3**: 54-56.

Alam, M., Khan, A.M. and Husain, A. 1986. Leaf blight and leaf spot diseases of Java citronella caused by *Curvularia andropogonis. Indian Phytopath.* **35**: 480-483.

Anandaraj M., Devasahayam, S., Krishnamoorthy, B., Mathew, P.A. and Rema, J. 2001. *Cinnamon.* Extn. Pamphlet, Indian Inst. of Spices Res., Calicut, Kerala, India.

Anandaraj, M. and Devasahayam, S. 2004. Pests and diseases of cinnamon and cassia, *In. Cinnamon and Cassia* (P.N. Ravindran, K.N. Babu and M. Shylajah, eds.), pp. 139-258. CRC Press, New York.

Anon. 1979. *Tropical Legumes: Resources for the Future.* National Acad. Sci., Washington DC, pp. 117-121.

Anon, 1987. Improvement of betel vine cultivation in Mahoba. *Progress Report 1986-87.* National Botanical Res. Inst., Lucknow, India.

Anon. 1990b. Biological suppression of pests and diseases in fruit crops. In *Proceedings of Indo-USSR Joint Workshop on Problems and Potentials of Biocontrol of Pests and Diseases*, Project Directorate of Biological Control, Bangalore, pp. 26-28.

Anon. 1996. *Cinnamon; Cultivation and Processing.* Tech. Bull. 5, Dept. of Export Agriculture, Ministry of Agriculture, Lands and Forestry, Sri Lanka, pp.7-8.

Bandare, J.M.R.S. 1981. *Puccinia* rust of citronella and lemon grass in Sri Lanka. *Plant Dis.* **65**: 164.

Barbercheck, M.E. and Broembsen, S.L. von. 1986. *Plant Disease* **70**: 945-950.

Baruah, A., Baruah, S.N. and Bora, K.N. 1980. A fungal disease of peach plant collected from the Gauhati University Campus, Assam (*Venturia carpophila*). *Sci. Cult.* **46**: 264.

Barua, A. and Bordoloi, D.N. 1983. Record of a new disease of lemon grass (*Cymbopogon flexuosus* Stapf.) caused by *Curvularia verruciformis* Agarwal and Sahni. *Curr. Sci.* **52**: 640-641.

Bedi, P.S. Singh, G. and Suryanarayana, D. 1969. Chemical control of plant diseases. *Hindustan Antibiotic Bull.* **11**: 251-253.

Bhardwaj, L.N. and Agarwala, R.K. (1986a). In *Advances in Research on Temperate Fruits* (T.R. Chadha *et al.*, eds.), pp. 397-401. Y.S. Parmar Univ. of Hort. & Forestry, Himachal Pradesh, Solan, India.

Bhardwaj, L.N. and Agarwala, R.K. 1986b. In *Advances in Research on Temperate Fruits* (T.R. Chadha *et al.*, eds.), pp. 403-405. Y.S. Parmar Univ. of Hort. & Forestry, Solan, Himachal Pradesh, India.

Bhargava, S.N. and Singh, A.P. 1975. *Aspergillus* rot of mango and its control. *Indian J. Hort.* **32**: 190.

Bhat, M.N., Hegde, R.K., Hiremath, P.C. and Naidu, K.S. 1988, A note on the occurrence of a die-back on cinnamon in Karnataka, *Curr. Res.,* Univ. Agric. Sci., Bangalore **17**(11): 13.

Bhattacharyya, S.K., Shekhawat, G.S. and Singh, B.P. 1990. *Potato Late Blight*. Tech. Bull. No. 27, Central Potato Res. Inst., Shimla, India, 40 pp.

Bijay Kumar 2007. *Indian Horticulture Database 2006*. National Hort. Board, Gurgaon, Haryana, India, 324 pp.

Bose, S.K., Sindnan, G.S. and Pandey, B.N. 1973. Studies on the die-back disease of mango in the Terai region of Kumaon. *Prog. Hort.* **5**: 41-53.

Butler, E.J. 1906. The wilt disease of pigeon-pea and pepper. *Agric. J. India* **1**: 25-36.

Capoor, S.P. and Sharma, D.C. 1965. Purple leaf disease of beet. *Indian Phytopath.* **18**: 88-89.

Chand, J.N., Rattan, B.K. and Suryanarayana, D. 1967. Epidemiology and control of fruit rot of citrus caused by *Alternaria citri* Ellis and Piece. *J. Res. PAU* **4**: 217-222.

Chang, T.T. 1992. Decline associated with brown *Phellinus noxius. Plant* **2**: 90-95.

Chattopadhyaya, S.B. and Sen Gupta, S.K. 1955. Studies on wilt of *Psidum guajava* in West Bengal. *Indian J. Hort.* **12**: 76-79.

Chaturvedi, A.N. 1985. *Firewood Farming on the Degraded Lands of the Gangetic Plain.* U.P. Forest Bull. No. 50, Lucknow, Govt. of India Press (I), 286 pp.

Coronel, R.E. 1991. *Tamarindus indica* L. In *Plant Resources of South East Asia, Wageningen; Pudoc. No. 2. Edible Fruits and Nuts* (E.W.M. Verheij and R.E. Coronel, eds.), pp. 298-301. PROSEA Foundation, Bogor, Indonesia.

Da Graca, J.V., Vuuren, S.P. and Van. 1980. Transmission of avocado sunblotch disease to cinnamon. *Plant Dis.* **64**: 475.

Durrani, M.K. 1976. *Studies on* Alternaria *Rot of Apple.* M. Sc. Thesis, Univ. of Udaipur, Jobner Campus, India.

Eckert, J.W. and Eaks, I.L. 1989. Post-harvest disorders and diseases of citrus fruits. In *The Citrus Industry* Vol. 5 (W. Reuther, E.C. Calavan and G.E. Carman, eds.), pp. 179-250. Univ. of Calif. Press, Oakland, USA.

Eckert, J.W. and Ogawa, J.M. 1985. The chemical control of post-harvest diseases: Subtropical and tropical fruits. *Ann. Rev. Phytopathol.* **23**: 421-454.

Fu and Chang, T. T. 1999. Anthracnose of cinnamon, *Taiwan J. Sci.* **14**: 513-515.

Gadewar, A.V., Trivedi, T.P. and Shekhawat, G.S. 1991. *Potato in Karnataka.* Tech. Bull. No. 17, Central Potato Res. Inst., Shimla, India, 72 pp.

Gardner, D.E. 1985. Lemon grass rust caused by *Puccinia makanishikii* in Hawaii. *Plant Dis.* **69**: 1100.

Gardner, P.D., Eckert, J.W., Baritelle, J.L. and Bancroft, M.N. 1986. Management strategies for control of *Pencillium* decay in lemon packing houses: Economic benefits. *Crop Prot.* **5**: 26-32.

Gowdar, S.B. and Kulkarni, S. 1999. Compatibility effect of antagonists and seed dressers against *Fusarium udum* – the causal agent of pigeon pea wilt. *Karnataka J. Agri. Sci.* **12**: 197-199.

Gunasena, H.P.M. and Hughes, A. 2000. *Tamarind*. International Centre for Underutilized Crops, Southampton, UK, 171 pp.

Gupta, G.K. 1985. *Apple Scab* (Venturia inaequalis). R. Merck (India) Ltd., Bombay, 36 pp.

Gupta, J.C. and Madaan, R.L. 1978. Occurrence of *Alternaria* leaf spot disease of ber (*Zizyphus mauritiana* Lamk.) and its control (Abstr.). *Symp. on Advancing Frontiers in Plant Biol.*, Banaras Hindu Univ., Varanasi, India, p. 45.

Gupta, J.C. and Madaan, R.L. 1979a. *Hansfordia pulvinata* – A mycoparasite on *Isariopsis indica* var. *zizyphi*. *Curr. Sci.* **48**: 121-122.

Gupta, J.C. and Madaan, R.L. 1980. Varietal resistance against leaf spot on ber. *Haryana Agri. Univ. J. Res.* **10**: 75-76.

Gupta, J.C. and Madaan, R.L. 1985. Report on pathological aspects of fruit crops (1975-83). Dept. of Hort., Haryana Agri. Univ., Hisar, India, 42 pp.

Gupta, J.C., Madaan, R.L. and Chauhan, K.S. 1980. Varietal resistance of ber against *Isariopsis* mouldy leaf spot disease. *Indian Phytopath.* **33**: 140-141.

Gupta, J.H. 1976. Reaction of mango varieties to powdery mildew (*Oidium mangiferae* Berthet) in Uttar Pradesh. *Prog. Hort.* **8**: 63-64.

Gupta, M.L. and Husaini, A. 1977. *Ellesiella* leaf blight of palmarosa. *Sci. Cult.* **43**: 412.

Gupta, M.L., Janardhan, K.K. and Husaini, A. 1977. *Curvularia* leaf blotch – A new disease of palmarosa in India. *Indian Phytopath.* **29**: 312-313.

Gupta, P.K., Agarwal, R.K. and Sharma, S.L.1986. Efficacy of soil amendments in managing 'Stalk rot' (*Sclerotinia sclerotiorum*) of cauliflower. *Indian J. Pl. Path.* **4**: 129-131.

Hare, W.W. 1951. *Phytopathology* **41**: 16.

Hocking, D. 1993. *Trees for Drylands*, Tamarindus indica *L. Family Leguminosae: Caesalpinodeae*. Oxford & IBH Publishing Co., New Delhi, pp. 305-307.

Janardhan, K.K., Gupta, M.L. and Husaini, A. 1980. Effect of *Curvularia* leaf blotch disease on essential oil contents of palmarosa. *Indian J. Expt. Biol.* **18**: 439-440.

Jeyarajan, R. and Cheema, S.S. 1972. Screening of ber varieties for resistance to leaf spot and powdery mildew diseases. *Indian J. Hort.* **29**: 353.

Kader, A.A., El-Goorani, M.A. and Sommer, N.F. 1982. Post harvest decay, respiration, ethylene production of quality peach held in controlled atmosphere with added carbon monoxide. *J. Amer. Soc. Hort. Sci.* **107**: 856-859.

Kanwar, Z.S. and Thakur, D.P. 1972. Controlling postharvest soft rot of pomegranate fruits by treatment of packing straw. *Sci. Cult.* **38**: 450-452.

Kanwar, Z.S. and Thakur, D.P. 1974. Control of soft rot of pomegranate fruits caused by *Rhizopus arhizus* Fisher. *Pesticides* **8**: 43-44.

Kapoor, S.P. and Chohan, J.S. 1974a. Fruit rot of papaya by *Macrophomina phaseoli*. *Indian Phytopath.* **27**: 251-252.

Kapoor, S.P. and Chohan, J.S. 1974b. Factors affecting infection and development of fruit rot of papaya by *Macrophomina phaseoli. Indian Phytopath.* **27**: 252-253.

Kapur, S.P., Cheema, S.S. and Singh, M.P. 1975. Occurrence and control of powdery mildew of ber (*Zizyphus maurutiana* Lame.) in Punjab. *J. Res. Punjab Agri. Univ.* **12**: 26-29.

Karunakaran, P. and Nair, M.C. 1980. Leaf spot and die back disease of *Cinnamomum zeylanicum* caused by *Colletotrichum gloeosporioides, Plant Dis.* **64**: 220-221.

Karunakaran, P., Nair, M.C. and Das, L. 1993. Grey blight disease of cinnamon *(Cinnamomum verum* Bercht. & Presl.) leaves. *J. Spices Aromatic Plants* **2**: 66-67.

Katan, J. 1981. Solar heating (solarization) of soil for control of soil borne pests. *Ann. Rev. Phytopath.* **19**: 211-236.

Kaul, J.L. and Munjal, R.L. 1982. Apple losses in Himachal Pradesh due to post-harvest fungal pathogens. *Indian J. Mycol. Pl. Path.* **12**: 209-210.

Kaushik, C.D., Thakur, D.P. and Chand, J.N. 1972. Parasitism and control of *Pestalotia psidii* causing cankerous disease of ripe guava fruits. *Indian Phytopath.* 25: *61-64.*

Khan, S.N., Misra, B.M. and Rehill, B.S. 1988. Two new species of *Cercospora. Indian J. Forestry* **11**: 188-190.

Khanna, K.K. and Chandra, S. 1976. Control of storage diseases of fruits. *Curr. Sci.* **45**: 70.

Kothari, K.L. 1968. Controlling fruit rot of guava. *Indian Hort.* **12**: 9-12.

Kumara, K.L.W. 1999a, Study of leaf galls, leaf spots and scab-like condition of cinnamon *(Cinnamomum zeylanicum* Blume) in Matara District, *Proc.* SLAAS, Colombo, Sri Lanka, 65 p.

Kumara, K.L.W. 1999b. Effect of some agronomic practices on the incidence of leaf spot disease *(Colletotrichum gloeosporioides)* in cinnamon *(Cinnamomum zeylanicum* Blume) in southern region of Sri Lanka. SLAAS-University of Ruhuna Seminar on "*Pest Control in the Next Millennium*", SLAAS, Colombo.

Lal, B., Arya, A., Rai, R.N. and Tewari, D.K. 1982. A new soft rot of aonla caused by *Phomopsis phyllanthi* Punith and its chemical control. *Nat. Acad. Sci. Letter* **5**: 183-185.

Lazarovits, G., Conn, K. and Tenuta, M. 2000. Control of *Verticillium dahliae* with soil amendments: Efficacy and mode of action. In *Advances in Verticillium Research and Disease Management* (E.C. Tjamos, R.C. Rowe, J.B. Heale, and D.R. Fravel, eds.), PP. 274-291. American Phytopathological Society, St. Paul, Minn.

Lele, V.C. and Butani, D.K. 1975. Trends in plant disease control in India with particular reference to fruit crops. *Pesticides* **9**: 75-95.

Leonian, L.H. 1922. Stem and fruit blight of chillies caused by *Phytophthora capsici* sp. nov. *Phytopathology* **12**:401-408.

Madaan, R.L. and Chand, J.N. 1985. Epidemiological studies on *Alternaria* leaf spot and fruit rot disease of ber (*Zizyphus mauritiana* Lamk.). *Indian J. Mycol. Pl. Path.* **15**: 26-27.

Mallikarjunaradhya, A.V., Bhat, A.V., Crown, J.K., Vijayendra Rao, A.R., Ramanna, K.V.R. and Narasimha, P. 1979. Control of fungal stem end rot during the transport of pineapples. *J. Food Sci. Tech.* **16**: 232-234.

Mammen, K.V. 1974. On a wilt disease of betel vine in Kerala caused by root-knot nematode. *Agric. Res. J. Kerala* **12**: 76.

Menge, J.A. 1993. Improved root rot control with mulches. *Calif. Citrog.* **79**: 16-17.

Menon, K.K. 1949. The survey of pollu and root diseases of pepper. *Indian J. Agric. Sci.* **19**: 89-136.

Mir, N.M. 1987. Screening of fungicides for field control of downy mildew (*Peronospora destructor*) on onions. *Indian J. Mycol. Pl. Path.* **17**: 321-322.

Mirakhur, R.K., Dhar, A.K. and Kaw, M.R. 1977. Downy mildew of *Allium cepa* and its control with fungicides in Kashmir valley. *Indian Phytopath.* **30**: 576-577.

Mishra, R.C. and Bhalla, O.P. 1974. Occurrence of *Eulecanium* sp. ? *tiliae* (L.) on plum and apple in Himachal Pradesh. *Curr. Sci.* **43**: 428.

Misra, A. and Shivpuri, A. 1983. Anthracnose, a new disease of aonla. *Indian Phytopath.* **36**: 406-407.

Misra, A.K. 1995. *Indian J. Mycol. Pl. Path.* **25**: 214-217.

Naik, S.L. and Joshi, L.K. 1973. Effect of relative humidity and temperature on the development of *Pencillium* rot of apple and its control. *Hindustan Antibiotics Bull.* **16**: 81-83.

Naqvi, S.A.M.H. 1992. Chemical control of *Aspergillus* rot of Nagpur mandarin. *Plant Dis. Res.* **9**: 95-97.

Naqvi, S.A.M.H. 1993a. Influence of pre and post harvest factors on export oriented production of Nagpur mandarin. In *Proc. of Vasantrao Naik Memorial National Seminar on Agri. Sci. – Export Oriented Hortil. Prodn.*, pp. 181-185.

Naqvi, S.A.M.H. 1993b. Pre harvest application of fungicides in Nagpur mandarin orchards to control post-harvest storage decay. *Indian Phytopath.* **46**: 190-193.

Naqvi, S.A.M.H. 1994. Efficacy of some fungicides in control of *Phytophthora* diseases of Nagpur mandarin in Central India. *Indian Phytopath.* **44**: 5-9.

Naqvi, S.A.M.H. 1997. Role of pre and post harvest applications of benzimidazole fungicides in control of post-harvest decay of Nagpur mandarin. *Plant Dis. Res.* **12**: 6-10.

Naqvi, S.A.M.H. 1999. Integrated management of fungal diseases of citrus. In *IPM System in Agriculture* (Rajjev K. Upadhyay, K.G. Mukerji and O.P. Dubey, eds.), pp. 489-503. Aditya Books Pvt. Ltd., New Delhi.

Naqvi, S.A.M.H. 2000. Managing *Phytophthora* diseases on citrus. *Indian Hort.* **47**: 430-434.

Naqvi, S.A.M.H. 2001. Diagnosis and management of fungal diseases of citrus. In *Citrus* (Shyam Singh & S.A.M.H. Naqvi eds.), pp. 375-391. Int. Book Distributing Co., Lucknow, India.

Naqvi, S.A.M.H. 2002. *Fungal Diseases of Citrus: Diagnosis and Management.* Tech. Bull. No. 5, National Res. Centre for Citrus, Nagpur, India, 61 pp.

Nanjan, K. and Kumar, S. 1983. Notes on the control of fruit borer in pomegranate. *South Ind.. Hort.* **31**: 104-105.

Narasimhan, V., Subramanian, K.S., Shanmugam, N. and Jeyarajan, H. 1984. Efficacy of certain fungicides in the control of powdery mildew of mandarin. *Pesticides* **18**: 61.

Nayak, G.M. 1986. *Studies on Papaya in Relation to Certain Plant Diseases.* M. Sc. (Agri.) thesis, Univ. of Agri. Sci., Bangalore, 106 pp.

Padwick, G.W. 1946. Notes on Indian fungi IV. *Mycol. Pap. Imp. Mycol. Inst.* **17**, 12 pp.

Paharia, K.D. 1963. *Proc. Second All India Potato Workers Conf.,* Central Potato Res. Inst., Shimla, India, pp. 198-202.

Pandey, K.K. and Pandey, P.K. 2001. Soil borne diseases in vegetable crops: Biological control approach. *Souvenir of AICRP on Vegetable Crops*, pp. 51-52.

Pandey, K.K. and Pandey, P.K. 2002a. Field evaluation of some fungal toxicants against fungal disease complex of cowpea. *Pestalogy* **24**(6): 20-22.

Pandey, K.K. and Pandey, P.K. 2002b. Study on soil solarization for nursery management of vegetable crops. *Asian Cong. Mycol. Pl. Path.*, Mysore.

Pandey, K.K., Pandey, P.K., Kalloo, G., Kumar, R. and Singh, B. 1999. Sources of resistance against powdery mildew (*Erysiphe pisi*) of pea and its pathogen reaction in natural and artificial conditions. *Veg. Sci.* **26**: 160-163.

Patel, M.K., Kulkarni, Y.S. and Moniz, L. 1948a. *Indian Phytopath.* **1**: 147-152.

Patel, M.K., Moniz, L. and Kulkarni, Y.S. 1948b. *Curr. Sci.* **17**: 180-190.

Pathak, V.N., Sharma, H.C. and Bhatnagar, L.G. 1972. Efficacy of fungicides and hot water for control of *Fusarium* and *Rhizopus* rot of papaya. *Third Intn. Symp. Subtrop. Trop. Hort.*, Bangalore, India, pp. 118-119.

Patwardhan, P.G. 1965. *Ellisiella caudata* (P.K.) Sacc. A new record from India. *Curr. Sci.* **34**: 88-89.

Peethambharam, G.K. and Aravindakhan, M. 1975. Varietal susceptibility of mangoes to sooty mold. *Agric. Res. J. Kerala* **16**: 260-261.

Prakash, Om. 1988. Sooty mould disease of mango and its control. *Int. J. Pl. Dis.* (In Press).

Prakash, Om and Raoof, M.A. 1989. Die-back disease of mango (*Mangifera indica*), its distribution, incidence, cause and management. *Fitopatol. Bras.* **14**: 207-215.

Prakash, Om and Srivastava, K.C. 1987. *Mango Diseases and their Management – A World Review.* Today and tomorrow's Printers and Publishers, New Delhi, 175 pp.

Prakasam, V. 1991. Red leaf spot of cinnamon in Lower Pulney hills of Tamil Nadu. *Indian Cocoa, Arecanut and Spices J.* **14**: 123 p.

Puranik, S.B. and Suryanarayana, D. 1967. A new record of *Gloeocercospora* on khas (*Vetiveria zizanoides* Stap.). *Indian Phytopath.* **19**: 394.

Raghunath, T. 1965. A new smut on lemon grass from India. *Pl. Dis. Reptr.* **49**: 916.

Raju, K.S. and Rao, G.S. 1984. Studies to fix up optimum spray intervals against bacterial leaf spot and fruit rot disease of chilli. *South Indian Hort.* **32**: 319-320.

Ramakrishnan, T.S. 1954. *Common Diseases of Citrus in Madras.* Govt. of Madras Publn.

Ramakrishnan, T.S. and Damodaran, S. 1956. *Proc. Indian Acad. Sci.* **43**: 213-222.

Rands, R.D. 1922. Stripe canker of cinnamon caused by *Phytophthora cinnamomi* n. sp. *Meded. Inst. Voor Plantenzieketen* **54**: 53.

Rao, V.G. and Mhaskar, D.N. 1973. Studies on a leaf blotch disease of pineapple. *Revista Di Pathol. Veg.* **9**: 59-60.

Rao, V.G. and Mhaskar, D.N. 1981. Physiological studies into *Pestalotia microspora* Speg. inciting leaf blotch of pineapple. *J. Univ. Poona Sci. Tech.* **54**: 163-172.

Rawal, R.D. 1986. Control of papaya powdery mildew through fungicidal sprays. *Indian J. Hort.* **43**: 295-298.

Rawal, R.D., Muniyappa, N.C. and Ullasa, B.A. 1983. Effect of preharvest field sprays of the fungicides on the control of storage rot of papaya. *Indian J. Agri. Sci.* **53**: 614-615.

Rawal, R.D., Sohi, H.S. and Manjunath, K.L. 1981. Evaluation of different systemic and non-systemic fungicides against powdery mildew of papaya. *Pestology* **5**:11-14.

Rawal, R.D. and Ullasa, B.A. 1985. Annual Report, Indian Inst. of Hort. Res., Bangalore, India, p. 131.

Rawal, R.D. and Ullasa, B.A. 1988. Occurrence of fungal diseases of citrus in Bangalore. *VII National Citrus Seminar*, New Delhi.

Rawal, R.D. and Ullasa, B.A. 1989. Management of anthracnose of banana through fungicidal sprays. *Pl. Dis. Reptr.* **4**: 24-26.

Reddy, G.S. and Murti, V.D. 1985. *Citrus Diseases and their Control.* ICAR Bull., pp. 31-57.

Reddy, S.M. and Reddy, C.K.R. 1983. *Greenaria* fruit rot – An endemic disease of grapes in India. *Indian Phytopath.* **36**: 110-11.

Rhoade, R.A. and Jenkins, N.R. 1957. *Phytopathology* **47**: 29.

Ristaino, J.B. and Johnston, S.A. 1999. Ecologically based approaches to management of *Phytophthora* blight on bell pepper. *Plant Dis.* **83**:1080-1089.

Rowe, R.C., Davis, J.R., Powelson, M.L. and Rouse, D.I. 1987. Potato early dying: causal agents and management strategies. *Plant Dis.* **71**: 482-489.

Roy, M.K. 1975. Radiation, heat and chemical combines in the extension of shelf life of apples infected with blue mould rot (*Pencillium expansum*). *Pl. Dis. Reptr.* **59**: 61-64.

Sarma, Y.R., Nambiar, K.R.N. and Nair, C.P.R. 1974. Brown rot of turmeric. *J. Plantation Crops* **2**: 33-34.

Sarwar, M. and Srinath, K.V. 1963. Occurrence of *Curvularia trifolii* (Kauf.) Boed. on khas - *Vetiveria zizanoides* Stapf. *Curr. Sci.* **34**: 668.

Sawant, I.S. and Sawant, S.D. 1989. Coffee fruit skin and cherry husk as substrates for mass multiplication of *Trichoderma harzianum* an antagonist of citrus *Phytophthora*. *Indian Phytopath.* **42**: 336.

Seaby, D. 1989. *The Mushroom J.* **197**: 147-151.

Sen, B. and Kapoor, I.J. 1974. Chemical control of *Fusarium* wilt of tomato. *Pesticides* **8**: 40-42.

Shah, A., Boru, S.S. and Roy, A.J. 1985. Effect of fungicidal sprays on the incidence of leaf spot disease of pear. *Prog. Hort.* **17**: 363-365.

Sharma, R.C., Gupta, V.K. and Garg, R.C. 1988. Dormant application of fungicides for the management of peach leaf curl. *Indian Phytopath.* **41**: 259-260.

Sharma, R.C., Jindal, K.K. and Gupta, V.K. 1987. Managing fungal peach leaf curl through chemical and host resistance. *Proc. XIth Intn. Cong. Pl. Prot.,* Manila, Philippines, p. 51.

Sharma, R.C., Kaul, J.L. and Sharma, R.L. 1989. Effect of post-inoculation fungicidal dips on post-harvest diseases of peaches. *Int. J. Trop. Pl. Dis.* **7**: 221-224.

Sharma, R.C. and Sharma, S.L. 1985. Correlation of meteorological factors with stalk rot of cauliflower caused by *Sclerotinia sclerotiorum. Indian J. Ecol.* **12**: 327-330.

Sharma, R.L. 1989a. Post-harvest diseases of China pear in Himachal Pradesh. *XI Annual Meet of Soc. of Mycol. and Pl. Path.*, Orissa Univ. Agri. & Tech., Bhubaneswar, India.

Sharma, R.L. 1989b. Control of storage decay of China pear fruit by pre-harvest application of fungicides. *XI Annual Meet of Soc. of Mycol. and Pl. Path.*, Orissa Univ. Agri. & Tech., Bhubaneswar, India.

Sharma, R.L. and Kaul, J.L. 1989. Efficacy of post harvest fungicidal treatments in controlling brown rot (*Monilinia* spp.) of apricot. *Plant Dis. Res.* **4**: 54-58.

Sheikh, K.M., Bajwa, M.N., Rashid, A. and Mahmood, T. 1995. *Pak. J. Phytopath.* **7**: 29-32.

Siddaramaiah, A.L. and Kulkarni, S. 1982. Control of powdery mildew disease of tamarind seedlings. *Indian Forester* **108**: 361-364.

Singh, B. and Lal, S.B. 1953. Wilt of guava. *Agric. Husb.* **111**(10, 11 & 12): 78-79.

Singh, D. and Agarwala, R.K. 1988. Incidence of apple frog eye leaf spot (*Sphaeropsis malorum*) and its correlation with weather conditions. *Indian J. Mycol. Pl. Path.* **17**: 169-172.

Singh, G. 1988. Research Reports. *Fruit Res. Workshop,* Rajendra Agri. Univ., Pusa, Bihar, India.

Singh, G. and Singh, R.N. 1986. Control of leaf spot disease of fig caused by *Ascochyta caricae. Indian Phytopath.* **39**: 462-463.

Singh, N. and Gangopadhyay, S. 1984. Control of white rot of seed cauliflower. *Pesticides* **18**(12): 23-24.

Singh, R., Trivedi, B.M., Gill, H.S. and Sen, B. 1987. Breeding for resistance to black rot, downy mildew and curd blight in Indian cauliflower. *Cruciferae Newsl.* **12**: 96-97.

Sohi, H.S. 1983. Diseases of tropical and sub-tropical fruits and their control. In *Recent Advances in Plant Pathology* (A. Husain, K. Singh, B.P. Singh and V.P. Agnihotri, eds.), pp. 73-76. Print House, Lucknow.

Sohi, H.S. and Sridhar, T.S. 1970. Relative resistance and susceptibility of different grape varieties against downy mildew (*Plasmopara viticola* Berk & Curt) and rust (*Phakopsora ampelopsides* Diet) diseases. *Indian. J. Hort.* **27**: 57-59.

Sohi, H.S. and Sridhar, T.S. 1971. Controlling fruit rot of guava. *Indian Hort.* **16**(2): 9-10, 23.

Sohi, H.S. and Sridhar, T.S. 1973. Chemical control of leaf spot disease of sapota caused by *Phaeophleospora indica. Indian J. Hort.* **30**: 353-357..

Smilanick, J.L. and Margosan, D.A. 1996. Sodaash to control post-harvest green mold. *Caif. Citrograph* **9**: 10-17.

Sridhar, T.S. 1975. Black rot of pineapple – A new record from South India. *Curr. Sci.* **44**: 869.

Swamy, K.R.M. and Sadashiva, A.T. 2004. *Vegetable Varieties/Hybrids of IIHR.* Tech. Bull. 2, Indian Inst. of Hort. Res., Bangalore, 30 pp.

Tandon, I.N. 1961. Control of green moulds in cold storage. *Hort. Adv.* **5**: 77-82.

Tandon, M.P., Jamaluddin and Bhargava, V. 1975. Chemical control of decay of fruits of *Vitis vinefera* caused by *Aspegillus niger* and *Pencillium* sp. *Curr. Sci.* **44**: 478.

Thakur, D.P. 1972. Factors influencing storage rots of certain fruits and vegetables caused by three species of *Rhizopus. Indian Phytopath.* **25**: 354-358.

Thakur, D.P. and Chenulu, V.V. 1970. Chemical control of soft rot of apple and mango caused by *Rhizopus arhizus. Indian Phytopath.* **23**: 58-61.

Thakur, D.P. and Mishra, A.P. 1966. Varietal resistance in banana against anthracnose fungus (*Gloeosporium musarum*). *Indian J. Hort.* **23**: 59-61.

Thakur, R.N. and Husain, A. 1975. A new leaf spot disease of lemon grass. *Indian Phytopath.* **28**: 100-102.

Timyan, J. 1996. *BWA Yo: Important Trees in Haiti.* Southeast Consortium for International development, Washington DC, USA.

Tuset, J.J., Hinarejos, C. and Garcia, J. 1990. *Phytophthora* foot rot control in citrus with *Myrothecium roridum. Bull. OEPP* **20**: 169-176.

Ullasa, B.A. and Rawal, R.D. 1982. Annual Report, Indian Inst. of Hort. Res., Bangalore, India, p. 207.

Ullasa, B.A. and Rawal, R.D. 1983. Annual Report, Indian Inst. of Hort. Res., Bangalore, India, p. 212.

Upadhyaya, D.N. and Bordoloi, D.N. 1975. New records of diseases of cultivated essential oil bearing plants from North-East India. *Indian Phytopath.* **28**: 532-534.

Verma, O.P. and Singh, R.D. 1973. A method of testing varietal reaction in mango against *Botryodiplodia theobromae. J. Mycol. Pl. Path.*.**3**: 110-111.

Vir, D. and Gaur, A. 1971. Post-harvest chemical treatment of apple fruits against rot and deterioration. *Curr. Sci.* **40**: 387.

Weiss, E.A. 2002. *Essential Oil Crops.* CAB International, 191 pp.

SUBJECT INDEX